Supersymmetry

"Symmetry beyond The Standard Model"

Edited by Paul F. Kisak

Contents

3   Minimal Supersymmetric Standard Model   17

Chapter 1

Supersymmetry

"SUSY" redirects here. For other uses, see Susy (disambiguation).

For the episode of the American TV series Angel, see Supersymmetry (Angel).

Supersymmetry (**SUSY**), a theory of particle physics, is a proposed type of spacetime symmetry that relates two basic classes of elementary particles: bosons, which have an integer-valued spin, and fermions, which have a half-integer spin.[1] Each particle from one group is associated with a particle from the other, known as its superpartner, the spin of which differs by a half-integer. In a theory with perfectly "unbroken" supersymmetry, each pair of superpartners would share the same mass and internal quantum numbers besides spin. For example, there would be a "selectron" (superpartner electron), a bosonic version of the electron with the same mass as the electron, that would be easy to find in a laboratory. Thus, since no superpartners have been observed, if supersymmetry exists it must be a spontaneously broken symmetry so that superpartners may differ in mass.[2][3] Spontaneously-broken supersymmetry could solve many mysterious problems in particle physics including the hierarchy problem. The simplest realization of spontaneously-broken supersymmetry, the so-called Minimal Supersymmetric Standard Model, is one of the best studied candidates for physics beyond the Standard Model.

There is only indirect evidence and motivation for the existence of supersymmetry. Direct confirmation would entail production of superpartners in collider experiments, such as the Large Hadron Collider (LHC). The first run of the LHC found no evidence for supersymmetry (all results were consistent with the Standard Model), and thus set limits on superpartner masses in supersymmetric theories. Whilst many remain enthusiastic about supersymmetry,[4] this first run at the LHC led some physicists to explore other ideas.[5] In any case, in 2015 the LHC resumed its search for supersymmetry and other new physics in its second run.

1.1 Motivations

There are numerous phenomenological motivations for supersymmetry close to the electroweak scale, as well as technical motivations for supersymmetry at any scale.

1.1.1 The hierarchy problem

Supersymmetry close to the electroweak scale ameliorates the hierarchy problem that afflicts the Standard Model. In the Standard Model, the electroweak scale receives enormous Planck-scale quantum corrections. The observed hierarchy between the electroweak scale and the Planck scale must be achieved with extraordinary fine tuning. In a supersymmetric theory, on the other hand, Planck-scale quantum corrections cancel between partners and superpartners (owing to a minus sign associated with fermionic loops). The hierarchy between the electroweak scale and the Planck sale is achieved in a natural manner, without miraculous fine-tuning.

1

1.1.2   Gauge coupling unification

The idea that the gauge symmetry groups unify at high-energy is called Grand unification theory. In the Standard Model, however, the weak, strong and electromagnetic couplings fail to unify at high energy. In a supersymmetry theory, the running of the gauge couplings are modified, and precise high-energy unification of the gauge couplings is achieved. The modified running also provides a natural mechanism for radiative electroweak symmetry breaking.

1.1.3   Dark matter

TeV-scale supersymmetry (augmented with a discrete symmetry) typically provides a candidate dark matter particle at a mass scale consistent with thermal relic abundance calculations.[6][7]

1.1.4   Other technical motivations

Supersymmetry is also motivated by solutions to several theoretical problems, for generally providing many desirable mathematical properties, and for ensuring sensible behavior at high energies. Supersymmetric quantum field theory is often much easier to analyze, as many more problems become exactly solvable. When supersymmetry is imposed as a *local* symmetry, Einstein's theory of general relativity is included automatically, and the result is said to be a theory of supergravity. It is also a necessary feature of the most popular candidate for a theory of everything, superstring theory.

Another theoretically appealing property of supersymmetry is that it offers the only "loophole" to the Coleman–Mandula theorem, which prohibits spacetime and internal symmetries from being combined in any nontrivial way, for quantum field theories like the Standard Model with very general assumptions. The Haag-Lopuszanski-Sohnius theorem demonstrates that supersymmetry is the only way spacetime and internal symmetries can be combined consistently.[8]

1.2   History

A supersymmetry relating mesons and baryons was first proposed, in the context of hadronic physics, by Hironari Miyazawa during 1966. This supersymmetry did not involve spacetime, that is, it concerned internal symmetry, and was broken badly. Miyazawa's work was largely ignored at the time.[9][10][11][12]

J. L. Gervais and B. Sakita (during 1971),[13] Yu. A. Golfand and E. P. Likhtman (also during 1971), and D.V. Volkov and V.P. Akulov (1972),[14] independently rediscovered supersymmetry in the context of quantum field theory, a radically new type of symmetry of spacetime and fundamental fields, which establishes a relationship between elementary particles of different quantum nature, bosons and fermions, and unifies spacetime and internal symmetries of microscopic phenomena. Supersymmetry with a consistent Lie-algebraic graded structure on which the Gervais–Sakita rediscovery was based directly first arose during 1971[15] in the context of an early version of string theory by Pierre Ramond, John H. Schwarz and André Neveu.

Finally, Julius Wess and Bruno Zumino (during 1974)[16] identified the characteristic renormalization features of four-dimensional supersymmetric field theories, which identified them as remarkable QFTs, and they and Abdus Salam and their fellow researchers introduced early particle physics applications. The mathematical structure of supersymmetry (Graded Lie superalgebras) has subsequently been applied successfully to other topics of physics, ranging from nuclear physics,[17][18] critical phenomena,[19] quantum mechanics to statistical physics. It remains a vital part of many proposed theories of physics.

The first realistic supersymmetric version of the Standard Model was proposed during 1977 by Pierre Fayet and is known as the Minimal Supersymmetric Standard Model or MSSM for short. It was proposed to solve, amongst other things, the hierarchy problem.

1.3   Applications

1.3.1 Extension of possible symmetry groups

One reason that physicists explored supersymmetry is because it offers an extension to the more familiar symmetries of quantum field theory. These symmetries are grouped into the Poincaré group and internal symmetries and the Coleman–Mandula theorem showed that under certain assumptions, the symmetries of the S-matrix must be a direct product of the Poincaré group with a compact internal symmetry group or if there is not any mass gap, the conformal group with a compact internal symmetry group. During 1971 Golfand and Likhtman were the first to show that the Poincaré algebra can be extended through introduction of four anticommuting spinor generators (in four dimensions), which later became known as supercharges. During 1975 the Haag-Lopuszanski-Sohnius theorem analyzed all possible superalgebras in the general form, including those with an extended number of the supergenerators and central charges. This extended super-Poincaré algebra paved the way for obtaining a very large and important class of supersymmetric field theories.

The supersymmetry algebra

Main article: Supersymmetry algebra

Traditional symmetries of physics are generated by objects that transform by the tensor representations of the Poincaré group and internal symmetries. Supersymmetries, however, are generated by objects that transform by the spinor representations. According to the spin-statistics theorem, bosonic fields commute while fermionic fields anticommute. Combining the two kinds of fields into a single algebra requires the introduction of a $\mathbf{Z}_2$-grading under which the bosons are the even elements and the fermions are the odd elements. Such an algebra is called a Lie superalgebra.

The simplest supersymmetric extension of the Poincaré algebra is the Super-Poincaré algebra. Expressed in terms of two Weyl spinors, has the following anti-commutation relation:

$$\{Q_\alpha, \bar{Q}_{\dot\beta}\} = 2(\sigma^\mu)_{\alpha\dot\beta} P_\mu$$

and all other anti-commutation relations between the Qs and commutation relations between the Qs and Ps vanish. In the above expression $P_\mu = -i\partial_\mu$ are the generators of translation and σ^μ are the Pauli matrices.

There are representations of a Lie superalgebra that are analogous to representations of a Lie algebra. Each Lie algebra has an associated Lie group and a Lie superalgebra can sometimes be extended into representations of a Lie supergroup.

1.3.2 The Supersymmetric Standard Model

Main article: Minimal Supersymmetric Standard Model

Incorporating supersymmetry into the Standard Model requires doubling the number of particles since there is no way that any of the particles in the Standard Model can be superpartners of each other. With the addition of new particles, there are many possible new interactions. The simplest possible supersymmetric model consistent with the Standard Model is the Minimal Supersymmetric Standard Model (MSSM) which can include the necessary additional new particles that are able to be superpartners of those in the Standard Model.

One of the main motivations for SUSY comes from the quadratically divergent contributions to the Higgs mass squared. The quantum mechanical interactions of the Higgs boson causes a large renormalization of the Higgs mass and unless there is an accidental cancellation, the natural size of the Higgs mass is the greatest scale possible. This problem is known as the hierarchy problem. Supersymmetry reduces the size of the quantum corrections by having automatic cancellations between fermionic and bosonic Higgs interactions. If supersymmetry is restored at the weak scale, then the Higgs mass is related to supersymmetry breaking which can be induced from small non-perturbative effects explaining the vastly different scales in the weak interactions and gravitational interactions.

In many supersymmetric Standard Models there is a heavy stable particle (such as neutralino) which could serve as a weakly interacting massive particle (WIMP) dark matter candidate. The existence of a supersymmetric dark matter candidate is related closely to R-parity.

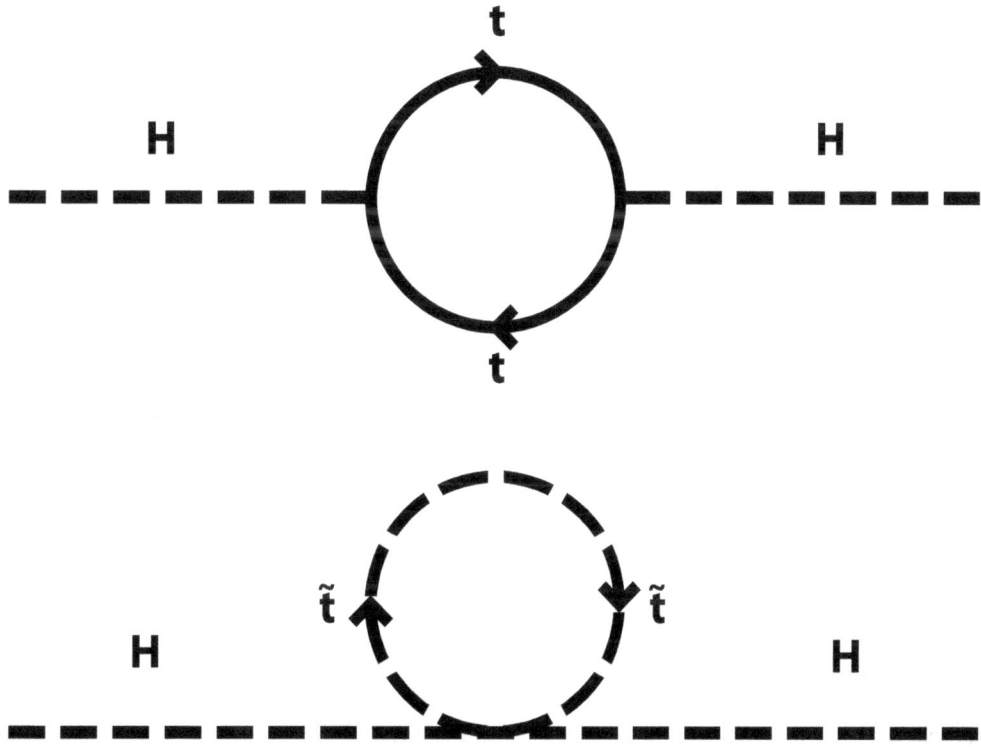

Cancellation of the Higgs boson quadratic mass renormalization between fermionic top quark loop and scalar stop squark tadpole Feynman diagrams in a supersymmetric extension of the Standard Model

The standard paradigm for incorporating supersymmetry into a realistic theory is to have the underlying dynamics of the theory be supersymmetric, but the ground state of the theory does not respect the symmetry and supersymmetry is broken spontaneously. The supersymmetry break can not be done permanently by the particles of the MSSM as they currently appear. This means that there is a new sector of the theory that is responsible for the breaking. The only constraint on this new sector is that it must break supersymmetry permanently and must give superparticles TeV scale masses. There are many models that can do this and most of their details do not matter. In order to parameterize the relevant features of supersymmetry breaking, arbitrary soft SUSY breaking terms are added to the theory which temporarily break SUSY explicitly but could never arise from a complete theory of supersymmetry breaking.

Gauge-coupling unification

Main article: Minimal Supersymmetric Standard Model § Gauge-coupling unification

One piece of evidence for supersymmetry existing is gauge coupling unification. The renormalization group evolution of the three gauge coupling constants of the Standard Model is somewhat sensitive to the present particle content of the theory. These coupling constants do not quite meet together at a common energy scale if we run the renormalization group using the Standard Model.[20] With the addition of minimal SUSY joint convergence of the coupling constants is projected at approximately 10^{16} GeV.[20]

1.3.3 Supersymmetric quantum mechanics

Main article: Supersymmetric quantum mechanics

Supersymmetric quantum mechanics adds the SUSY superalgebra to quantum mechanics as opposed to quantum field theory. Supersymmetric quantum mechanics often becomes relevant when studying the dynamics of supersymmetric solitons, and due to the simplified nature of having fields which are only functions of time (rather than space-time), a great deal of progress has been made in this subject and it is now studied in its own right.

SUSY quantum mechanics involves pairs of Hamiltonians which share a particular mathematical relationship, which are called *partner Hamiltonians*. (The potential energy terms which occur in the Hamiltonians are then known as *partner potentials*.) An introductory theorem shows that for every eigenstate of one Hamiltonian, its partner Hamiltonian has a corresponding eigenstate with the same energy. This fact can be exploited to deduce many properties of the eigenstate spectrum. It is analogous to the original description of SUSY, which referred to bosons and fermions. We can imagine a "bosonic Hamiltonian", whose eigenstates are the various bosons of our theory. The SUSY partner of this Hamiltonian would be "fermionic", and its eigenstates would be the theory's fermions. Each boson would have a fermionic partner of equal energy.

1.3.4 Supersymmetry: Applications to condensed matter physics

SUSY concepts have provided useful extensions to the WKB approximation. Additionally, SUSY has been applied to disorder averaged systems both quantum and non-quantum (through statistical mechanics), the Fokker-Planck equation being an example of a non-quantum theory. The `supersymmetry' in all these systems arises from the fact that one is modelling one particle and as such the`statistics' don't matter. The use of the supersymmetry method provides a mathematical rigorous alternative to the replica trick, but only in non-interacting systems, which attempts to address the so-called `problem of the denominator' under disorder averaging. For more on the applications of supersymmetry in condensed matter physics see the book[21]

1.3.5 Supersymmetry in optics

Integrated optics was recently found[22] to provide a fertile ground on which certain ramifications of SUSY can be explored in readily-accessible laboratory settings. Making use of the analogous mathematical structure of the quantum-mechanical Schrödinger equation and the wave equation governing the evolution of light in one-dimensional settings, one may interpret the refractive index distribution of a structure as a potential landscape in which optical wave packets propagate. In this manner, a new class of functional optical structures with possible applications in phase matching, mode conversion[23] and space-division multiplexing becomes possible. SUSY transformations have been also proposed as a way to address inverse scattering problems in optics and as a one-dimensional transformation optics [24]

1.3.6 Mathematics

SUSY is also sometimes studied mathematically for its intrinsic properties. This is because it describes complex fields satisfying a property known as holomorphy, which allows holomorphic quantities to be exactly computed. This makes supersymmetric models useful "toy models" of more realistic theories. A prime example of this has been the demonstration of S-duality in four-dimensional gauge theories[25] that interchanges particles and monopoles.

The proof of the Atiyah-Singer index theorem is much simplified by the use of supersymmetric quantum mechanics.

1.4 General supersymmetry

Supersymmetry appears in many related contexts of theoretical physics. It is possible to have multiple supersymmetries and also have supersymmetric extra dimensions.

1.4.1   Extended supersymmetry

Main article: Extended supersymmetry

It is possible to have more than one kind of supersymmetry transformation. Theories with more than one supersymmetry transformation are known as extended supersymmetric theories. The more supersymmetry a theory has, the more constrained are the field content and interactions. Typically the number of copies of a supersymmetry is a power of 2, i.e. 1, 2, 4, 8. In four dimensions, a spinor has four degrees of freedom and thus the minimal number of supersymmetry generators is four in four dimensions and having eight copies of supersymmetry means that there are 32 supersymmetry generators.

The maximal number of supersymmetry generators possible is 32. Theories with more than 32 supersymmetry generators automatically have massless fields with spin greater than 2. It is not known how to make massless fields with spin greater than two interact, so the maximal number of supersymmetry generators considered is 32. This corresponds to an $N = 8$ supersymmetry theory. Theories with 32 supersymmetries automatically have a graviton.

For four dimensions there are the following theories, with the corresponding multiplets[26](CPT adds a copy, whenever they are not invariant under such symmetry)

- $N = 1$

Chiral multiplet: $(0, \frac{1}{2})$ Vector multiplet: $(\frac{1}{2}, 1)$ Gravitino multiplet: $(1, \frac{3}{2})$ Graviton multiplet: $(\frac{3}{2}, 2)$

- $N = 2$

hypermultiplet: $(-\frac{1}{2}, 0^2, \frac{1}{2})$ vector multiplet: $(0, \frac{1}{2}^2, 1)$ supergravity multiplet: $(1, \frac{3}{2}^2, 2)$

- $N = 4$

Vector multiplet: $(-1, -\frac{1}{2}^4, 0^6, \frac{1}{2}^4, 1)$ Supergravity multiplet: $(0, \frac{1}{2}^4, 1^6, \frac{3}{2}^4, 2)$

- $N = 8$

Supergravity multiplet: $(-2, -\frac{3}{2}^8, -1^{28}, -\frac{1}{2}^{56}, 0^{70}, \frac{1}{2}^{56}, 1^{28}, \frac{3}{2}^8, 2)$

1.4.2   Supersymmetry in alternate numbers of dimensions

It is possible to have supersymmetry in dimensions other than four. Because the properties of spinors change drastically between different dimensions, each dimension has its characteristic. In d dimensions, the size of spinors is approximately $2^{d/2}$ or $2^{(d-1)/2}$. Since the maximum number of supersymmetries is 32, the greatest number of dimensions in which a supersymmetric theory can exist is eleven.

1.5   Supersymmetry as a quantum group

Main article: Supersymmetry as a quantum group

Supersymmetry can be reinterpreted in the language of noncommutative geometry and quantum groups. In particular, it involves a mild form of noncommutativity, namely supercommutativity. See the main article for more details.

1.6 Supersymmetry in quantum gravity

Supersymmetry is part of a larger enterprise of theoretical physics to unify everything we know about the universe into a single consistent set of physical principles, known as the quest for a Theory of Everything (TOE). A significant part of this larger enterprise is the quest for a theory of quantum gravity, which would unify the classical theory of general relativity and the Standard Model, which explains the other three basic forces in physics (electromagnetism, the strong interaction, and the weak interaction), and provides a palette of fundamental particles upon which all four forces act. Two of the most active methods of forming a theory of quantum gravity are string theory and loop quantum gravity (LQG), although in theory, supersymmetry could be a component of other theories as well.

For string theory to be consistent, supersymmetry seems to be required at some level (although it may be a strongly broken symmetry). In particle theory, supersymmetry is recognized as a way to stabilize the hierarchy between the unification scale and the electroweak scale (or the Higgs boson mass), and can also provide a natural dark matter candidate. String theory also requires extra spatial dimensions which have to be compactified as in Kaluza–Klein theory.

Loop quantum gravity (LQG) predicts no additional spatial dimensions, nor anything else about particle physics. These theories can be formulated in three spatial dimensions and one dimension of time, although in some LQG theories dimensionality is an emergent property of the theory, rather than a fundamental assumption of the theory. Also, LQG is a theory of quantum gravity which does not require supersymmetry. Lee Smolin, one of the originators of LQG, has proposed that a loop quantum gravity theory incorporating either supersymmetry or extra dimensions, or both, be called "loop quantum gravity II".

If experimental evidence confirms supersymmetry in the form of supersymmetric particles such as the neutralino that is often believed to be the lightest superpartner, some people believe this would be a major boost to string theory. Since supersymmetry is a required component of string theory, any discovered supersymmetry would be consistent with string theory. If the Large Hadron Collider and other major particle physics experiments fail to detect supersymmetric partners or evidence of extra dimensions, many versions of string theory which had predicted certain low mass superpartners to existing particles may need to be significantly revised. The failure of experiments to discover either supersymmetric partners or extra spatial dimensions, as of 2013, has encouraged loop quantum gravity researchers.

1.7 Current status

Supersymmetric models are constrained by a variety of experiments, including measurements of low-energy observables – for example, the anomalous magnetic moment of the muon at Brookhaven; the WMAP dark matter density measurement and direct detection experiments – for example, XENON–100 and LUX; and by particle collider experiments, including B-physics, Higgs phenomenology and direct searches for superpartners (sparticles), at the Large Electron–Positron Collider, Tevatron and the LHC.

Historically, the tightest limits were from direct production at colliders. The first mass limits for squarks and gluinos were made at CERN by the UA1 experiment and the UA2 experiment at the Super Proton Synchrotron. LEP later set very strong limits.,[27] which in 2006 were extended by the D0 experiment at the Tevatron.[28][29] From 2003, WMAP's and Planck's dark matter density measurements have strongly constrained supersymmetry models, which, if they explain dark matter, have to be tuned to invoke a particular mechanism to sufficiently reduce the neutralino density.

Prior to the beginning of the LHC, in 2009 fits of available data to CMSSM and NUHM1 indicated that squarks and gluinos were most likely to have masses in the 500 to 800 GeV range, though values as high as 2.5 TeV were allowed with low probabilities. Neutralinos and sleptons were expected to be quite light, with the lightest neutralino and the lightest stau most likely to be found between 100 to 150 GeV.[30]

The first run of the LHC found no evidence for supersymmetry, and, as a result, surpassed existing experimental limits from the Large Electron–Positron Collider and Tevatron and partially excluded the aforementioned expected ranges.[31]

During 2011 and 2012, the LHC discovered a Higgs boson with a mass of about 125 GeV, and with couplings to fermions and bosons which are consistent with the Standard Model. The MSSM predicts that the mass of the lightest Higgs boson should not be much higher than the mass of the Z boson, and, in the absence of fine tuning (with the supersymmetry breaking scale on the order of 1 TeV), should not exceed 130 GeV. Furthermore, for values of the MSSM parameter *tan* β ≤ 3, it predicts a Higgs mass below 114 GeV over most of the parameter space.[32] This region of Higgs mass was excluded by LEP by 2000. The LHC result is somewhat problematic for the minimal supersymmetric model, as the value of 125 GeV is relatively large for the model and can only be achieved with large radiative loop corrections from top squarks, which many theorists consider to be "unnatural" (see naturalness and fine tuning).[33] On the other hand, the lightest Higgs boson in the MSSM is Standard Model-like, which is consistent with measurements of the Higgs boson couplings at the LHC.

In spite of the null searches and the heavy Higgs, a recent analysis of the constrained minimal supersymmetric Standard Model, the CMSSM, suggests that the model is still compatible with all present experimental constraints.[34][35] The preferred masses for squarks and gluinos is about 2 TeV. The resulting fine-tuning of the electroweak scale, however, is considered "unnatural" (see little hierarchy problem), and some theorists now favor extended supersymmetry models – for example, the NMSSM.

1.8   See also

- Supersymmetric gauge theory

- Wess–Zumino model

- Minimal Supersymmetric Standard Model

- Supersymmetry as a quantum group

- Quantum group

- Supercharge

- Superfield

- Supergeometry

- Supergravity

- Supergroup

- Superspace

1.9   References

[1] Haber, Howie. "SUPERSYMMETRY, PART I (THEORY)" (PDF). *Reviews, Tables and Plots*. Particle Data Group (PDG). Retrieved 8 July 2015.

[2] Martin, Stephen P. (1997). "A Supersymmetry Primer". arXiv:hep-ph/9709356.

[3] Dine, Michael (2007). *Supersymmetry and String Theory: Beyond the Standard Model*. p. 169.

[4] Ellis, John. "The Physics Landscape after the Higgs Discovery at the LHC". *arXiv*. Invited plenary talk at SILAFAE 2014. Retrieved 8 July 2015.

[5] Wolchover, Natalie (November 20, 2012). "Supersymmetry Fails Test, Forcing Physics to Seek New Ideas". *Quanta Magazine*.

[6] Jonathan Feng: Supersymmetric Dark Matter *(pdf)*, University of California, Irvine, 11 May 2007

[7] Torsten Bringmann: The WIMP "Miracle" *(pdf)* University of Hamburg

[8] R. Haag, J. T. Lopuszanski and M. Sohnius, "All Possible Generators Of Supersymmetries Of The S Matrix", Nucl. Phys. B 88 (1975) 257

[9] H. Miyazawa (1966). "Baryon Number Changing Currents".*Prog. Theor. Phys.***36**(6): 1266–1276.Bibcode:1966PThPh..36.1M. doi:10.1143/PTP.36.1266.

[10] H. Miyazawa (1968). "Spinor Currents and Symmetries of Baryons and Mesons".*Phys. Rev.* **170**(5): 15:1968PhRv..170.1586M. doi:10.1103/PhysRev.170.1586.

[11] Michio Kaku, *Quantum Field Theory*, ISBN 0-19-509158-2, pg 663.

[12] Peter Freund, *Introduction to Supersymmetry*, ISBN 0-521-35675-X, pages 26-27, 138.

[13] Gervais, J. -L.; Sakita, B. (1971). "Field theory interpretation of supergauges in dual models". *Nuclear Physics B* **34** (2): 632. Bibcode:1971NuPhB..34..632G. doi:10.1016/0550-3213(71)90351-8.

[14] D.V. Volkov, V.P. Akulov, Pisma Zh.Eksp.Teor.Fiz. 16 (1972) 621; Phys.Lett. B46 (1973) 109; V.P. Akulov, D.V. Volkov, Teor.Mat.Fiz. 18 (1974) 39

[15] Ramond, P. (1971). "Dual Theory for Free Fermions". *Physical Review D* **3** (10): 2415. Bibcode:1971PhRvD...3.2415R. doi:10.1103/PhysRevD.3.2415.

[16] Wess, J.; Zumino, B. (1974). "Supergauge transformations in four dimensions".*Nuclear Physics B***70**: 39.Bibcode:1974NuPh39W. doi:10.1016/0550-3213(74)90355-1.

[17] http://users.physik.fu-berlin.de/~{}kleinert/kleinert/?p=supersym suggested here

[18] Iachello, F. (1980). "Dynamical Supersymmetries in Nuclei".*Physical Review Letters***44**(12): 772.Bibcode:1980PhRvL..4.772I. doi:10.1103/PhysRevLett.44.772.

[19] Friedan, D.; Qiu, Z.; Shenker, S. (1984). "Conformal Invariance, Unitarity, and Critical Exponents in Two Dimensions". *Physical Review Letters* **52** (18): 1575. Bibcode:1984PhRvL..52.1575F. doi:10.1103/PhysRevLett.52.1575.

[20] Gordon L. Kane, *The Dawn of Physics Beyond the Standard Model*, Scientific American, June 2003, page 60 and *The frontiers of physics*, special edition, Vol 15, #3, page 8

[21] *Supersymmetry in Disorder and Chaos*, Konstantin Efetov, Cambridge university press, 1997.

[22] Miri, M.-A.; Heinrich, M.; El-Ganainy, R.; Christodoulides, D. N. (2013). "Superymmetric optical structures". *Physical Review Letters* (APS) **110** (23): 233902. arXiv:1304.6646. Bibcode:2013PhRvL.110w3902M. doi:10.1103/PhysRevLett.110.233902. PMID 25167493. Retrieved April 2014.

[23] Heinrich, M.; Miri, M.-A.; Stützer, S.; El-Ganainy, R.; Nolte, S.; Szameit, A.; Christodoulides, D. N. (2014). "Superymmetric mode converters".*Nature Communications*(NPG)**5**: 3698.arXiv:1401.5734.Bibcod98.PMID24739256.RetrievedApril2014.

[24] Miri, M.-A.; Heinrich, Matthias; Christodoulides, D. N. (2014). "SUSY-inspired one-dimensional transformation optics". *Optica* (OSA) **1** (2): 89. arXiv:1408.0832. doi:10.1364/OPTICA.1.000089. Retrieved August 2014.

[25] Krasnitz, Michael (2002). *Correlation functions in supersymmetric gauge theories from supergravity fluctuafluctuations hHKtions* (PDF). Princeton University Department of Physics: Princeton University Department of Physics. p. 91.

[26] Polchinski,J. *String theory. Vol. 2: Superstring theory and beyond*, Appendix B

[27] LEPSUSYWG, ALEPH, DELPHI, L3 and OPAL experiments, charginos, large m0 LEPSUSYWG/01-03.1

[28] The D0-Collaboration (2009). "Search for associated production of charginos and neutralinos in the trilepton final state using 2.3 fb^{-1} of data". arXiv:0901.0646. Bibcode:2009PhLB..680...34D. doi:10.1016/j.physletb.2009.08.011.

[29] The D0 Collaboration (2006). "Search for squarks and gluinos in events with jets and missing transverse energy using 2.1 fb-1 of pp⁻ collision data at s=1.96 TeV". arXiv:0712.3805. Bibcode:2008PhLB..660..449D. doi:10.1016/j.physletb.2008.01.042.

[30] O. Buchmueller et al. (2009). "Likelihood Functions for Supersymmetric Observables in Frequentist Analyses of the CMSSM and NUHM1". *The European Physical Journal C* **64** (3): 391–415. arXiv:0907.5568. Bibcode:2009EPJC...64..391B. doi: /s-009-1159-z.

[31] Roszkowski, Leszek; Sessolo, Enrico Maria; Williams, Andrew J. (11 August 2014). "What next for the CMSSM and the NUHM: improved prospects for superpartner and dark matter detection". *Journal of High Energy Physics* **2014** (8). doi:10P08(.

[32] Marcela Carena and Howard E. Haber; Haber (1970). "Higgs Boson Theory and Phenomenology". *Progress in Particle and Nuclear Physics* **50**: 63. arXiv:hep-ph/0208209v3. Bibcode:2003PrPNP..50...63C. doi:10.1016/S0146-6410(02)00177-1.

[33] Patrick Draper et al. (December 2011). "Implications of a 125 GeV Higgs for the MSSM and Low-Scale SUSY Breaking". *Physical Review D* **85** (9): 095007. arXiv:1112.3068. Bibcode:2012PhRvD..85i5007D. doi:10.1103/PhysRevD.85.095007.

[34] Bechtle, Philip. "How alive is constrained SUSY really?". *arXiv*. Retrieved 8 July 2015.

[35] Jan de Vries, Kees. "SUSY fits with full LHC Run I data". *arXiv*. Retrieved 8 July 2015.

1.10   Further reading

- Supersymmetry and Supergravity page in String Theory Wiki lists more books and reviews.

1.10.1   Theoretical introductions, free and online

- S. Martin (2011). "A Supersymmetry Primer". arXiv:hep-ph/9709356.

- Joseph D. Lykken (1996). "Introduction to Supersymmetry". arXiv:hep-th/9612114.

- Manuel Drees (1996). "An Introduction to Supersymmetry". arXiv:hep-ph/9611409.

- Adel Bilal (2001). "Introduction to Supersymmetry". arXiv:hep-th/0101055.

- An Introduction to Global Supersymmetry by Philip Arygres, 2001

1.10.2   Monographs

- Weak Scale Supersymmetry by Howard Baer and Xerxes Tata, 2006.

- Cooper, F.; Khare, A.; Sukhatme, U. (1995). "Supersymmetry and quantum mechanics". *Physics Reports* **251** (5–6): 267. doi:10.1016/0370-1573(94)00080-M. (arXiv:hep-th/9405029).

- Junker, G. (1996). "Supersymmetric Methods in Quantum and Statistical Physics". doi:10.1007/978-3-642-61194-0. ISBN 978-3-540-61591-0..

- Gordon L. Kane.*Supersymmetry: Unveiling the Ultimate Laws of Nature* Basic Books, New York (2001). ISBN 0-7382-0489-7.

- Gordon L. Kane and Shifman, M., eds. *The Supersymmetric World: The Beginnings of the Theory*, World Scientific, Singapore (2000). ISBN 981-02-4522-X.

- Weinberg, Steven, *The Quantum Theory of Fields, Volume 3: Supersymmetry*, Cambridge University Press, Cambridge, (1999). ISBN 0-521-66000-9.

- Wess, Julius, and Jonathan Bagger, *Supersymmetry and Supergravity*, Princeton University Press, Princeton, (1992). ISBN 0-691-02530-4.

- "Concise Encyclopedia of Supersymmetry". 2003. doi:10.1007/1-4020-4522-0. ISBN 978-1-4020-1338-6.

1.10.3   On experiments

- Bennett GW; Muon (g−2) Collaboration; Bousquet; Brown; Bunce; Carey; Cushman; Danby; Debevec; Deile; Deng; Dhawan; Druzhinin; Duong; Farley; Fedotovich; Gray; Grigoriev; Grosse-Perdekamp; Grossmann; Hare; Hertzog; Huang; Hughes; Iwasaki; Jungmann; Kawall; Khazin; Krienen; Kronkvist et al. (2004). "Measurement of the negative muon anomalous magnetic moment to 0.7 ppm". *Physical Review Letters* **92** (16): 161802. arXiv:hep-ex/0401008. Bibcode:2004PhRvL..92p1802B. doi:10.1103/PhysRevLett.92.161802. PMID 15169217.

- Brookhaven National Laboratory (Jan. 8, 2004). *New g−2 measurement deviates further from Standard Model.* Press Release.

- Fermi National Accelerator Laboratory (Sept 25, 2006). *Fermilab's CDF scientists have discovered the quick-change behavior of the B-sub-s meson.* Press Release.

1.11   External links

- Supersymmetry (physics) at *Encyclopædia Britannica*

- What do current LHC results (mid-August 2011) imply about supersymmetry? Matt Strassler

- ATLAS Experiment Supersymmetry search documents

- CMS Experiment Supersymmetry search documents

- "Particle wobble shakes up supersymmetry", *Cosmos* magazine, September 2006

- LHC results put supersymmetry theory 'on the spot' BBC news 27/8/2011

- SUSY running out of hiding places BBC news 12/11/2012

- Supersymmetry in optics? "Skulls in the Stars" blog 22/08/2013

Chapter 2

Spacetime symmetries

For the notation, see Ricci calculus.

Spacetime symmetries are features of spacetime that can be described as exhibiting some form of symmetry. The role of symmetry in physics is important in simplifying solutions to many problems, spacetime symmetries are used in the study of exact solutions of Einstein's field equations of general relativity.

2.1 Physical motivation

Physical problems are often investigated and solved by noticing features which have some form of symmetry. For example, in the Schwarzschild solution, the role of spherical symmetry is important in deriving the Schwarzschild solution and deducing the physical consequences of this symmetry (such as the non-existence of gravitational radiation in a spherically pulsating star). In cosmological problems, symmetry finds a role to play in the cosmological principle which restricts the type of universes that are consistent with large-scale observations (e.g. the Friedmann-Lemaître-Robertson-Walker (FLRW) metric). Symmetries usually require some form of preserving property, the most important of which in general relativity include the following:

- preserving geodesics of the spacetime
- preserving the metric tensor
- preserving the curvature tensor

These and other symmetries will be discussed in more detail later. This preservation feature can be used to motivate a useful definition of symmetries.

2.2 Mathematical definition

A rigorous definition of symmetries in general relativity has been given by Hall (2004). In this approach, the idea is to use (smooth) vector fields whose local flow diffeomorphisms preserve some property of the spacetime. This preserving property of the diffeomorphisms is made precise as follows. A smooth vector field X on a spacetime M is said to *preserve* a smooth tensor T on M (or T is **invariant** under X) if, for each smooth local flow diffeomorphism ϕt associated with X, the tensors T and $\phi t^*(T)$ are equal on the domain of ϕt. This statement is equivalent to the more usable condition that the Lie derivative of the tensor under the vector field vanishes:

$$\mathcal{L}_X T = 0$$

on M. This has the consequence that, given any two points p and q on M, the coordinates of T in a coordinate system around p are equal to the coordinates of T in a coordinate system around q. A *symmetry on the spacetime* is a smooth vector field whose local flow diffeomorphisms preserve some (usually geometrical) feature of the spacetime. The (geometrical) feature may refer to specific tensors (such as the metric, or the energy-momentum tensor) or to other aspects of the spacetime such as its geodesic structure. The vector fields are sometimes referred to as *collineations*, *symmetry vector fields* or just *symmetries*. The set of all symmetry vector fields on M forms a Lie algebra under the Lie bracket operation as can be seen from the identity:

$$\mathcal{L}_{[X,Y]}T = \mathcal{L}_X(\mathcal{L}_Y T) - \mathcal{L}_Y(\mathcal{L}_X T)$$

the term on the right usually being written, with an abuse of notation, as $[\mathcal{L}_X, \mathcal{L}_Y]T$.

2.3 Killing symmetry

Main article: Killing vector field

A Killing vector field is one of the most important types of symmetries and is defined to be a smooth vector field that preserves the metric tensor:

$$\mathcal{L}_X g_{ab} = 0$$

This is usually written in the expanded form as:

$$X_{a;b} + X_{b;a} = 0$$

Killing vector fields find extensive applications (including in classical mechanics) and are related to conservation laws.

2.4 Homothetic symmetry

Main article: Homothetic vector field

A homothetic vector field is one which satisfies:

$$\mathcal{L}_X g_{ab} = 2c g_{ab}$$

where c is a real constant. Homothetic vector fields find application in the study of singularities in general relativity.

2.5 Affine symmetry

Main article: Affine vector field

An affine vector field is one that satisfies:

$$(\mathcal{L}_X g_{ab})_{;c} = 0$$

An affine vector field preserves geodesics and preserves the affine parameter.

The above three vector field types are special cases of projective vector fields which preserve geodesics without necessarily preserving the affine parameter.

2.6 Conformal symmetry

Main article: Conformal vector field

A conformal vector field is one which satisfies:

$$\mathcal{L}_X g_{ab} = \phi g_{ab}$$

where ϕ is a smooth real-valued function on M.

2.7 Curvature symmetry

Main article: Curvature collineation

A curvature collineation is a vector field which preserves the Riemann tensor:

$$\mathcal{L}_X R^a{}_{bcd} = 0$$

where $R^a bcd$ are the components of the Riemann tensor. The set of all smooth curvature collineations forms a Lie algebra under the Lie bracket operation (if the smoothness condition is dropped, the set of all curvature collineations need not form a Lie algebra). The Lie algebra is denoted by $CC(M)$ and may be infinite-dimensional. Every affine vector field is a curvature collineation.

2.8 Matter symmetry

Main article: Matter collineation

A less well-known form of symmetry concerns vector fields that preserve the energy-momentum tensor. These are variously referred to as matter collineations or matter symmetries and are defined by:

$$\mathcal{L}_X T_{ab} = 0$$

where *Tab* are the energy-momentum tensor components. The intimate relation between geometry and physics may be highlighted here, as the vector field X is regarded as preserving certain physical quantities along the flow lines of X, this being true for any two observers. In connection with this, it may be shown that *every Killing vector field is a matter collineation* (by the Einstein field equations, with or without cosmological constant). Thus, given a solution of the EFE, *a vector field that preserves the metric necessarily preserves the corresponding energy-momentum tensor*. When the energy-momentum tensor represents a perfect fluid, every Killing vector field preserves the energy density, pressure and the fluid flow vector field. When the energy-momentum tensor represents an electromagnetic field, a Killing vector field does *not necessarily* preserve the electric and magnetic fields.

2.9 Local and global symmetries

Main articles: Local symmetry and Global symmetry

2.10 Applications

As mentioned at the start of this article, the main application of these symmetries occur in general relativity, where solutions of Einstein's equations may be classified by imposing some certain symmetries on the spacetime.

2.10.1 Spacetime classifications

Classifying solutions of the EFE constitutes a large part of general relativity research. Various approaches to classifying spacetimes, including using the Segre classification of the energy-momentum tensor or the Petrov classification of the Weyl tensor have been studied extensively by many researchers, most notably Stephani et al. (2003). They also classify spacetimes using symmetry vector fields (especially Killing and homothetic symmetries). For example, Killing vector fields may be used to classify spacetimes, as there is a limit to the number of global, smooth Killing vector fields that a spacetime may possess (the maximum being 10 for 4-dimensional spacetimes). Generally speaking, the higher the dimension of the algebra of symmetry vector fields on a spacetime, the more symmetry the spacetime admits. For example, the Schwarzschild solution has a Killing algebra of dimension 4 (3 spatial rotational vector fields and a time translation), whereas the Friedmann-Lemaître-Robertson-Walker (FLRW) metric (excluding the Einstein static subcase) has a Killing algebra of dimension 6 (3 translations and 3 rotations). The Einstein static metric has a Killing algebra of dimension 7 (the previous 6 plus a time translation).

The assumption of a spacetime admitting a certain symmetry vector field can place restrictions on the spacetime.

2.11 See also

- Field (physics)

- Killing tensor

- Lie groups

- Noether's theorem

- Ricci decomposition

- Symmetry in physics

- Symmetry in quantum mechanics

- Derivations of the Lorentz transformations

2.12 References

- Hall, Graham (2004). *Symmetries and Curvature Structure in General Relativity (World Scientific Lecture Notes in Physics)*. Singapore: World Scientific Pub. Co. ISBN 981-02-1051-5. See *Section 10.1* for a definition of symmetries.

- Stephani, Hans; Kramer, Dietrich; MacCallum, Malcolm; Hoenselaers, Cornelius & Herlt, Eduard (2003). *Exact Solutions of Einstein's Field Equations*. Cambridge: Cambridge University Press. ISBN 0-521-46136-7.

- Schutz, Bernard (1980). *Geometrical Methods of Mathematical Physics*. Cambridge: Cambridge University Press. ISBN 0-521-29887-3. See *Chapter 3* for properties of the Lie derivative and *Section 3.10* for a definition of invariance.

Chapter 3

Minimal Supersymmetric Standard Model

The **Minimal Supersymmetric Standard Model** (**MSSM**) is an extension to the Standard Model that realizes supersymmetry. MSSM is the minimal supersymmetrical model as it considers only "the [minimum] number of new particle states and new interactions consistent with phenomenology".[1]Supersymmetry pair sbosons with fermions; therefore every StandardModel particle has a partner that has yet to be discovered. If the superparticles are found, it maybe analogous to discovering darkmatter[2] and depending on the details of what might be found, it could provide evidence for grandunification and might even, in principle, provide hints as to whether string theory describes nature. The failure to find evidence for supersymmetry using the Large Hadron Collider since 2010 has led to suggestions that the theory should be abandoned.[3]

3.1 Background

The MSSM was originally proposed in 1981 to stabilize the weak scale, solving the hierarchy problem.[4] The Higgs boson mass of the Standard Model is unstable to quantum corrections and the theory predicts that weak scale should be much weaker than what is observed to be. In the MSSM, the Higgs boson has a fermionic superpartner, the Higgsino, that has the same mass as it would if supersymmetry were an exact symmetry. Because fermion masses are radiatively stable, the Higgs mass inherits this stability. However, in MSSM there is a need for more than one Higgs field, as described below.

The only unambiguous way to claim discovery of supersymmetry is to produce superparticles in the laboratory. Because superparticles are expected to be 100 to 1000 times heavier than the proton, it requires a huge amount of energy to make these particles that can only be achieved at particle accelerators. The Tevatron was actively looking for evidence of the production of supersymmetric particles before it was shut down on 30 September 2011. Most physicists believe that supersymmetry must be discovered at the LHC if it is responsible for stabilizing the weak scale. There are five classes of particle that superpartners of the Standard Model fall into: squarks, gluinos, charginos, neutralinos, and sleptons. These superparticles have their interactions and subsequent decays described by the MSSM and each has characteristic signatures.

The MSSM imposes R-parity to explain the stability of the proton. It adds supersymmetry breaking by introducing explicit soft supersymmetry breaking operators into the Lagrangian that is communicated to it by some unknown (and unspecified) dynamics. This means that there are 120 new parameters in the MSSM. Most of these parameters lead to unacceptable phenomenology such as large flavor changing neutral currents or large electric dipole moments for the neutron and electron. To avoid these problems, the MSSM takes all of the soft supersymmetry breaking to be diagonal in flavor space and for all of the new CP violating phases to vanish.

3.2 Theoretical motivations

There are three principal motivations for the MSSM over other theoretical extensions of the Standard Model, namely:

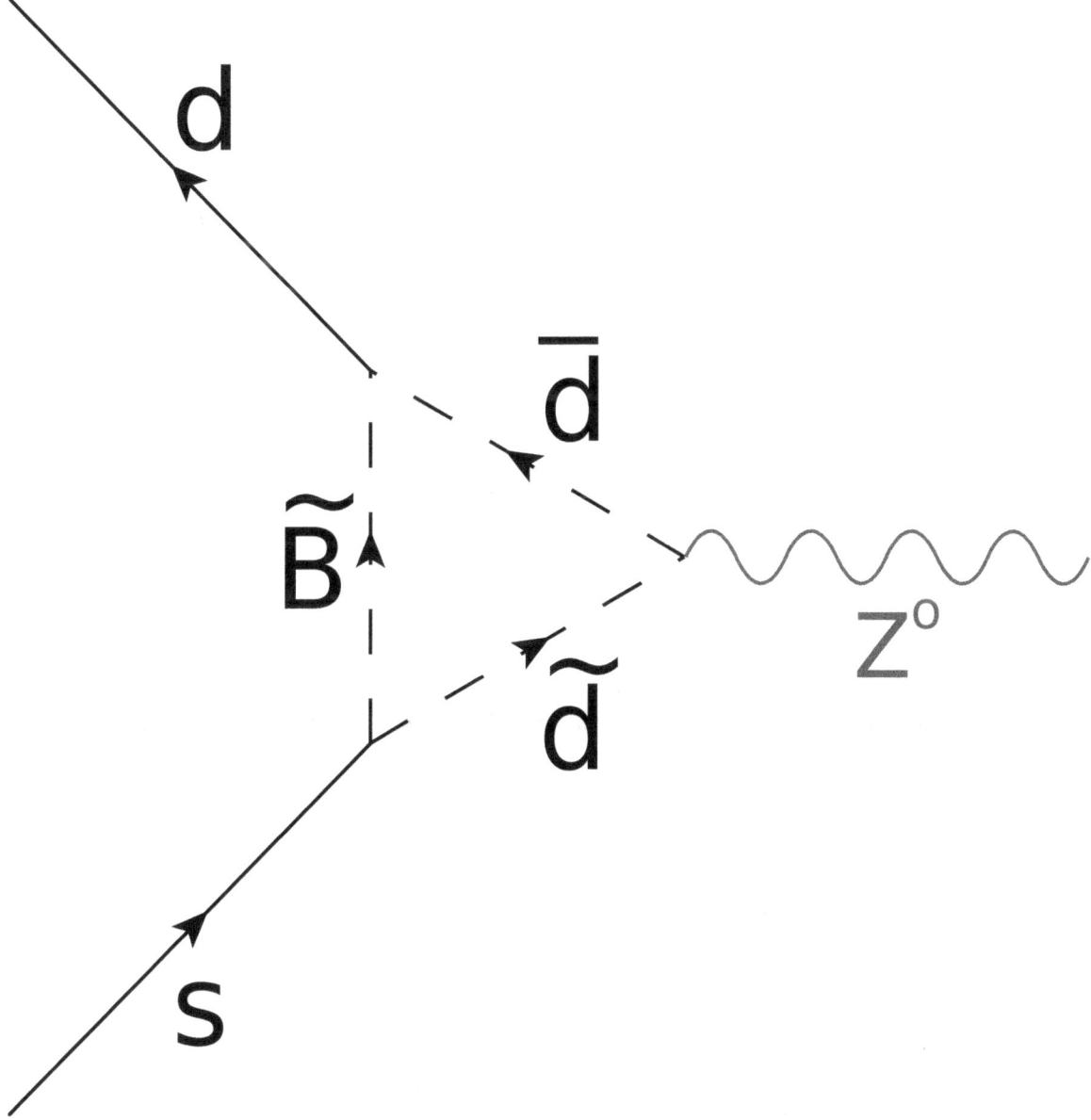

An example of a flavor changing neutral current process in MSSM. A strange quark emits a bino, turning into a sdown-type quark, which then emits a Z boson and reabsorbs the bino, turning into a down quark. If the MSSM squark masses are flavor violating, such a process can occur.

- Naturalness

- Gauge coupling unification

- Dark Matter

These motivations come out without much effort and they are the primary reasons why the MSSM is the leading candidate for a new theory to be discovered at collider experiments such as the Tevatron or the LHC.

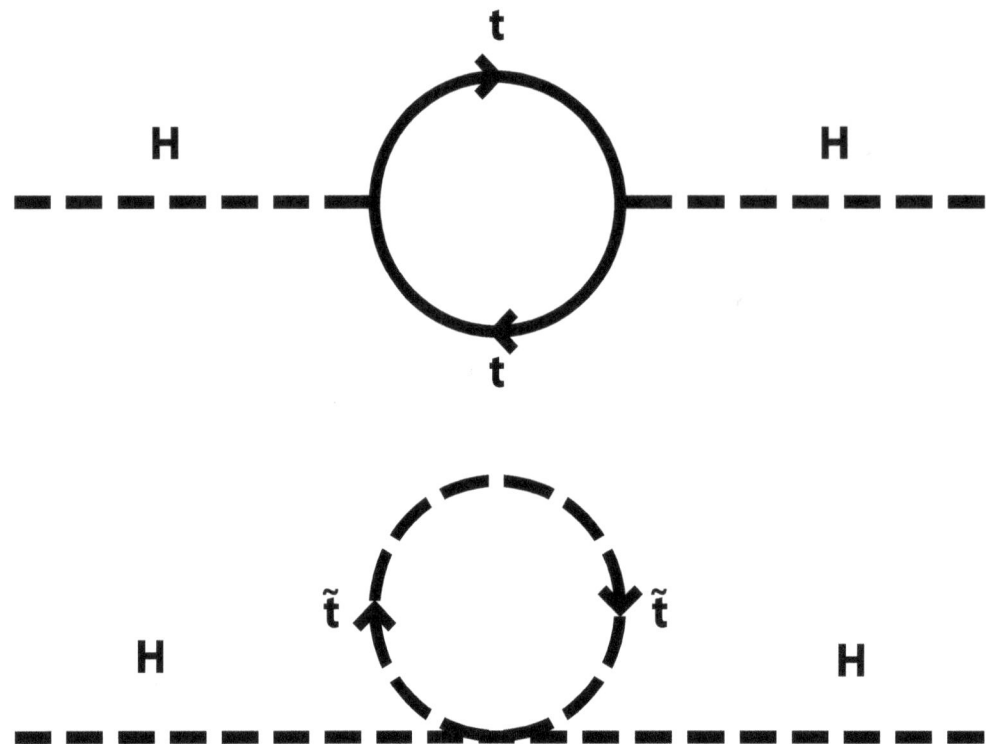

Cancellation of the Higgs boson quadratic mass renormalization between fermionic top quark loop and scalar top squark Feynman diagrams in a supersymmetric extension of the Standard Model

3.2.1 Naturalness

The original motivation for proposing the MSSM was to stabilize the Higgs mass to radiative corrections that are quadratically divergent in the Standard Model (hierarchy problem). In supersymmetric models, scalars are related to fermions and have the same mass. Since fermion masses are logarithmically divergent, scalar masses inherit the same radiative stability. The Higgs vacuum expectation value is related to the negative scalar mass in the Lagrangian. In order for the radiative corrections to the Higgs mass to not be dramatically larger than the actual value, the mass of the superpartners of the Standard Model should not be significantly heavier than the Higgs VEV—roughly 100 GeV. In 2012, the Higgs particle was discovered at the LHC, and its mass was found to be 125-126 GeV.

3.2.2 Gauge-coupling unification

If the superpartners of the Standard Model are near the TeV scale, then measured gauge couplings of the three gauge groups unify at high energies.[5] [6] [7] The beta-functions for the MSSM gauge couplings are given by

where α_1^{-1} is measured in SU(5) normalization—a factor of $\frac{3}{5}$ different than the Standard Model's normalization and predicted by Georgi–Glashow SU(5) .

The condition for gauge coupling unification at one loop is whether the following expression is satisfied $\frac{\alpha_3^{-1}-\alpha_2^{-1}}{\alpha_2^{-1}-\alpha_1^{-1}} = \frac{b_{0\,3}-b_{0\,2}}{b_{0\,2}-b_{0\,1}}$.

Remarkably, this is precisely satisfied to experimental errors in the values of $\alpha^{-1}(M_{Z^0})$. There are two loop corrections and both TeV-scale and GUT-scale threshold corrections that alter this condition on gauge coupling unification, and the results of more extensive calculations reveal that gauge coupling unification occurs to an accuracy of 1%, though this is about 3 standard deviations from the theoretical expectations.

This prediction is generally considered as indirect evidence for both the MSSM and SUSY GUTs.[8] It should be noted that gauge coupling unification does not necessarily imply grand unification and there exist other mechanisms to reproduce gauge coupling unification. However, if superpartners are found in the near future, the apparent success of gauge coupling unification would suggest that a supersymmetric grand unified theory is a promising candidate for high scale physics.

3.2.3   Dark matter

If R-parity is preserved, then the lightest superparticle (LSP) of the MSSM is stable and is a Weakly interacting massive particle (WIMP) — i.e. it does not have electromagnetic or strong interactions. This makes the LSP a good dark matter candidate and falls into the category of cold dark matter (CDM) particle.

3.3   Predictions of the MSSM regarding hadron colliders

The Tevatron and LHC have active experimental programs searching for supersymmetric particles. Since both of these machines are hadron colliders — proton antiproton for the Tevatron and proton proton for the LHC — they search best for strongly interacting particles. Therefore most experimental signature involve production of squarks or gluinos. Since the MSSM has R-parity, the lightest supersymmetric particle is stable and after the squarks and gluinos decay each decay chain will contain one LSP that will leave the detector unseen. This leads to the generic prediction that the MSSM will produce a 'missing energy' signal from these particles leaving the detector.

3.3.1   Neutralinos

There are four neutralinos that are fermions and are electrically neutral, the lightest of which is typically stable. They are typically labeled N0
1, N0
2, N0
3, N0
4 (although sometimes $\tilde{\chi}_1^0, \ldots, \tilde{\chi}_4^0$ is used instead). These four states are mixtures of the Bino and the neutral Wino (which are the neutral electroweak Gauginos), and the neutral Higgsinos. As the neutralinos are Majorana fermions, each of them is identical with its antiparticle. Because these particles only interact with the weak vector bosons, they are not directly produced at hadron colliders in copious numbers. They primarily appear as particles in cascade decays of heavier particles usually originating from colored supersymmetric particles such as squarks or gluinos.

In R-parity conserving models, the lightest neutralino is stable and all supersymmetric cascades decays end up decaying into this particle which leaves the detector unseen and its existence can only be inferred by looking for unbalanced momentum in a detector.

The heavier neutralinos typically decay through a Z0 to a lighter neutralino or through a W± to chargino. Thus a typical decay is

The mass splittings between the different Neutralinos will dictate which patterns of decays are allowed.

3.3.2   Charginos

There are two Charginos that are fermions and are electrically charged. They are typically labeled C $\tilde{\chi}\pm$
1 and C $\tilde{\chi}\pm$
2 (although sometimes $\tilde{\chi}_1^\pm$ and $\tilde{\chi}_2^\pm$ is used instead). The heavier chargino can decay through Z0 to the lighter chargino. Both can decay through a W± to neutralino.

3.3.3 Squarks

The squarks are the scalar superpartners of the quarks and there is one version for each Standard Model quark. Due to phenomenological constraints from flavor changing neutral currents, typically the lighter two generations of squarks have to be nearly the same in mass and therefore are not given distinct names. The superpartners of the top and bottom quark can be split from the lighter squarks and are called *stop* and *sbottom*.

On the other way, there may be a remarkable left-right mixing of the stops $\tilde{t}$ and of the sbottoms $\tilde{b}$ because of the high masses of the partner quarks top and bottom: [9]

- $\tilde{t}_1 = e^{+i\phi} \cos(\theta)\tilde{t}_L + \sin(\theta)\tilde{t}_R$

- $\tilde{t}_2 = e^{-i\phi} \cos(\theta)\tilde{t}_R - \sin(\theta)\tilde{t}_L$

Same holds for bottom $\tilde{b}$ with its own parameters ϕ and θ .

Squarks can be produced through strong interactions and therefore are easily produced at hadron colliders. They decay to quarks and neutralinos or charginos which further decay. In R-parity conserving scenarios, squarks are pair produced and therefore a typical signal is

- $\tilde{q}\tilde{\bar{q}} \to q\tilde{N}_1^0\bar{q}\tilde{N}_1^0 \to 2$ jets + missing energy

- $\tilde{q}\tilde{\bar{q}} \to q\tilde{N}_2^0\bar{q}\tilde{N}_1^0 \to q\tilde{N}_1^0\ell\bar{\ell}\bar{q}\tilde{N}_1^0 \to 2$ jets + 2 leptons + missing energy

3.3.4 Gluinos

Gluinos are Majorana fermionic partners of the gluon which means that they are their own antiparticles. They interact strongly and therefore can be produced significantly at the LHC. They can only decay to a quark and a squark and thus a typical gluino signal is

- $\tilde{g}\tilde{g} \to (q\tilde{\bar{q}})(\bar{q}\tilde{q}) \to (q\bar{q}\tilde{N}_1^0)(\bar{q}q\tilde{N}_1^0) \to 4$ jets + Missing energy

Because gluinos are Majorana, gluinos can decay to either a quark+anti-squark or an anti-quark+squark with equal probability. Therefore pairs of gluinos can decay to

- $\tilde{g}\tilde{g} \to (\bar{q}\tilde{q})(\bar{q}\tilde{q}) \to (q\bar{q}\tilde{C}_1^+)(q\bar{q}\tilde{C}_1^+) \to (q\bar{q}W^+)(q\bar{q}W^+) \to 4$ jets+ $\ell^+\ell^+$ + Missing energy

This is a distinctive signature because it has same-sign di-leptons and has very little background in the Standard Model.

3.3.5 Sleptons

Sleptons are the scalar partners of the leptons of the Standard Model. They are not strongly interacting and therefore are not produced very often at hadron colliders unless they are very light.

Because of the high mass of the tau lepton there will be left-right mixing of the stau similar to that of stop and sbottom (see above).

Sfermions will typically be found in decays of a charginos and neutralinos if they are light enough to be a decay product

- $\tilde{C}^+ \to \tilde{\ell}^+\nu$

- $\tilde{N}^0 \to \tilde{\ell}^+\ell^-$

3.4 MSSM fields

Fermions have bosonic superpartners (called sfermions), and bosons have fermionic superpartners (called bosinos). For most of the Standard Model particles, doubling is very straightforward. However, for the Higgs boson, it is more complicated.

A single Higgsino (the fermionic superpartner of the Higgs boson) would lead to a gauge anomaly and would cause the theory to be inconsistent. However, if two Higgsinos are added, there is no gauge anomaly. The simplest theory is one with two Higgsinos and therefore two scalar Higgs doublets. Another reason for having two scalar Higgs doublets rather than one is in order to have Yukawa couplings between the Higgs and both down-type quarks and up-type quarks; these are the terms responsible for the quarks' masses. In the Standard Model the down-type quarks couple to the Higgs field (which has Y=−1/2) and the up-type quarks to its complex conjugate (which has Y=+1/2). However in a supersymmetric theory this is not allowed, so two types of Higgs fields are needed.

3.4.1 MSSM superfields

In supersymmetric theories, every field and its superpartner can be written together as a superfield. The superfield formulation of supersymmetry is very convenient to write down manifestly supersymmetric theories (i.e. one does not have to tediously check that the theory is supersymmetric term by term in the Lagrangian). The MSSM contains vector superfields associated with the Standard Model gauge groups which contain the vector bosons and associated gauginos. It also contains chiral superfields for the Standard Model fermions and Higgs bosons (and their respective superpartners).

3.4.2 MSSM Higgs Mass

The MSSM Higgs Mass is a prediction of the Minimal Supersymmetric Standard Model. The mass of the lightest Higgs boson is set by the Higgs *quartic coupling*. Quartic couplings are not soft supersymmetry-breaking parameters since they lead to a quadratic divergence of the Higgs mass. Furthermore, there are no supersymmetric parameters to make the Higgs mass a free parameter in the MSSM (though not in non-minimal extensions). This means that Higgs mass is a prediction of the MSSM. The LEP II and the IV experiments placed a lower limit on the Higgs mass of 114.4 GeV. This lower limit is significantly above where the MSSM would typically predict it to be, and while it does not rule out the MSSM, the discovery of the Higgs with a mass of 125 GeV makes proponents of the MSSM nervous.[10][11]

Formulas

The only susy-preserving operator that creates a quartic coupling for the Higgs in the MSSM arise for the D-terms of the SU(2) and U(1) gauge sector and the magnitude of the quartic coupling is set by the size of the gauge couplings.

This leads to the prediction that the Standard Model-like Higgs mass (the scalar that couples approximately to the vev) is limited to be less than the Z mass

$$m_{h^0}^2 \leq m_{Z^0}^2 \cos^2 2\beta \ .$$

Since supersymmetry is broken, there are radiative corrections to the quartic coupling that can increase the Higgs mass. These dominantly arise from the 'top sector'

$$m_{h^0}^2 \leq m_{Z^0}^2 \cos^2 2\beta + \frac{3}{\pi^2} \frac{m_t^4 \sin^4 \beta}{v^2} \log \frac{m_{\tilde{t}}}{m_t}$$

where m_t is the top mass and $m_{\tilde{t}}$ is the mass of the top squark. This result can be interpreted as the RG running of the Higgs quartic coupling from the scale of supersymmetry to the top mass—however since the top squark mass should be relatively close to the top mass, this is usually a fairly modest contribution and increases the Higgs mass to roughly the LEP II bound of 114 GeV before the top squark becomes too heavy.

Finally there is a contribution from the top squark A-terms

$$\mathcal{L} = y_t \, m_{\tilde{t}} \, a \, h_u \tilde{q}_3 \tilde{u}_3^c$$

where a is a dimensionless number. This contributes an additional term to the Higgs mass at loop level, but is not logarithmically enhanced

$$m_{h^0}^2 \leq m_{Z^0}^2 \cos^2 2\beta + \frac{3}{\pi^2} \frac{m_t^4 \sin^4 \beta}{v^2} \left(\log \frac{m_{\tilde{t}}}{m_t} + a^2 (1 - a^2/12) \right)$$

by pushing $a \to \sqrt{6}$ (known as 'maximal mixing') it is possible to push the Higgs mass to 125 GeV without decoupling the top squark or adding new dynamics to the MSSM.

As the Higgs was found at around 125 GeV (along with no other superparticles) at the LHC, this strongly hints at new dynamics beyond the MSSM, such as the 'Next to Minimal Supersymmetric Standard Model' (NMSSM); and suggests some correlation to the little hierarchy problem.

3.5 The MSSM Lagrangian

The Lagrangian for the MSSM contains several pieces.

- The first is the Kähler potential for the matter and Higgs fields which produces the kinetic terms for the fields.

- The second piece is the gauge field superpotential that produces the kinetic terms for the gauge bosons and gauginos.

- The next term is the superpotential for the matter and Higgs fields. These produce the Yukawa couplings for the Standard Model fermions and also the mass term for the Higgsinos. After imposing R-parity, the renormalizable, gauge invariant operators in the superpotential are

$$W = \mu H_u H_d + y_u H_u Q U^c + y_d H_d Q D^c + y_l H_d L E^c$$

The constant term is unphysical in global supersymmetry (as opposed to supergravity).

3.5.1 Soft Susy breaking

Main article: Soft SUSY breaking

The last piece of the MSSM Lagrangian is the soft supersymmetry breaking Lagrangian. The vast majority of the parameters of the MSSM are in the susy breaking Lagrangian. The soft susy breaking are divided into roughly three pieces.

- The first are the gaugino masses

$$\mathcal{L} \supset m_{\frac{1}{2}} \tilde{\lambda}\tilde{\lambda} + \text{h.c.}$$

Where $\tilde{\lambda}$ are the gauginos and $m_{\frac{1}{2}}$ is different for the wino, bino and gluino.

- The next are the soft masses for the scalar fields

$$\mathcal{L} \supset m_0^2 \phi^\dagger \phi$$

where ϕ are any of the scalars in the MSSM and m_0 are 3×3 hermitean matrices for the squarks and sleptons of a given set of gauge quantum numbers. The eigenvalues of these matrices are actually the masses squared, rather than the masses.

- There are the A and B terms which are given by

$$\mathcal{L} \supset B_\mu h_u h_d + A h_u \tilde{q}\tilde{u^c} + A h_d \tilde{q}\tilde{d^c} + A h_d \tilde{l}\tilde{e^c} + \text{h.c.}$$

The A terms are 3×3 complex matrices much as the scalar masses are.

- Although not often mentioned with regard to soft terms, to be consistent with observation, one must also include Gravitino and Goldstino soft masses given by

$$\mathcal{L} \supset m_{3/2}\Psi_\mu^\alpha (\sigma^{\mu\nu})_\alpha^\beta \Psi_\beta + m_{3/2}G^\alpha G_\alpha + \text{h.c.}$$

The reason these soft terms are not often mentioned are that they arise through local supersymmetry and not global supersymmetry, although they are required otherwise if the Goldstino were massless it would contradict observation. The Goldstino mode is eaten by the Gravitino to become massive, through a gauge shift, which also absorbs the would-be "mass" term of the Goldstino.

3.6 Problems with the MSSM

There are several problems with the MSSM — most of them falling into understanding the parameters.

- The mu problem: The Higgsino mass parameter μ appears as the following term in the superpotential: $\mu H_u H_d$. It should have the same order of magnitude as the electroweak scale, many orders of magnitude smaller than that of the Planck scale, which is the natural cutoff scale. The soft supersymmetry breaking terms should also be of the same order of magnitude as the electroweak scale. This brings about a problem of naturalness: why are these scales so much smaller than the cutoff scale yet happen to fall so close to each other?

- Flavor universality of soft masses and A-terms: since no flavor mixing additional to that predicted by the standard model has been discovered so far, the coefficients of the additional terms in the MSSM Lagrangian must be, at least approximately, flavor invariant (i.e. the same for all flavors).

- Smallness of CP violating phases: since no CP violation additional to that predicted by the standard model has been discovered so far, the additional terms in the MSSM Lagrangian must be, at least approximately, CP invariant, so that their CP violating phases are small.

3.7 Theories of supersymmetry breaking

A large amount of theoretical effort has been spent trying to understand the mechanism for soft supersymmetry breaking that produces the desired properties in the superpartner masses and interactions. The three most extensively studied mechanisms are:

3.7.1 Gravity-mediated supersymmetry breaking

Gravity-mediated supersymmetry breaking is a method of communicating supersymmetry breaking to the supersymmetric Standard Model through gravitational interactions. It was the first method proposed to communicate supersymmetry breaking. In gravity-mediated supersymmetry-breaking models, there is a part of the theory that only interacts with the MSSM through gravitational interaction. This hidden sector of the theory breaks supersymmetry. Through the supersymmetric version of the Higgs mechanism, the gravitino, the supersymmetric version of the graviton, acquires a mass. After the gravitino has a mass, gravitational radiative corrections to soft masses are incompletely cancelled beneath the gravitino's mass.

It is currently believed that it is not generic to have a sector completely decoupled from the MSSM and there should be higher dimension operators that couple different sectors together with the higher dimension operators suppressed by the Planck scale. These operators give as large of a contribution to the soft supersymmetry breaking masses as the gravitational loops; therefore, today people usually consider gravity mediation to be gravitational sized direct interactions between the hidden sector and the MSSM.

mSUGRA stands for minimal supergravity. The construction of a realistic model of interactions within $N = 1$ supergravity framework where supersymmetry breaking is communicated through the supergravity interactions was carried out by Ali

Chamseddine, Richard Arnowitt, and Pran Nath in 1982.[12] mSUGRA is one of the most widely investigated models of particle physics due to its predictive power requiring only 4 input parameters and a sign, to determine the low energy phenomenology from the scale of Grand Unification. The most widely used set of parameters is:

Gravity-Mediated Supersymmetry Breaking was assumed to be flavor universal because of the universality of gravity; however, in 1986 Hall, Kostelecky, and Raby [13] showed that Planck-scale physics that are necessary to generate the Standard-Model Yukawa couplings spoil the universality of the supersymmetry breaking.

3.7.2 Gauge-mediated supersymmetry breaking (GMSB)

Gauge-mediated supersymmetry breaking is method of communicating supersymmetry breaking to the supersymmetric Standard Model through the Standard Model's gauge interactions. Typically a hidden sector breaks supersymmetry and communicates it to massive messenger fields that are charged under the Standard Model. These messenger fields induce a gaugino mass at one loop and then this is transmitted on to the scalar superpartners at two loops. Requiring stop squarks below 2 TeV, the maximum Higg's boson mass predicted is just 121.5GeV.[14] With the Higgs being discovered at 125GeV - this model requires stops above 2 TeV.

3.7.3 Anomaly-mediated supersymmetry breaking (AMSB)

Anomaly-mediated supersymmetry breaking is a special type of gravity mediated supersymmetry breaking that results in supersymmetry breaking being communicated to the supersymmetric Standard Model through the conformal anomaly.[15][16] Requiring stop squarks below 2 TeV, the maximum Higg's boson mass predicted is just 121.0GeV.[14] With the Higgs being discovered at 125GeV - this scenario requires stops heavier than 2 TeV.

3.8 Phenomenological MSSM (pMSSM)

The unconstrained MSSM has more than 100 parameters in addition to the Standard Model parameters. This makes any phenomenological analysis (e.g. finding regions in parameter space consistent with observed data) impractical. Under the following three assumptions:

- no new source of CP-violation
- no Flavour Changing Neutral Currents
- first and second generation universality

one can reduce the number of additional parameters to the following 19 quantities of the phenomenological MSSM (pMSSM):[17] The large parameter space of pMSSM makes searches in pMSSM extremely challenging and makes pMSSM difficult to exclude.

3.9 See also

- MSSM Higgs Mass
- Desert (particle physics)

3.10 References

[1] Howard Baer; Xerxes Tata (2006). "8 - The Minimal Supersymmetric Standard Model". *Weak Scale Supersymmetry From Superfields to Scattering Events*. Cambridge: Cambridge University Press. p. 127. ISBN 9780511617270. It is minimal in the sense that it contains the smallest number of new particle states and new interactions consistent with phenomenology.

[2] Murayama, Hitoshi (2000). "Supersymmetry phenomenology". arXiv:hep-ph/0002232.

[3] Wolchover, Natalie (November 29, 2012). "Supersymmetry Fails Test, Forcing Physics to Seek New Ideas". *Scientific American*.

[4] S. Dimopoulos, H. Georgi; Georgi (1981). "Softly Broken Supersymmetry and SU(5)".*Nuclear Physics B***193**: 1hB.193..150D. doi:10.1016/0550-3213(81)90522-8.

[5] S. Dimopoulos, S. Raby and F. Wilczek; Raby; Wilczek (1981). "Supersymmetry and the Scale of Unification". *Physical Review D* **24** (6): 1681–1683. Bibcode:1981PhRvD..24.1681D. doi:10.1103/PhysRevD.24.1681.

[6] L.E. Ibanez and G.G. Ross; Ross (1981). "Low-energy predictions in supersymmetric grand unified theories". *Physics Letters B* **105** (6): 439. Bibcode:1981PhLB..105..439I. doi:10.1016/0370-2693(81)91200-4.

[7] W.J. Marciano and G. Senjanovic; Senjanović (1982). "Predictions of supersymmetric grand unified theories". *Physical Review D* **25** (11): 3092. Bibcode:1982PhRvD..25.3092M. doi:10.1103/PhysRevD.25.3092.

[8] Gordon Kane, "The Dawn of Physics Beyond the Standard Model", *Scientific American*, June 2003, page 60 and *The frontiers of physics*, special edition, Vol 15, #3, page 8 "Indirect evidence for supersymmetry comes from the extrapolation of interactions to high energies."

[9] Bartl, A.; Hesselbach, S.; Hidaka, K.; Kernreiter, T.; Porod, W. (2003). "Impact of SUSY CP Phases on Stop and Sbottom Decays in the MSSM". arXiv:hep-ph/0306281 [hep-ph].

[10] Heinemeyer, S.; Stål, O.; Weiglein, G. (2012). "Interpreting the LHC Higgs search results in the MSSM". *Physics Letters B* **710**: 201. arXiv:1112.3026v3. Bibcode:2012PhLB..710..201H. doi:10.1016/j.physletb.2012.02.084.

[11] Carena, M.; Heinemeyer, S.; Wagner, C. E. M.; Weiglein, G. (2006). "MSSM Higgs boson searches at the evatron and the LHC: Impact of different benchmark scenarios" (PDF). *The European Physical Journal C* **45** (3): 797. arXiv:hep-ph/0511023. Bibcode:2006EPJC...45..797C. doi:10.1140/epjc/s2005-02470-y.

[12] A. Chamseddine, R. Arnowitt, P. Nath; Arnowitt; Nath (1982). "Locally Supersymmetric Grand Unification". *Physical Review Letters* **49** (14): 970–974. Bibcode:1982PhRvL..49..970C. doi:10.1103/PhysRevLett.49.970.

[13] L.J. Hall, V.A. Kostelecky, S. Raby; Kostelecky; Raby (1986). "New Flavor Violations in Supergravity Models". *Nuclear Physics B* **267** (2): 415. Bibcode:1986NuPhB.267..415H. doi:10.1016/0550-3213(86)90397-4.

[14] Arbey, A.; Battaglia, M.; Djouadi, A.; Mahmoudi, F.; Quevillon, J. (2011). "Implications of a 125 GeV Higgs for supersymmetric models". *Physics Letters B*. 3 **708** (2012): 162–169. arXiv:1112.3028. Bibcode:2012PhLB..708..162A. doi:

[15] L. Randall, R. Sundrum; Sundrum (1999). "Out of this world supersymmetry breaking". *Nuclear Physics B* **557**: 79–118. arXiv:hep-th/9810155. Bibcode:1999NuPhB.557...79R. doi:10.1016/S0550-3213(99)00359-4.

[16] G. Giudice, M. Luty, H. Murayama, R. Rattazzi; Rattazzi; Luty; Murayama (1998). "Gaugino mass without singlets". *Journal of High Energy Physics* **9812** (12): 027. arXiv:hep-ph/9810442.Bibcode:1998JHEP...12..027G.doi:10.1088/1126-6708/1998/1.

[17] Djouadi, A.; Rosier-Lees, S.; Bezouh, M.; Bizouard, M. A.; Boehm, C.; Borzumati, F.; Briot, C.; Carr, J.; Causse, M. B.; Charles, F.; Chereau, X.; Colas, P.; Duflot, L.; Dupperin, A.; Ealet, A.; El-Mamouni, H.; Ghodbane, N.; Gieres, F.; Gonzalez-Pineiro, B.; Gourmelen, S.; Grenier, G.; Gris, Ph.; Grivaz, J. -F.; Hebrard, C.; Ille, B.; Kneur, J. -L.; Kostantinidis, N.; Layssac, J.; Lebrun, P. et al. (1999). "The Minimal Supersymmetric Standard Model: Group Summary Report". arXiv:hep-ph/9901246.

3.11   External links

- MSSM on arxiv.org

- Stephen P. Martin (1997). "A Supersymmetry Primer". arXiv:hep-ph/9709356.

- Particle Data Group review of MSSM and search for MSSM predicted particles

- Ian J R Aitchison (2005). "Supersymmetry and the MSSM: An Elementary Introduction". arXiv:hep-ph/0505105.

Chapter 4

Physics beyond the Standard Model

Physics beyond the Standard Model (**BSM**) refers to the theoretical developments needed to explain the deficiencies of the Standard Model, such as the origin of mass, the strong CP problem, neutrino oscillations, matter–antimatter asymmetry, and the nature of dark matter and dark energy.[1] Another problem lies within the mathematical framework of the Standard Model itself – the Standard Model is inconsistent with that of general relativity, to the point that one or both theories break down under certain conditions (for example within known space-time singularities like the Big Bang and black hole event horizons).

Theories that lie beyond the Standard Model include various extensions of the standard model through supersymmetry, such as the Minimal Supersymmetric Standard Model (MSSM) and Next-to-Minimal Supersymmetric Standard Model (NMSSM), or entirely novel explanations, such as string theory, M-theory and extra dimensions. As these theories tend to reproduce the entirety of current phenomena, the question of which theory is the right one, or at least the "best step" towards a Theory of Everything, can only be settled via experiments, and is one of the most active areas of research in both theoretical and experimental physics.

4.1 Problems with the Standard Model

Despite being the most successful theory of particle physics to date, the Standard Model is not perfect.[2] A large share of the published output of theoretical physicists consists of proposals for various forms of "Beyond the Standard Model" new physics proposals that would modify the Standard Model in ways subtle enough to be consistent with existing data, yet address its imperfections materially enough to predict non-Standard Model outcomes of new experiments that can be proposed.

4.1.1 Phenomena not explained

The Standard Model is inherently an incomplete theory. There are fundamental physical phenomena in nature that the Standard Model does not adequately explain:

- *Gravity.* The standard model does not explain gravity. The approach of simply adding a "graviton" (whose properties are the subject of considerable consensus among physicists if it exists) to the Standard Model does not recreate what is observed experimentally without other modifications, as yet undiscovered, to the Standard Model. Moreover, instead, the Standard Model is widely considered to be incompatible with the most successful theory of gravity to date, general relativity.[3]

- *Dark matter and dark energy.* Cosmological observations tell us the standard model explains about 5% of the energy present in the universe. About 26% should be dark matter, which would behave just like other matter, but which only interacts weakly (if at all) with the Standard Model fields. Yet, the Standard Model does not supply

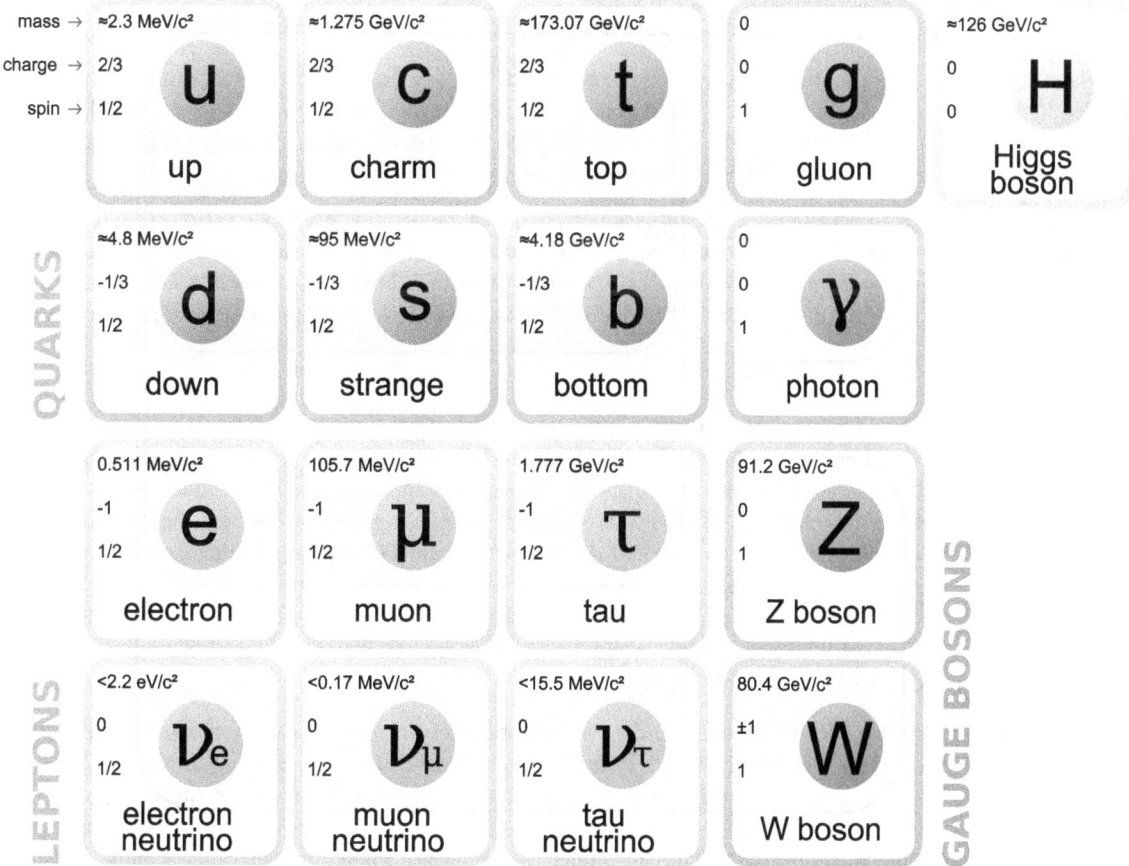

The Standard Model of elementary particles

any fundamental particles that are good dark matter candidates. The rest (69%) should be dark energy, a constant energy density for the vacuum. Attempts to explain dark energy in terms of vacuum energy of the standard model lead to a mismatch of 120 orders of magnitude.[4]

- *Neutrino masses.* According to the standard model, neutrinos are massless particles. However, neutrino oscillation experiments have shown that neutrinos do have mass. Mass terms for the neutrinos can be added to the standard model by hand, but these lead to new theoretical problems. For example, the mass terms need to be extraordinarily small and it is not clear if the neutrino masses would arise in the same way that the masses of other fundamental particles do in the Standard Model.

- *Matter-antimatter asymmetry.* The universe is made out of mostly matter. However, the standard model predicts that matter and antimatter should have been created in (almost) equal amounts if the initial conditions of the universe did not involve disproportionate matter relative to antimatter. Yet, no mechanism sufficient to explain this asymmetry exists in the Standard Model.

4.1.2 Experimental results not explained

No experimental result is widely accepted as contradicting the Standard Model at a level that definitively contradicts it at the "five sigma" (i.e. five standard deviation) level widely considered to be the threshold of a "discovery" in particle physics. But, because every experiment contains some degree of statistical and systemic uncertainty, and the theoretical predictions themselves are also almost never calculated exactly and are subject to uncertainties in measurements of the fundamental constants of the Standard Model (some of which are tiny and others of which are substantial,) it is mathematically expected that some of the hundreds of experimental tests of the Standard Model will deviate to some extent from the Standard Model even if there were no "new physics" beyond the Standard Model to be discovered.

At any given time there are a number of experimental results that are significantly different from the Standard Model expectation, although many of these have been found to be statistical flukes or experimental errors as more data has been collected. On the other hand, any "beyond the Standard Model" physics would necessarily first manifest experimentally as a statistically significant difference between an experiment and a Standard Model theoretical prediction.

In each case, physicists seek to determine if a result is a mere statistical fluke or experimental error on the one hand, or a sign of new physics on the other. More statistically significant results cannot be mere statistical flukes but can still result from experimental error or inaccurate estimates of experimental precision. Frequently, experiments are tailored to be more sensitive to experimental results that would distinguish the Standard Model from theoretical alternatives.

Some of the most notable examples include the following:

- *Muonic hydrogen* – the Standard Model makes precise theoretical predictions regarding the atomic radius size of ordinary hydrogen (a proton-electron system) and that of muonic hydrogen (a proton-muon system in which a muon is a "heavy" variant of an electron). However, the measured atomic radius of muonic hydrogen differs significantly from that of the radius predicted by the Standard Model using existing physical constant measurements by what appears to be as many as seven standard deviations.[5] Doubts about the accuracy of the error estimates in earlier experiments, which are still within 4% of each other in measuring a truly tiny distance, and a lack of a well motivated theory that could explain the discrepancy, have caused physicists to be hesitant to describe these results as contradicting the Standard Model despite the apparent statistical significance of the result and a lack of any clearly identified possible source of experimental error in the results.

- *BaBar Data Suggests Possible Flaws in the Standard Model* – results from a BaBar experiment may suggest a surplus over Standard Model predictions of a type of particle decay called "B to D-star-tau-nu." In this, an electron and positron collide, resulting in a B meson and an antimatter B-bar meson, which then decays into a D meson and a tau lepton as well as a smaller antineutrino. While the level of certainty of the excess (3.4 sigma in statistical language) is not enough to claim a break from the Standard Model, the results are a potential sign of something amiss and are likely to affect existing theories, including those attempting to deduce the properties of Higgs bosons.[6] However, results at LHCb have demonstrated no significant deviation from the Standard Model prediction of very nearly zero asymmetry.[7][8]

- Proton radius - radius measured using electrons is different from radius measured using muons[9]

4.1.3 Theoretical predictions not observed

Observation at particle colliders of all of the fundamental particles predicted by the Standard Model has been confirmed. The Higgs boson is predicted by the Standard Model's explanation of the Higgs mechanism, which describes how the weak SU(2) gauge symmetry is broken and how fundamental particles obtain mass; it was the last particle predicted by the Standard Model to be observed. On July 4, 2012, CERN scientists using the Large Hadron Collider announced the discovery of a particle consistent with the Higgs boson, with a mass of about 126 GeV/c^2. A Higgs boson was confirmed to exist on March 14, 2013, although efforts to confirm that it has all of the properties predicted by the Standard Model are ongoing.[10]

A few hadrons (i.e. composite particles made of quarks) whose existence is predicted by the Standard Model, which can be produced only at very high energies in very low frequencies have not yet been definitively observed, and "glueballs"[11] (i.e. composite particles made of gluons) have also not yet been definitively observed. Some very low frequency particle decays predicted by the Standard Model have also not yet been definitively observed because insufficient data is available to make a statistically significant observation.

4.1.4 Theoretical problems

Some features of the standard model are added in an ad hoc way. These are not problems per se (i.e. the theory works fine with these ad hoc features), but they imply a lack of understanding. These ad hoc features have motivated theorists

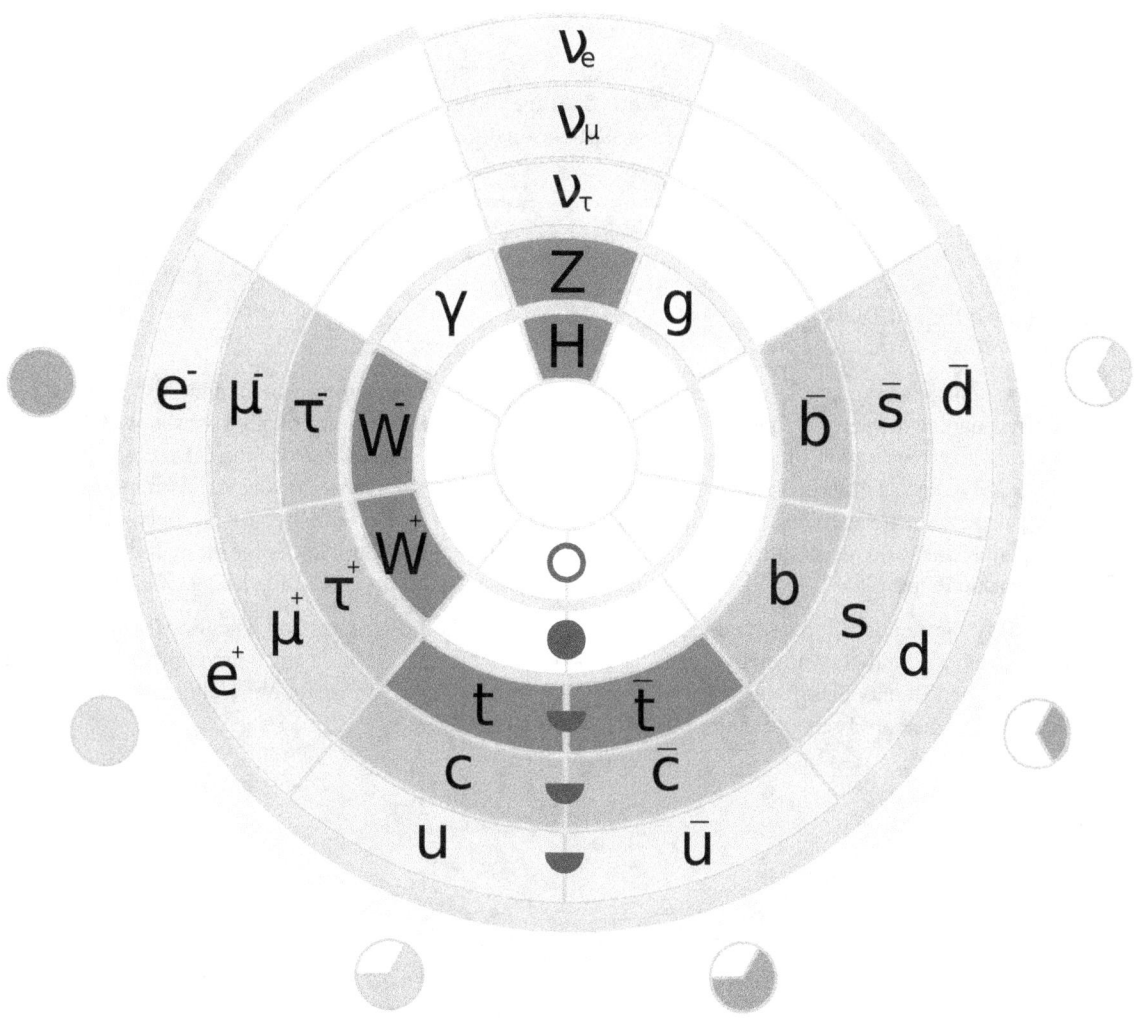

Masses of fundamental particles ----
more than 80 GeV/c²
1-5 GeV/c²
90-110 MeV/c²
less than 16 MeV/c²
Massless

to look for more fundamental theories with fewer parameters. Some of the ad hoc features are:

- *Hierarchy problem* – the standard model introduces particle masses through a process known as spontaneous symmetry breaking caused by the Higgs field. Within the standard model, the mass of the Higgs gets some very large quantum corrections due to the presence of virtual particles (mostly virtual top quarks). These corrections are much larger than the actual mass of the Higgs. This means that the bare mass parameter of the Higgs in the standard model must be fine tuned in such a way that almost completely cancels the quantum corrections. This level of fine-tuning is deemed unnatural by many theorists.There are also issues of Quantum triviality, which suggests that it may not be possible to create a consistent quantum field theory involving elementary scalar particles.

- *Strong CP problem* – theoretically it can be argued that the standard model should contain a term that breaks CP symmetry —relating matter to antimatter— in the strong interaction sector. Experimentally, however, no such violation has been found, implying that the coefficient of this term is very close to zero. This fine tuning is also considered unnatural.

- *Number of parameters* – the standard model depends on 19 numerical parameters. Their values are known from experiment, but the origin of the values is unknown. Some theorists have tried to find relations between different parameters, for example, between the masses of particles in different generations.

4.2 Grand unified theories

Main article: Grand Unified Theory

The standard model has three gauge symmetries; the colour SU(3), the weak isospin SU(2), and the hypercharge U(1) symmetry, corresponding to the three fundamental forces. Due to renormalization the coupling constants of each of these symmetries vary with the energy at which they are measured. Around 10^{16} GeV these couplings become approximately equal. This has led to speculation that above this energy the three gauge symmetries of the standard model are unified in one single gauge symmetry with a simple group gauge group, and just one coupling constant. Below this energy the symmetry is spontaneously broken to the standard model symmetries.[12] Popular choices for the unifying group are the special unitary group in five dimensions SU(5) and the special orthogonal group in ten dimensions SO(10).[13]

Theories that unify the standard model symmetries in this way are called Grand Unified Theories (or GUTs), and the energy scale at which the unified symmetry is broken is called the GUT scale. Generically, grand unified theories predict the creation of magnetic monopoles in the early universe,[14] and instability of the proton.[15] Neither of which have been observed, and this absence of observation puts limits on the possible GUTs.

4.3 Supersymmetry

Main article: Supersymmetry

Supersymmetry extends the Standard Model by adding another class of symmetries to the Lagrangian. These symmetries exchange fermionic particles with bosonic ones. Such a symmetry predicts the existence of *supersymmetric particles*, abbreviated as *sparticles*, which include the sleptons, squarks, neutralinos and charginos. Each particle in the Standard Model would have a superpartner whose spin differs by 1/2 from the ordinary particle. Due to the breaking of supersymmetry, the sparticles are much heavier than their ordinary counterparts; they are so heavy that existing particle colliders may not be powerful enough to produce them.

4.4 Neutrinos

In the standard model, neutrinos have exactly zero mass. This is a consequence of the standard model containing only left-handed neutrinos. With no suitable right-handed partner, it is impossible to add a renormalizable mass term to the standard model.[16] Measurements however indicated that neutrinos spontaneously change flavour, which implies that neutrinos have a mass. These measurements only give the relative masses of the different flavours. The best constraint on the absolute mass of the neutrinos comes from precision measurements of tritium decay, providing an upper limit 2 eV, which makes them at least five orders of magnitude lighter than the other particles in the standard model.[17] This necessitates an extension of the standard model, which not only needs to explain how neutrinos get their mass, but also why the mass is so small.[18]

One approach to add masses to the neutrinos, the so-called seesaw mechanism, is to add right-handed neutrinos and have these couple to left-handed neutrinos with a Dirac mass term. The right-handed neutrinos have to be sterile, meaning that they do not participate in any of the standard model interactions. Because they have no charges, the right-handed neutrinos can act as their own anti-particles, and have a Majorana mass term. Like the other Dirac masses in the standard model, the neutrino Dirac mass is expected to be generated through the Higgs mechanism, and is therefore unpredictable. The standard model fermion masses differ by many orders of magnitude; the Dirac neutrino mass has at least the same uncertainty. On the other hand, the Majorana mass for the right-handed neutrinos does not arise from the Higgs mechanism, and is therefore expected to be tied to some energy scale of new physics beyond the standard model, for example the Planck scale.[19] Therefore, any process involving right-handed neutrinos will be suppressed at low energies. The correction due to these suppressed processes effectively gives the left-handed neutrinos a mass that is inversely proportional to the right-handed Majorana mass, a mechanism known as the see-saw.[20] The presence of heavy right-handed neutrinos thereby explains both the small mass of the left-handed neutrinos and the absence of the right-handed neutrinos in observations. However, due to the uncertainty in the Dirac neutrino masses, the right-handed neutrino masses can lie anywhere. For example, they could be as light as keV and be dark matter,[21] they can have a mass in the LHC energy range[22][23] and lead to observable lepton number violation,[24] or they can be near the GUT scale, linking the right-handed neutrinos to the possibility of a grand unified theory.[25][26]

The mass terms mix neutrinos of different generations. This mixing is parameterized by the PMNS matrix, which is the neutrino analogue of the CKM quark mixing matrix. Unlike the quark mixing, which is almost minimal, the mixing of the neutrinos appears to be almost maximal. This has led to various speculations of symmetries between the various generations that could explain the mixing patterns.[27] The mixing matrix could also contain several complex phases that break CP invariance, although there has been no experimental probe of these. These phases could potentially create a surplus of leptons over anti-leptons in the early universe, a process known as leptogenesis. This asymmetry could then at a later stage be converted in an excess of baryons over anti-baryons, and explain the matter-antimatter asymmetry in the universe.[13]

The light neutrinos are disfavored as an explanation for the observation of dark matter, due to considerations of large-scale structure formation in the early universe. Simulations of structure formation show that they are too hot—i.e. their kinetic energy is large compared to their mass—while formation of structures similar to the galaxies in our universe requires cold dark matter. The simulations show that neutrinos can at best explain a few percent of the missing dark matter. The heavy sterile right-handed neutrinos are however a possible candidate for a dark matter WIMP.[28]

4.5   Preon Models

Several preon models have been proposed to address the unsolved problem concerning the fact that there are three generations of quarks and leptons. Preon models generally postulate some additional new particles which are further postulated to be able to combine to form the quarks and leptons of the standard model. One of the earliest preon models was the Rishon model.[29][30][31]

To date, no preon model is widely accepted or fully verified.

4.6   Theories of everything

4.6.1   Theory of everything

Main article: Theory of everything

Theoretical physics continues to strive toward a theory of everything, a theory that fully explains and links together all known physical phenomena, and predicts the outcome of any experiment that could be carried out in principle. In practical terms the immediate goal in this regard is to develop a theory which would unify the Standard Model with General Relativity in a theory of quantum gravity. Additional features, such as overcoming conceptual flaws in either theory or accurate prediction of particle masses, would be desired. The challenges in putting together such a theory are not just conceptual - they include the experimental aspects of the very high energies needed to probe exotic realms.

Several notable attempts in this direction are supersymmetry, string theory, and loop quantum gravity.

4.6.2 String theory

Main article: String theory

Extensions, revisions, replacements, and reorganizations of the Standard Model exist in attempt to correct for these and other issues. String theory is one such reinvention, and many theoretical physicists think that such theories are the next theoretical step toward a true Theory of Everything. Theories of quantum gravity such as loop quantum gravity and others are thought by some to be promising candidates to the mathematical unification of quantum field theory and general relativity, requiring less drastic changes to existing theories.[32] However recent work places stringent limits on the putative effects of quantum gravity on the speed of light, and disfavours some current models of quantum gravity.[33]

Among the numerous variants of string theory, M-theory, whose mathematical existence was first proposed at a String Conference in 1995, is believed by many to be a proper "ToE" candidate, notably by physicists Brian Greene and Stephen Hawking. Though a full mathematical description is not yet known, solutions to the theory exist for specific cases.[34] Recent works have also proposed alternate string models, some of which lack the various harder-to-test features of M-theory (e.g. the existence of Calabi–Yau manifolds, many extra dimensions, etc.) including works by well-published physicists such as Lisa Randall.[35][36]

4.7 See also

- *A New Kind of Science*
- Antimatter tests of Lorentz violation
- Fundamental physical constants in the standard model
- Higgsless model
- Holographic principle
- Little Higgs
- Lorentz-violating neutrino oscillations
- Minimal Supersymmetric Standard Model
- Peccei–Quinn theory
- Preon
- Standard-Model Extension
- Supergravity
- Seesaw mechanism
- Supersymmetry
- Superfluid vacuum theory
- String theory
- Technicolor (physics)
- Theory of everything
- Unsolved problems in physics
- Unparticle physics

4.8 References

[1] Womersley, J. (February 2005). "Beyond the Standard Model" (PDF). *Symmetry Magazine*. Retrieved 2010-11-23.

[2] Lykken, J. D. (2010). "Beyond the Standard Model". *CERN Yellow Report*. CERN. pp. 101–109. arXiv:1005.1676. CERN-2010-002.

[3] Sushkov, A. O.; Kim, W. J.; Dalvit, D. A. R.; Lamoreaux, S. K. (2011). "New Experimental Limits on Non-Newtonian Forces in the Micrometer Range". *Physical Review Letters* **107** (17): 171101. arXiv:1108.2547. Bibcode:2011PhRvL.107q1101S. doi:10.1103/PhysRevLett.107.171101. It is remarkable that two of the greatest successes of 20th century physics, general relativity and the standard model, appear to be fundamentally incompatible. But see also Donoghue, John F. (2012). "The effective field theory treatment of quantum gravity". *AIP Conference Proceedings* **1473**: 73. arXiv:1209.3511. doi:10.1063/1.4756964. One can find thousands of statements in the literature to the effect that "general relativity and quantum mechanics are incompatible". These are completely outdated and no longer relevant. Effective field theory shows that general relativity and quantum mechanics work together perfectly normally over a range of scales and curvatures, including those relevant for the world that we see around us. However, effective field theories are only valid over some range of scales. General relativity certainly does have problematic issues at extreme scales. There are important problems which the effective field theory does not solve because they are beyond its range of validity. However, this means that the issue of quantum gravity is not what we thought it to be. Rather than a fundamental incompatibility of quantum mechanics and gravity, we are in the more familiar situation of needing a more complete theory beyond the range of their combined applicability. The usual marriage of general relativity and quantum mechanics is fine at ordinary energies, but we now seek to uncover the modifications that must be present in more extreme conditions. This is the modern view of the problem of quantum gravity, and it represents progress over the outdated view of the past."

[4] Krauss, L. (2009). *A Universe from Nothing*. AAI Conference.

[5] Randolf Pohl, Ronald Gilman, Gerald A. Miller, Krzysztof Pachucki, "Muonic hydrogen and the proton radius puzzle" (May 30, 2013) http://arxiv.org/abs/1301.0905 in print Annu. Rev. Nucl. Part. Sci. Vol 63 (2013) 10.1146/annurev-nucl-102212-170627 ("The recent determination of the proton radius using the measurement of the Lamb shift in the muonic hydrogen atom startled the physics world. The obtained value of 0.84087(39) fm differs by about 4% or 7 standard deviations from the CODATA value of 0.8775(51) fm. The latter is composed from the electronic hydrogenate atom value of 0.8758(77) fm and from a similar value with larger uncertainties determined by electron scattering.")

[6] Lees, J. P.; et al. (BaBar Collaboration) (1970). "Evidence for an excess of B → D$^{(*)}$τ$^-$τν decays". *Physical Review Letters* **109** (10). arXiv:1205.5442. Bibcode:2012PhRvL.109j1802L. doi:10.1103/PhysRevLett.109.101802.

[7] Article on LHCb results

[8] 2012 LHCb paper

[9] http://arxiv.org/pdf/1502.05314.pdf

[10] O'Luanaigh, C. (14 March 2013). "New results indicate that new particle is a Higgs boson". CERN.

[11] Marco Frasca, "What is a Glueball?" (March 31, 2009) http://marcofrasca.wordpress.com/2009/03/31/what-is-a-glueball-2/

[12] Peskin, M. E.; Schroeder, D. V. (1995). *An introduction to quantum field theory*. Addison-Wesley. pp. 786–791. ISBN 978-0-201-50397-5.

[13] Buchmüller, W. (2002). "Neutrinos, Grand Unification and Leptogenesis". arXiv:hep-ph/0204288 [hep-ph].

[14] Milstead, D.; Weinberg, E.J. (2009). "Magnetic Monopoles" (PDF). Particle Data Group. Retrieved 2010-12-20.

[15] P., Nath; P. F., Perez (2006). "Proton stability in grand unified theories, in strings, and in branes". *Physics Reports* **441** (5–6): 191–317. arXiv:hep-ph/0601023. Bibcode:2007PhR...441..191N. doi:10.1016/j.physrep.2007.02.010.

[16] Peskin, M. E.; Schroeder, D. V. (1995). *An introduction to quantum field theory*. Addison-Wesley. pp. 713–715. ISBN 978-0-201-50397-5.

[17] Nakamura, K.; et al. (Particle Data Group) (2010). "Neutrino Properties". Particle Data Group. Retrieved 2010-12-20.

[18] Mohapatra, R. N.; Pal, P. B. (2007). *Massive neutrinos in physics and astrophysics*. Lecture Notes in Physics **72** (3rd ed.). World Scientific. ISBN 978-981-238-071-5.

[19] Senjanovic, G. (2011). "Probing the Origin of Neutrino Mass: from GUT to LHC". arXiv:1107.5322 [hep-ph].

[20] Grossman, Y. (2003). "TASI 2002 lectures on neutrinos". arXiv:hep-ph/0305245v1 [hep-ph].

[21] Dodelson, S.; Widrow, L. M. (1993). "Sterile neutrinos as dark matter". *Physical Review Letters* **72**: 17. arXiv:hep-ph/9303287. Bibcode:1994PhRvL..72...17D. doi:10.1103/PhysRevLett.72.17.

[22] Minkowski, P. (1977). "$\mu \to e\, \gamma$ at a Rate of One Out of 10₉Muon Decays?".*Physics Letters B***67**(4): 421.Bibcode:1977PhLB...67 ..42 IdMi:10.1016/0370-2693(77)90435-X.

[23] Mohapatra, R. N.; Senjanovic, G. (1980). "Neutrino mass and spontaneous parity nonconservation". *Physical Review Letters* **44** (14): 912. Bibcode:1980PhRvL..44..912M. doi:10.1103/PhysRevLett.44.912.

[24] Keung, W.-Y.; Senjanovic, G. (1983). "Majorana Neutrinos And The Production Of The Right-handed Charged Gauge Boson". *Physical Review Letters* **50** (19): 1427. Bibcode:1983PhRvL..50.1427K. doi:10.1103/PhysRevLett.50.1427.

[25] Gell-Mann, M.; Ramond, P.; Slansky, R. (1979). P. van Nieuwenhuizen; D. Freedman, eds. *Supergravity*. North Holland.

[26] Glashow, S. L. (1979). M. Levy, ed. *Proceedings of the 1979 Cargèse Summer Institute on Quarks and Leptons*. Plenum Press.

[27] Altarelli, G. (2007). "Lectures on Models of Neutrino Masses and Mixings". arXiv:0711.0161 [hep-ph].

[28] Murayama, H. (2007). "Physics Beyond the Standard Model and Dark Matter". arXiv:0704.2276 [hep-ph].

[29] Harari, H. (1979). "A Schematic Model of Quarks and Leptons". Physics Letters B 86 (1): 83-86.

[30] Shupe, M. A. (1979). "A Composite Model of Leptons and Quarks". Physics Letters B 86 (1): 87-92.

[31] Zenczykowski, P. (2008). "The Harari-Shupe preon model and nonrelativistic quantum phase space". Physics Letters B 660 (5): 567-572.

[32] Smolin, L. (2001). *Three Roads to Quantum Gravity*. Basic Books. ISBN 0-465-07835-4.

[33] Abdo, A. A.; et al. (Fermi GBM/LAT Collaborations) (2009). "A limit on the variation of the speed of light arising from quantum gravity effects". *Nature* **462** (7271): 331–4. arXiv:0908.1832. Bibcode:2009Natur.462..331A. doi:10.1038/nature08574. PMID 19865083.

[34] Maldacena, J.; Strominger, A.; Witten, E. (1997). "Black hole entropy in M-Theory". *Journal of High Energy Physics* **1997** (12): 2. arXiv:hep-th/9711053. Bibcode:1997JHEP...12..002M. doi:10.1088/1126-6708/1997/12/002.

[35] Randall, L.; Sundrum, R. (1999). "Large Mass Hierarchy from a Small Extra Dimension". *Physical Review Letters* **83** (17): 3370. arXiv:hep-ph/9905221. Bibcode:1999PhRvL..83.3370R. doi:10.1103/PhysRevLett.83.3370.

[36] Randall, L.; Sundrum, R. (1999). "An Alternative to Compactification". *Physical Review Letters* **83** (23): 4690. arXiv:hep-th/9906064. Bibcode:1999PhRvL..83.4690R. doi:10.1103/PhysRevLett.83.4690.

4.9   Further reading

- Lisa Randall (2005). *Warped Passages: Unraveling the Mysteries of the Universe's Hidden Dimensions*. HarperCollins. ISBN 0-06-053108-8.

4.10   External resources

- Standard Model Theory @ SLAC
- Scientific American Apr 2006
- LHC. Nature July 2007
- Open Questions
- Working group - schedule
- Les Houches Conference, Summer 2005

Chapter 5

Hierarchy problem

In theoretical physics, the **hierarchy problem** is the large discrepancy between aspects of the weak force and gravity.[1] There is no scientific consensus on, for example, why the weak force is 10^{32} times stronger than gravity.

5.1 Technical definition

A hierarchy problem occurs when the fundamental value of some physical parameter, such as a coupling constant or a mass, in some Lagrangian is vastly different from its effective value, which is the value that gets measured in an experiment. This happens because the effective value is related to the fundamental value by a prescription known as renormalization, which applies quantum corrections to it. Typically the renormalized value of parameters are close to their fundamental values, but in some cases, it appears that there has been a delicate cancellation between the fundamental quantity and the quantum corrections. Hierarchy problems are related to fine-tuning problems and problems of naturalness.

Studying renormalization in hierarchy problems is difficult, because such quantum corrections are usually power-law divergent, which means that the shortest-distance physics are most important. Because we do not know the precise details of the shortest-distance theory of physics, we cannot even address how this delicate cancellation between two large terms occurs. Therefore, researchers are led to postulate new physical phenomena that resolve hierarchy problems without fine tuning.

5.2 The Higgs mass

In particle physics, the most important **hierarchy problem** is the question that asks why the weak force is 10^{32} times stronger than gravity. Both of these forces involve constants of nature, Fermi's constant for the weak force and Newton's constant for gravity. Furthermore if the Standard Model is used to calculate the quantum corrections to Fermi's constant, it appears that Fermi's constant is surprisingly large and is expected to be closer to Newton's constant, unless there is a delicate cancellation between the bare value of Fermi's constant and the quantum corrections to it.

More technically, the question is why the Higgs boson is so much lighter than the Planck mass (or the grand unification energy, or a heavy neutrino mass scale): one would expect that the large quantum contributions to the square of the Higgs boson mass would inevitably make the mass huge, comparable to the scale at which new physics appears, unless there is an incredible fine-tuning cancellation between the quadratic radiative corrections and the bare mass.

It should be remarked that the problem cannot even be formulated in the strict context of the Standard Model, for the Higgs mass cannot be calculated. In a sense, the problem amounts to the worry that a future theory of fundamental particles, in which the Higgs boson mass will be calculable, should not have excessive fine-tunings.

One proposed solution, popular amongst many physicists, is that one may solve the hierarchy problem via supersymmetry. Supersymmetry can explain how a tiny Higgs mass can be protected from quantum corrections. Supersymmetry removes

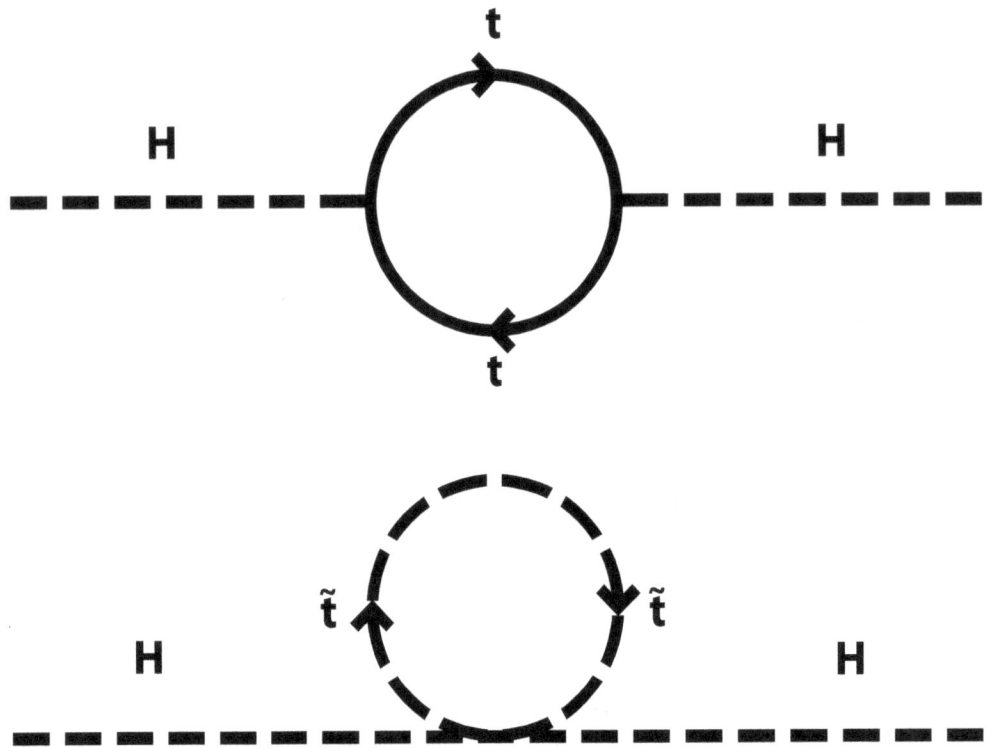

Cancellation of the Higgs boson quadratic mass renormalization between fermionic top quark loop and scalar stop squark tadpole Feynman diagrams in a supersymmetric extension of the Standard Model

the power-law divergences of the radiative corrections to the Higgs mass and solves the hierarchy problem as long as the supersymmetric particles are light enough to satisfy the Barbieri–Giudice criterion.[2] This still leaves open the mu problem, however. Currently the tenets of supersymmetry are being tested at the LHC, although no evidence has been found so far for supersymmetry.

5.3 Theoretical solutions

5.3.1 Supersymmetric solution

Each particle that couples to the Higgs field has a Yukawa coupling λ_f. The coupling with the Higgs field for fermions gives an interaction term $\mathcal{L}_{\text{Yukawa}} = -\lambda_f \bar{\psi} H \psi$, with ψ being the Dirac Field and H the Higgs Field. Also, the mass of a fermion is proportional to its Yukawa coupling, meaning that the Higgs boson will couple most to the most massive particle. This means that the most significant corrections to the Higgs mass will originate from the heaviest particles, most prominently the top quark. By applying the Feynman rules, one gets the quantum corrections to the Higgs mass squared from a fermion to be:

$$\Delta m_H^2 = -\frac{|\lambda_f|^2}{8\pi^2}[\Lambda_{\text{UV}}^2 + ...].$$

The Λ_{UV} is called the ultraviolet cutoff and is the scale up to which the Standard Model is valid. If we take this scale to be the Planck scale, then we have the quadratically diverging Lagrangian. However, suppose there existed two complex scalars (taken to be spin 0) such that:

$\lambda S = |\lambda_f|^2$ (the couplings to the Higgs are exactly the same).

Then by the Feynman rules, the correction (from both scalars) is:

$$\Delta m_H^2 = 2 \times \frac{\lambda_S}{16\pi^2}[\Lambda_{\text{UV}}^2 + ...].$$

(Note that the contribution here is positive. This is because of the spin-statistics theorem, which means that fermions will have a negative contribution and bosons a positive contribution. This fact is exploited.)

This gives a total contribution to the Higgs mass to be zero if we include both the fermionic and bosonic particles. Supersymmetry is an extension of this that creates 'superpartners' for all Standard Model particles.

This section adapted from Stephen P. Martin's "A Supersymmetry Primer" on arXiv.[3]

5.3.2   Conformal solution

Without supersymmetry, a solution to the hierarchy problem has been proposed using just the Standard Model. The idea can be traced back to the fact that the term in the Higgs field that produces the uncontrolled quadratic correction upon renormalization is the quadratic one. If the Higgs field had no mass term, then no hierarchy problem arises. But by missing a quadratic term in the Higgs field, one must find a way to recover the breaking of electroweak symmetry through a non-null vacuum expectation value. This can be obtained using the Weinberg–Coleman mechanism with terms in the Higgs potential arising from quantum corrections. Mass obtained in this way is far too small with respect to what is seen in accelerator facilities and so a conformal Standard Model needs more than one Higgs particle. This proposal has been put forward in 2006 by Krzysztof Meissner and Hermann Nicolai[4] and is currently under scrutiny. But if no further excitation is observed beyond the one seen so far at LHC, this model would have to be abandoned.

5.3.3   Solution via extra dimensions

If we live in a 3+1 dimensional world, then we calculate the Gravitational Force via Gauss' law for gravity:

$$\mathbf{g}(\mathbf{r}) = -Gm\frac{\mathbf{e_r}}{r^2}$$

which is simply Newton's law of gravitation. Note that Newton's constant G can be rewritten in terms of the Planck mass.

$$\frac{1}{M_{\text{Pl}}^2}$$

If we extend this idea to δ extra dimensions, then we get:

$$\mathbf{g}(\mathbf{r}) = -m\frac{\mathbf{e_r}}{M_{\text{Pl}_{3+1+\delta}}^{2+\delta} r^{2+\delta}}$$

where $M_{\text{Pl}_{3+1+\delta}}$ is the $3+1+\delta$ dimensional Planck mass. However, we are assuming that these extra dimensions are the same size as the normal 3+1 dimensions. Let us say that the extra dimensions are of size $n \lll$ than normal dimensions. If we let $r \ll n$, then we get (2). However, if we let $r \gg n$, then we get our usual Newton's law. However, when $r \gg n$, the flux in the extra dimensions becomes a constant, because there is no extra room for gravitational flux to flow through. Thus the flux will be proportional to n^δ because this is the flux in the extra dimensions. The formula is:

$$\mathbf{g}(\mathbf{r}) = -m \frac{\mathbf{e_r}}{M_{\mathrm{Pl}_{3+1+\delta}}^{2+\delta}} r^2 n^\delta$$

$$-m \frac{\mathbf{e_r}}{M_{\mathrm{Pl}}^2 r^2} = -m \frac{\mathbf{e_r}}{M_{\mathrm{Pl}_{3+1+\delta}}^{2+\delta}} r^2 n^\delta$$

which gives:

$$\frac{1}{M_{\mathrm{Pl}}^2 r^2} = \frac{1}{M_{\mathrm{Pl}_{3+1+\delta}}^{2+\delta} r^2 n^\delta} \Rightarrow$$

$$M_{\mathrm{Pl}}^2 = M_{\mathrm{Pl}_{3+1+\delta}}^{2+\delta} n^\delta.$$

Thus the fundamental Planck mass (the extra dimensional one) could actually be small, meaning that gravity is actually strong, but this must be compensated by the number of the extra dimensions and their size. Physically, this means that gravity is weak because there is a loss of flux to the extra dimensions.

This section adapted from "Quantum Field Theory in a Nutshell" by A. Zee.[5]

Braneworld models

Main article: Brane cosmology

In 1998 Nima Arkani-Hamed, Savas Dimopoulos, and Gia Dvali proposed the **ADD model**, also known as the model with large extra dimensions, an alternative scenario to explain the weakness of gravity relative to the other forces.[6][7] This theory requires that the fields of the Standard Model are confined to a four-dimensional membrane, while gravity propagates in several additional spatial dimensions that are large compared to the Planck scale.[8]

In 1998/99 Merab Gogberashvili published on the arXiv (and subsequently in peer-reviewed journals) a number of articles where he showed that if the Universe is considered as a thin shell (a mathematical synonym for "brane") expanding in 5-dimensional space then it is possible to obtain one scale for particle theory corresponding to the 5-dimensional cosmological constant and Universe thickness, and thus to solve the hierarchy problem.[9][10][11] It was also shown that four-dimensionality of the Universe is the result of stability requirement since the extra component of the Einstein field equations giving the localized solution for matter fields coincides with the one of the conditions of stability.

Subsequently, there were proposed the closely related Randall–Sundrum scenarios which offered their solution to the hierarchy problem.

Finite Groups

It has also been noted that the group order of the Baby Monster group is of the right order of magnitude, 4×10^{33}. It is known that the Monster Group is related to the symmetries of a particular bosonic string theory on the Leech lattice. However, there's no physical reason for why the size of the Monster Group or its subgroups should appear in the Lagrangian. Most physicists think this is merely a coincidence. Another coincidence is that in *reduced* Plank units, the Higgs mass is approximately $48.|M|^{-1/3} = 125.5$ GeV where $|M|$ is the order of the Monster group. This suggests that the smallness of the Higgs mass may be due to a redundancy caused by a symmetry of the extra dimensions, which must be divided out. There are other groups that are also of the right order of magnitude for example $Weyl(E_8 \times E_8)$.

Extra dimensions

Until now, no experimental or observational evidence of extra dimensions has been officially reported. Analyses of results from the Large Hadron Collider severely constrain theories with large extra dimensions.[12] However, extra dimensions could explain why the gravity force is so weak, and why the expansion of the universe is faster than expected.[13]

5.4   The cosmological constant

In physical cosmology, current observations in favor of an accelerating universe imply the existence of a tiny, but nonzero cosmological constant. This is a hierarchy problem very similar to that of the Higgs boson mass problem, since the cosmological constant is also very sensitive to quantum corrections. It is complicated, however, by the necessary involvement of general relativity in the problem and may be a clue that we do not understand gravity on long distance scales (such as the size of the universe today). While quintessence has been proposed as an explanation of the acceleration of the Universe, it does not actually address the cosmological constant hierarchy problem in the technical sense of addressing the large quantum corrections. Supersymmetry does not address the cosmological constant problem, since supersymmetry cancels the $M^4 P_{lanck}$ contribution, but not the $M^2 P_{lanck}$ one (quadratically diverging).

5.5   See also

- CP violation

- Little hierarchy problem

- Quantum triviality

5.6   References

[1] http://profmattstrassler.com/articles-and-posts/particle-physics-basics/the-hierarchy-problem/

[2] R. Barbieri, G. F. Giudice (1988). "Upper Bounds on Supersymmetric Particle Masses". *Nucl. Phys.* **B306**: 63. Bibcode:19. doi:10.1016/0550-3213(88)90171-X.

[3] Stephen P. Martin, A Supersymmetry Primer

[4] K. Meissner, H. Nicolai (2006). "Conformal Symmetry and the Standard Model". *Physics Letters* **B648**: 312–317. arXiv:hep-th/0612165. Bibcode:2007PhLB..648..312M. doi:10.1016/j.physletb.2007.03.023.

[5] Zee, A. (2003). "Quantum field theory in a nutshell". Princeton University Press. Bibcode:2003qftn.book.....Z.

[6] N. Arkani-Hamed, S. Dimopoulos, G. Dvali (1998). "The Hierarchy problem and new dimensions at a millimeter". *Physics Letters* **B429**: 263–272. arXiv:hep-ph/9803315. Bibcode:1998PhLB..429..263A. doi:10.1016/S0370-2693(98)00466-3.

[7] N. Arkani-Hamed, S. Dimopoulos, G. Dvali (1999). "Phenomenology, astrophysics and cosmology of theories with submillimeter dimensions and TeV scale quantum gravity". *Physical Review* **D59**: 086004. arXiv:hep-ph/9807344. Bibcode:1999PA. doi:10.1103/PhysRevD.59.086004.

[8] For a pedagogical introduction, see M. Shifman (2009). *Large Extra Dimensions: Becoming acquainted with an alternative paradigm*. Crossing the boundaries: Gauge dynamics at strong coupling. Singapore: World Scientific. arXiv:0907.3074.

[9] M. Gogberashvili, *Hierarchy problem in the shell universe model*, Arxiv:hep-ph/9812296.

[10] M. Gogberashvili, *Our world as an expanding shell*, Arxiv:hep-ph/9812365.

[11] M. Gogberashvili, *Four dimensionality in noncompact Kaluza-Klein model*, Arxiv:hep-ph/9904383.

[12] ATLAS Collaboration, "Search for Quantum Black Hole Production in High-Invariant-Mass Lepton+Jet Final States Using pp Collisions at s√=8 TeV and the ATLAS Detector" http://arxiv.org/abs/1311.2006

[13] CERN (20 January 2012). "Extra dimensions, gravitons, and tiny black holes".

Chapter 6

Higgs mechanism

In the Standard Model of particle physics, the **Higgs mechanism** is essential to explain the generation mechanism of the property "mass" for gauge bosons. Without the Higgs mechanism, or some other effect like it, all bosons (a type of fundamental particle) would be massless, but measurements show that the W^+, W^-, and Z bosons actually have relatively large masses of around 80 GeV/c^2. The Higgs field resolves this conundrum. The simplest description of the mechanism adds a quantum field (the Higgs field) that permeates all space, to the Standard Model. Below some extremely high temperature, the field causes spontaneous symmetry breaking during interactions. The breaking of symmetry triggers the Higgs mechanism, causing the bosons it interacts with to have mass. In the Standard Model, the phrase "Higgs mechanism" refers specifically to the generation of masses for the $W^\pm$, and Z weak gauge bosons through electroweak symmetry breaking.[1] The Large Hadron Collider at CERN announced results consistent with the Higgs particle on March 14, 2013, making it extremely likely that the field, or one like it, exists, and explaining how the Higgs mechanism takes place in nature.

The mechanism was proposed in 1962 by Philip Warren Anderson,[2] following work in the late 1950s on symmetry breaking in superconductivity and a 1960 paper by Yoichiro Nambu that discussed its application within particle physics. A theory able to finally explain mass generation without "breaking" gauge theory was published almost simultaneously by three independent groups in 1964: by Robert Brout and François Englert;[3] by Peter Higgs;[4] and by Gerald Guralnik, C. R. Hagen, and Tom Kibble.[5][6][7] The Higgs mechanism is therefore also called the **Brout–Englert–Higgs mechanism** or **Englert–Brout–Higgs–Guralnik–Hagen–Kibble mechanism**,[8] **Anderson–Higgs mechanism**,[9] **Anderson–Higgs-Kibble mechanism**,[10] **Higgs–Kibble mechanism** by Abdus Salam[11] and **ABEGHHK'tH mechanism** [for Anderson, Brout, Englert, Guralnik, Hagen, Higgs, Kibble and 't Hooft] by Peter Higgs.[11]

On October 8, 2013, following the discovery at CERN's Large Hadron Collider of a new particle that appeared to be the long-sought Higgs boson predicted by the theory, it was announced that Peter Higgs and François Englert had been awarded the 2013 Nobel Prize in Physics (Englert's co-author Robert Brout had died in 2011 and the Nobel Prize is not usually awarded posthumously).[12]

6.1   Standard model

The Higgs mechanism was incorporated into modern particle physics by Steven Weinberg and Abdus Salam, and is an essential part of the standard model.

In the standard model, at temperatures high enough that electroweak symmetry is unbroken, all elementary particles are massless. At a critical temperature the Higgs field becomes tachyonic, the symmetry is spontaneously broken by condensation, and the W and Z bosons acquire masses. (EWSB, ElectroWeak Symmetry Breaking, is an abbreviation used for this.)

Fermions, such as the leptons and quarks in the Standard Model, can also acquire mass as a result of their interaction with the Higgs field, but not in the same way as the gauge bosons.

6.1.1 Structure of the Higgs field

In the standard model, the Higgs field is an **SU**(2) doublet, a complex scalar with four real components (or equivalently with two complex components). Its (weak hypercharge) **U**(1) charge is 1. That means that it transforms as a spinor under **SU**(2). Under **U**(1) rotations, it is multiplied by a phase, which thus mixes the real and imaginary parts of the complex spinor into each other—so this is *not the same* as two complex spinors mixing under **U**(1) (which would have eight real components between them), but instead is the spinor representation of the group **U**(2).

The Higgs field, through the interactions specified (summarized, represented, or even simulated) by its potential, induces spontaneous breaking of three out of the four generators ("directions") of the gauge group **SU**(2) × **U**(1): three out of its four components would ordinarily amount to Goldstone bosons, if they were not coupled to gauge fields.

However, after symmetry breaking, these three of the four degrees of freedom in the Higgs field mix with the three W and Z bosons (W+, W− and Z), and are only observable as spin components of these weak bosons, which are now massive; while the one remaining degree of freedom becomes the Higgs boson—a new scalar particle.

6.1.2 The photon as the part that remains massless

The gauge group of the electroweak part of the standard model is **SU**(2) × **U**(1). The group **SU**(2) is the group of all 2-by-2 unitary matrices with unit determinant; all the orthonormal changes of coordinates in a complex two dimensional vector space.

Rotating the coordinates so that the second basis vector points in the direction of the Higgs boson makes the vacuum expectation value of H the spinor $(0, v)$. The generators for rotations about the x, y, and z axes are by half the Pauli matrices σx, σy, and σz, so that a rotation of angle θ about the z-axis takes the vacuum to

$$(0, ve^{-i\theta/2}).$$

While the T_x and T_y generators mix up the top and bottom components of the spinor, the T_z rotations only multiply each by opposite phases. This phase can be undone by a **U**(1) rotation of angle $1/2\theta$. Consequently, under both an **SU**(2) T_z-rotation and a **U**(1) rotation by an amount $1/2\theta$, *the vacuum is invariant.*

This combination of generators

$$Q = T_z + \frac{Y}{2}$$

defines the unbroken part of the gauge group, where Q is the electric charge, T_z is the generator of rotations around the z-axis in the **SU**(2) and Y is the hypercharge generator of the **U**(1). This combination of generators (a z rotation in the **SU**(2) and a simultaneous **U**(1) rotation by half the angle) preserves the vacuum, and defines the unbroken gauge group in the standard model, namely *the electric charge* group. The part of the gauge field in this direction stays massless, and amounts to the physical photon.

6.1.3 Consequences for fermions

In spite of the introduction of spontaneous symmetry breaking, the mass terms oppose the chiral gauge invariance. For these fields the mass terms should always be replaced by a gauge-invariant "Higgs" mechanism. One possibility is some kind of "Yukawa coupling" (see below) between the fermion field ψ and the Higgs field Φ, with unknown couplings $G\psi$, which after symmetry breaking (more precisely: after expansion of the Lagrange density around a suitable ground state) again results in the original mass terms, which are now, however (i.e. by introduction of the Higgs field) written in a gauge-invariant way. The Lagrange density for the "Yukawa" interaction of a fermion field ψ and the Higgs field Φ is

$$\mathcal{L}_{\text{Fermion}}(\phi, A, \psi) = \overline{\psi}\gamma^{\mu}D_{\mu}\psi + G_{\psi}\overline{\psi}\phi\psi,$$

where again the gauge field A only enters $D\mu$ (i.e., it is only indirectly visible). The quantities γ^{μ} are the Dirac matrices, and $G\psi$ is the already-mentioned "Yukawa" coupling parameter. Already now the mass-generation follows the same principle as above, namely from the existence of a finite expectation value $|\langle\phi\rangle|$, as described above. Again, this is crucial for the existence of the property "mass".

6.2 History of research

6.2.1 Background

Spontaneous symmetry breaking offered a framework to introduce bosons into relativistic quantum field theories. However, according to Goldstone's theorem, these bosons should be massless.[13] The only observed particles which could be approximately interpreted as Goldstone bosons were the pions, which Yoichiro Nambu related to chiral symmetry breaking.

A similar problem arises with Yang–Mills theory (also known as non-abelian gauge theory), which predicts massless spin-1 gauge bosons. Massless weakly interacting gauge bosons lead to long-range forces, which are only observed for electromagnetism and the corresponding massless photon. Gauge theories of the weak force needed a way to describe massive gauge bosons in order to be consistent.

6.2.2 Discovery

The mechanism was proposed in 1962 by Philip Warren Anderson,[2] who discussed its consequences for particle physics but did not work out an explicit relativistic model. The relativistic model was developed in 1964 by three independent groups – Robert Brout and François Englert;[3] Peter Higgs;[4] and Gerald Guralnik, Carl Richard Hagen, and Tom Kibble.[5][6][7] Slightly later, in 1965, but independently from the other publications[14][15][16][17][18][19] the mechanism was also proposed by Alexander Migdal and Alexander Polyakov,[20] at that time Soviet undergraduate students. However, the paper was delayed by the Editorial Office of JETP, and was published only in 1966.

The mechanism is closely analogous to phenomena previously discovered by Yoichiro Nambu involving the "vacuum structure" of quantum fields in superconductivity.[21] A similar but distinct effect (involving an affine realization of what is now recognized as the Higgs field), known as the Stueckelberg mechanism, had previously been studied by Ernst Stueckelberg.

These physicists discovered that when a gauge theory is combined with an additional field that spontaneously breaks the symmetry group, the gauge bosons can consistently acquire a nonzero mass. In spite of the large values involved (see below) this permits a gauge theory description of the weak force, which was independently developed by Steven Weinberg and Abdus Salam in 1967. Higgs's original article presenting the model was rejected by Physics Letters. When revising the article before resubmitting it to Physical Review Letters, he added a sentence at the end,[22] mentioning that it implies the existence of one or more new, massive scalar bosons, which do not form complete representations of the symmetry group; these are the Higgs bosons.

The three papers by Brout and Englert; Higgs; and Guralnik, Hagen, and Kibble were each recognized as "milestone letters" by *Physical Review Letters* in 2008.[23] While each of these seminal papers took similar approaches, the contributions and differences among the 1964 PRL symmetry breaking papers are noteworthy. All six physicists were jointly awarded the 2010 J. J. Sakurai Prize for Theoretical Particle Physics for this work.[24]

Benjamin W. Lee is often credited with first naming the "Higgs-like" mechanism, although there is debate around when this first occurred.[25][26][27] One of the first times the *Higgs* name appeared in print was in 1972 when Gerardus 't Hooft and Martinus J. G. Veltman referred to it as the "Higgs–Kibble mechanism" in their Nobel winning paper.[28][29]

Five of the six 2010 APS Sakurai Prize Winners – (L to R) Tom Kibble, Gerald Guralnik, Carl Richard Hagen, François Englert, and Robert Brout

6.3   Examples

The Higgs mechanism occurs whenever a charged field has a vacuum expectation value. In the nonrelativistic context, this is the Landau model of a charged Bose–Einstein condensate, also known as a superconductor. In the relativistic condensate, the condensate is a scalar field, and is relativistically invariant.

6.3.1   Landau model

The Higgs mechanism is a type of superconductivity which occurs in the vacuum. It occurs when all of space is filled with a sea of particles which are charged, or, in field language, when a charged field has a nonzero vacuum expectation value. Interaction with the quantum fluid filling the space prevents certain forces from propagating over long distances (as it does in a superconducting medium; e.g., in the Ginzburg–Landau theory).

A superconductor expels all magnetic fields from its interior, a phenomenon known as the Meissner effect. This was mysterious for a long time, because it implies that electromagnetic forces somehow become short-range inside the superconductor. Contrast this with the behavior of an ordinary metal. In a metal, the conductivity shields electric fields by rearranging charges on the surface until the total field cancels in the interior. But magnetic fields can penetrate to any distance, and if a magnetic monopole (an isolated magnetic pole) is surrounded by a metal the field can escape without collimating into a string. In a superconductor, however, electric charges move with no dissipation, and this allows for permanent surface currents, not just surface charges. When magnetic fields are introduced at the boundary of a superconductor, they produce surface currents which exactly neutralize them. The Meissner effect is due to currents in a thin surface layer, whose thickness, the London penetration depth, can be calculated from a simple model (the Ginzburg–Landau theory).

This simple model treats superconductivity as a charged Bose–Einstein condensate. Suppose that a superconductor con-

Number six: Peter Higgs 2009

tains bosons with charge q. The wavefunction of the bosons can be described by introducing a quantum field, ψ, which obeys the Schrödinger equation as a field equation (in units where the reduced Planck constant, $\hbar$, is set to 1):

$$i\frac{\partial}{\partial t}\psi = \frac{(\nabla - iqA)^2}{2m}\psi.$$

The operator $\psi(x)$ annihilates a boson at the point x, while its adjoint $\psi^\dagger$ creates a new boson at the same point. The wavefunction of the Bose–Einstein condensate is then the expectation value ψ of $\psi(x)$, which is a classical function that obeys the same equation. The interpretation of the expectation value is that it is the phase that one should give to a newly created boson so that it will coherently superpose with all the other bosons already in the condensate.

When there is a charged condensate, the electromagnetic interactions are screened. To see this, consider the effect of a gauge transformation on the field. A gauge transformation rotates the phase of the condensate by an amount which changes from point to point, and shifts the vector potential by a gradient:

$$\psi \to e^{iq\phi(x)}\psi$$
$$A \to A + \nabla\phi.$$

When there is no condensate, this transformation only changes the definition of the phase of ψ at every point. But when there is a condensate, the phase of the condensate defines a preferred choice of phase.

The condensate wave function can be written as

$$\psi(x) = \rho(x)\, e^{i\theta(x)},$$

where ρ is real amplitude, which determines the local density of the condensate. If the condensate were neutral, the flow would be along the gradients of θ, the direction in which the phase of the Schrödinger field changes. If the phase θ changes slowly, the flow is slow and has very little energy. But now θ can be made equal to zero just by making a gauge transformation to rotate the phase of the field.

The energy of slow changes of phase can be calculated from the Schrödinger kinetic energy,

$$H = \frac{1}{2m}|(qA + \nabla)\psi|^2,$$

and taking the density of the condensate ρ to be constant,

$$H \approx \frac{\rho^2}{2m}(qA + \nabla\theta)^2.$$

Fixing the choice of gauge so that the condensate has the same phase everywhere, the electromagnetic field energy has an extra term,

$$\frac{q^2\rho^2}{2m}A^2.$$

When this term is present, electromagnetic interactions become short-ranged. Every field mode, no matter how long the wavelength, oscillates with a nonzero frequency. The lowest frequency can be read off from the energy of a long wavelength A mode,

$$E \approx \frac{\dot{A}^2}{2} + \frac{q^2\rho^2}{2m}A^2.$$

This is a harmonic oscillator with frequency

$$\sqrt{\frac{1}{m}q^2\rho^2}.$$

The quantity |ψ|² (=ρ²) is the density of the condensate of superconducting particles.

In an actual superconductor, the charged particles are electrons, which are fermions not bosons. So in order to have superconductivity, the electrons need to somehow bind into Cooper pairs. The charge of the condensate q is therefore twice the electron charge e. The pairing in a normal superconductor is due to lattice vibrations, and is in fact very weak; this means that the pairs are very loosely bound. The description of a Bose–Einstein condensate of loosely bound pairs is actually more difficult than the description of a condensate of elementary particles, and was only worked out in 1957 by Bardeen, Cooper and Schrieffer in the famous BCS theory.

6.3.2 Abelian Higgs mechanism

Gauge invariance means that certain transformations of the gauge field do not change the energy at all. If an arbitrary gradient is added to A, the energy of the field is exactly the same. This makes it difficult to add a mass term, because a mass term tends to push the field toward the value zero. But the zero value of the vector potential is not a gauge invariant idea. What is zero in one gauge is nonzero in another.

So in order to give mass to a gauge theory, the gauge invariance must be broken by a condensate. The condensate will then define a preferred phase, and the phase of the condensate will define the zero value of the field in a gauge-invariant way. The gauge-invariant definition is that a gauge field is zero when the phase change along any path from parallel transport is equal to the phase difference in the condensate wavefunction.

The condensate value is described by a quantum field with an expectation value, just as in the Ginzburg-Landau model.

In order for the phase of the vacuum to define a gauge, the field must have a phase (also referred to as 'to be charged'). In order for a scalar field Φ to have a phase, it must be complex, or (equivalently) it should contain two fields with a symmetry which rotates them into each other. The vector potential changes the phase of the quanta produced by the field when they move from point to point. In terms of fields, it defines how much to rotate the real and imaginary parts of the fields into each other when comparing field values at nearby points.

The only renormalizable model where a complex scalar field Φ acquires a nonzero value is the Mexican-hat model, where the field energy has a minimum away from zero. The action for this model is

$$S(\phi) = \int \frac{1}{2}|\partial\phi|^2 - \lambda\left(|\phi|^2 - \Phi^2\right)^2,$$

which results in the Hamiltonian

$$H(\phi) = \frac{1}{2}|\dot{\phi}|^2 + |\nabla\phi|^2 + V(|\phi|).$$

The first term is the kinetic energy of the field. The second term is the extra potential energy when the field varies from point to point. The third term is the potential energy when the field has any given magnitude.

This potential energy, $V(z, \Phi) = \lambda(|z|^2 - \Phi^2)^2$,[30] has a graph which looks like a Mexican hat, which gives the model its name. In particular, the minimum energy value is not at $z = 0$, but on the circle of points where the magnitude of z is Φ.

When the field $\Phi(x)$ is not coupled to electromagnetism, the Mexican-hat potential has flat directions. Starting in any one of the circle of vacua and changing the phase of the field from point to point costs very little energy. Mathematically, if

$$\phi(x) = \Phi e^{i\theta(x)}$$

with a constant prefactor, then the action for the field $\theta(x)$, i.e., the "phase" of the Higgs field $\Phi(x)$, has only derivative terms. This is not a surprise. Adding a constant to $\theta(x)$ is a symmetry of the original theory, so different values of $\theta(x)$ cannot have different energies. This is an example of Goldstone's theorem: spontaneously broken continuous symmetries normally produce massless excitations.

The Abelian Higgs model is the Mexican-hat model coupled to electromagnetism:

$$S(\phi, A) = \int -\frac{1}{4}F^{\mu\nu}F_{\mu\nu} + |(\partial - iqA)\phi|^2 - \lambda(|\phi|^2 - \Phi^2)^2.$$

The classical vacuum is again at the minimum of the potential, where the magnitude of the complex field φ is equal to Φ. But now the phase of the field is arbitrary, because gauge transformations change it. This means that the field $\theta(x)$ can be set to zero by a gauge transformation, and does not represent any actual degrees of freedom at all.

Furthermore, choosing a gauge where the phase of the vacuum is fixed, the potential energy for fluctuations of the vector field is nonzero. So in the abelian Higgs model, the gauge field acquires a mass. To calculate the magnitude of the mass, consider a constant value of the vector potential A in the x direction in the gauge where the condensate has constant phase. This is the same as a sinusoidally varying condensate in the gauge where the vector potential is zero. In the gauge where A is zero, the potential energy density in the condensate is the scalar gradient energy:

$$E = \frac{1}{2}\left|\partial\left(\Phi e^{iqAx}\right)\right|^2 = \frac{1}{2}q^2\Phi^2 A^2.$$

This energy is the same as a mass term $1/2m^2 A^2$ where $m = q\Phi$.

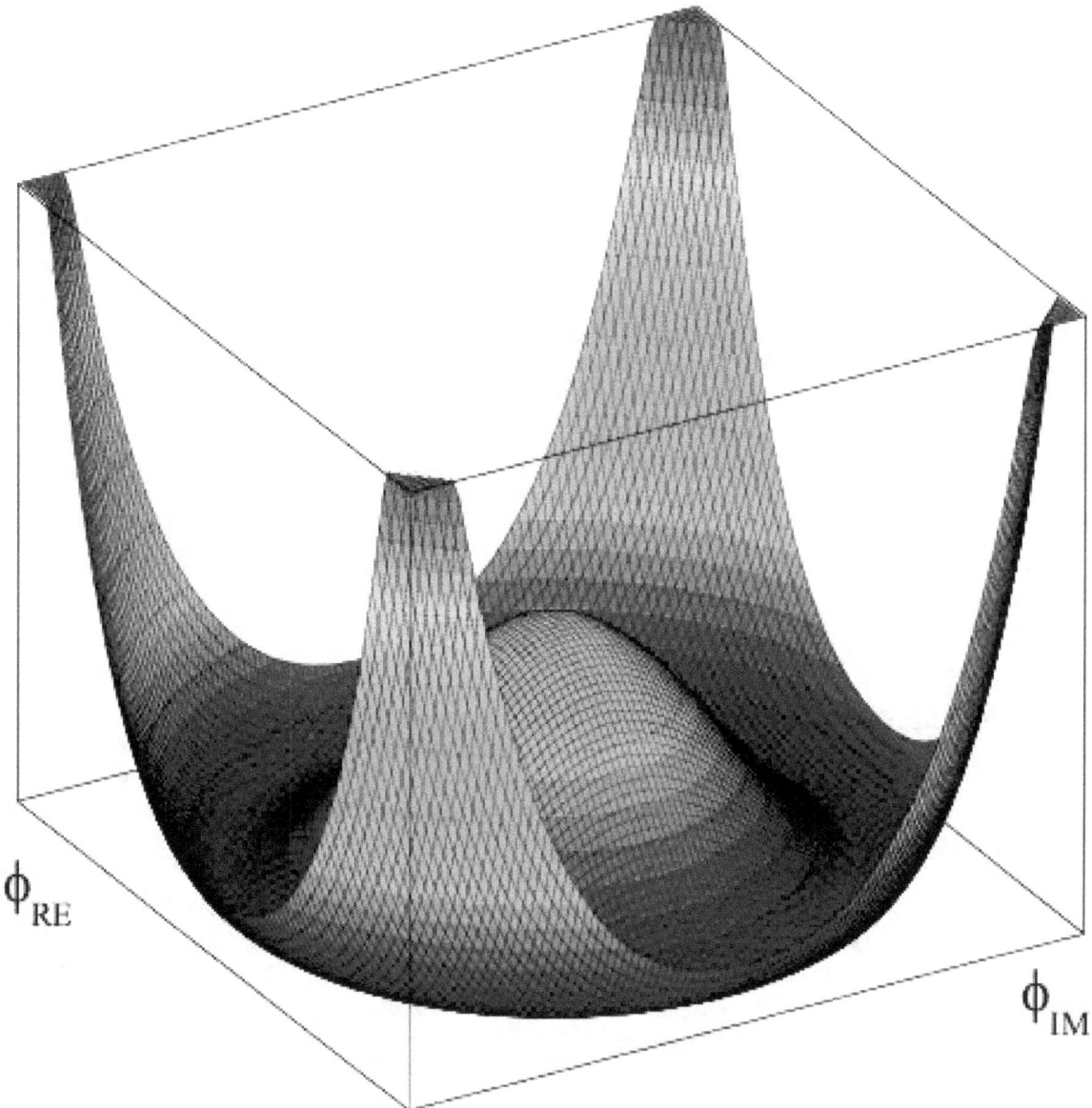

Higgs potential V. *For a fixed value of λ the potential is presented upwards against the real and imaginary parts of Φ. The* Mexican-hat *or* champagne-bottle *profile at the ground should be noted.*

6.3.3   Nonabelian Higgs mechanism

The Nonabelian Higgs model has the following action:

$$S(\phi, \mathbf{A}) = \int \frac{1}{4g^2} \mathrm{tr}(F^{\mu\nu} F_{\mu\nu}) + |D\phi|^2 + V(|\phi|)$$

where now the nonabelian field $\mathbf{A}$ is contained in D and in the tensor components $F^{\mu\nu}$ and $F_{\mu\nu}$ (the relation between $\mathbf{A}$ and those components is well-known from the Yang–Mills theory).

It is exactly analogous to the Abelian Higgs model. Now the field φ is in a representation of the gauge group, and the gauge covariant derivative is defined by the rate of change of the field minus the rate of change from parallel transport using the gauge field A as a connection.

$$D\phi = \partial\phi - iA^k t_k \phi$$

Again, the expectation value of Φ defines a preferred gauge where the vacuum is constant, and fixing this gauge, fluctuations in the gauge field A come with a nonzero energy cost.

Depending on the representation of the scalar field, not every gauge field acquires a mass. A simple example is in the renormalizable version of an early electroweak model due to Julian Schwinger. In this model, the gauge group is **SO**(3) (or **SU**(2) – there are no spinor representations in the model), and the gauge invariance is broken down to **U**(1) or **SO**(2) at long distances. To make a consistent renormalizable version using the Higgs mechanism, introduce a scalar field φ^a which transforms as a vector (a triplet) of **SO**(3). If this field has a vacuum expectation value, it points in some direction in field space. Without loss of generality, one can choose the z-axis in field space to be the direction that φ is pointing, and then the vacuum expectation value of φ is $(0, 0, A)$, where A is a constant with dimensions of mass ($c = \hbar = 1$).

Rotations around the z-axis form a **U**(1) subgroup of **SO**(3) which preserves the vacuum expectation value of φ, and this is the unbroken gauge group. Rotations around the x and y-axis do not preserve the vacuum, and the components of the **SO**(3) gauge field which generate these rotations become massive vector mesons. There are two massive **W** mesons in the Schwinger model, with a mass set by the mass scale A, and one massless **U**(1) gauge boson, similar to the photon.

The Schwinger model predicts magnetic monopoles at the electroweak unification scale, and does not predict the Z meson. It doesn't break electroweak symmetry properly as in nature. But historically, a model similar to this (but not using the Higgs mechanism) was the first in which the weak force and the electromagnetic force were unified.

6.3.4 Affine Higgs mechanism

Ernst Stueckelberg discovered[31] a version of the Higgs mechanism by analyzing the theory of quantum electrodynamics with a massive photon. Effectively, Stueckelberg's model is a limit of the regular Mexican hat Abelian Higgs model, where the vacuum expectation value H goes to infinity and the charge of the Higgs field goes to zero in such a way that their product stays fixed. The mass of the Higgs boson is proportional to H, so the Higgs boson becomes infinitely massive and decouples, so is not present in the discussion. The vector meson mass, however, equals to the product eH, and stays finite.

The interpretation is that when a **U**(1) gauge field does not require quantized charges, it is possible to keep only the angular part of the Higgs oscillations, and discard the radial part. The angular part of the Higgs field θ has the following gauge transformation law:

$$\theta \to \theta + e\alpha$$

$$A \to A + \alpha.$$

The gauge covariant derivative for the angle (which is actually gauge invariant) is:

$$D\theta = \partial\theta - eA.$$

In order to keep θ fluctuations finite and nonzero in this limit, θ should be rescaled by H, so that its kinetic term in the action stays normalized. The action for the theta field is read off from the Mexican hat action by substituting $\phi = He^{\frac{1}{H}i\theta}$
.

$$S = \int \frac{1}{4}F^2 + \frac{1}{2}(D\theta)^2 = \int \frac{1}{4}F^2 + \frac{1}{2}(\partial\theta - HeA)^2 = \int \frac{1}{4}F^2 + \frac{1}{2}(\partial\theta - mA)^2$$

since eH is the gauge boson mass. By making a gauge transformation to set $\theta = 0$, the gauge freedom in the action is eliminated, and the action becomes that of a massive vector field:

$$S = \int \frac{1}{4}F^2 + \frac{1}{2}m^2A^2.$$

To have arbitrarily small charges requires that the $\mathbf{U}(1)$ is not the circle of unit complex numbers under multiplication, but the real numbers $\mathbf{R}$ under addition, which is only different in the global topology. Such a $\mathbf{U}(1)$ group is *non-compact*. The field θ transforms as an affine representation of the gauge group. Among the allowed gauge groups, only non-compact $\mathbf{U}(1)$ admits affine representations, and the $\mathbf{U}(1)$ of electromagnetism is experimentally known to be compact, since charge quantization holds to extremely high accuracy.

The Higgs condensate in this model has infinitesimal charge, so interactions with the Higgs boson do not violate charge conservation. The theory of quantum electrodynamics with a massive photon is still a renormalizable theory, one in which electric charge is still conserved, but magnetic monopoles are not allowed. For nonabelian gauge theory, there is no affine limit, and the Higgs oscillations cannot be too much more massive than the vectors.

6.4   See also

- Electromagnetic mass

- Higgs bundle

- Mass generation

- QCD vacuum

- Quantum triviality

- Top quark condensate

- Yang–Mills–Higgs equations

6.5   References

[1] G. Bernardi, M. Carena, and T. Junk: "Higgs bosons: theory and searches", Reviews of Particle Data Group: Hypothetical particles and Concepts, 2007, http://pdg.lbl.gov/2008/reviews/higgs_s055.pdf

[2] P. W. Anderson (1962). "Plasmons, Gauge Invariance, and Mass".*Physical Review***130**(1): 439–442.Bibcode:1963PhRv..130..A. doi:10.1103/PhysRev.130.439.

[3] F. Englert and R. Brout (1964). "Broken Symmetry and the Mass of Gauge Vector Mesons". *Physical Review Letters* **13** (9): 321–323. Bibcode:1964PhRvL..13..321E. doi:10.1103/PhysRevLett.13.321.

[4] Peter W. Higgs (1964). "Broken Symmetries and the Masses of Gauge Bosons". *Physical Review Letters* **13** (16): 508–509. Bibcode:1964PhRvL..13..508H. doi:10.1103/PhysRevLett.13.508.

[5] G. S. Guralnik, C. R. Hagen, and T. W. B. Kibble (1964). "Global Conservation Laws and Massless Particles". *Physical Review Letters* **13** (20): 585–587. Bibcode:1964PhRvL..13..585G. doi:10.1103/PhysRevLett.13.585.

[6] Gerald S. Guralnik (2009). "The History of the Guralnik, Hagen and Kibble development of the Theory of Spontaneous Symmetry Breaking and Gauge Particles". *International Journal of Modern Physics* **A24** (14): 2601–2627. arXiv:0907.3466. Bibcode:2009IJMPA..24.2601G. doi:10.1142/S0217751X09045431.

[7] History of Englert–Brout–Higgs–Guralnik–Hagen–Kibble Mechanism. Scholarpedia.

[8] "Englert–Brout–Higgs–Guralnik–Hagen–Kibble Mechanism". Scholarpedia. Retrieved 2012-06-16.

[9] Liu, G. Z.; Cheng, G. (2002). "Extension of the Anderson-Higgs mechanism". *Physical Review B* **65** (13): 132513. arXiv:cond-mat/0106070. Bibcode:2002PhRvB..65m2513L. doi:10.1103/PhysRevB.65.132513.

[10] Matsumoto, H.; Papastamatiou, N. J.; Umezawa, H.; Vitiello, G. (1975). "Dynamical rearrangement in the Anderson-Higgs-Kibble mechanism". *Nuclear Physics B* **97**: 61. doi:10.1016/0550-3213(75)90215-1.

[11] Close, Frank (2011). *The Infinity Puzzle: Quantum Field Theory and the Hunt for an Orderly Universe.* Oxford: Oxford University Press. ISBN 978-0-19-959350-7.

[12] "Press release from Royal Swedish Academy of Sciences" (PDF). 8 October 2013. Retrieved 8 October 2013.

[13] "Guralnik, G S; Hagen, C R and Kibble, T W B (1967). Broken Symmetries and the Goldstone Theorem. Advances in Physics, vol. 2" (PDF).

[14] A.M. Polyakov, A View From The Island, 1992

[15] Farhi, E., & Jackiw, R. W. (1982). *Dynamical Gauge Symmetry Breaking: A Collection Of Reprints.* Singapore: World Scientific Pub. Co.

[16] Frank Close. "The Infinity Puzzle." 2011, p.158

[17] Norman Dombey, "Higgs Boson: Credit Where It's Due". The Guardian, July 6, 2012

[18] Cern Courier, Mar 1, 2006

[19] Sean Carrol, "The Particle At The End Of The Universe: The Hunt For The Higgs And The Discovery Of A New World", 2012, p.228

[20] A. A. Migdal and A. M. Polyakov, "Spontaneous Breakdown of Strong Interaction Symmetry and Absence of Massless Particles", *JETP* **51**, 135, July 1966 (English translation: *Soviet Physics JETP*, **24**, 1, January 1967)

[21] Nambu, Y (1960). "Quasiparticles and Gauge Invariance in the Theory of Superconductivity". *Physical Review* **117** (3): 648–663. Bibcode:1960PhRv..117..648N. doi:10.1103/PhysRev.117.648.

[22] Higgs, Peter (2007). "Prehistory of the Higgs boson". *Comptes Rendus Physique* **8** (9): 970–972. Bibcode:2007CRPhy...8..970H. doi:10.1016/j.crhy.2006.12.006.

[23] "Physical Review Letters – 50th Anniversary Milestone Papers". Prl.aps.org. Retrieved 2012-06-16.

[24] "American Physical Society – J. J. Sakurai Prize Winners". Aps.org. Retrieved 2012-06-16.

[25] Department of Physics and Astronomy. "Rochester's Hagen Sakurai Prize Announcement". Pas.rochester.edu. Retrieved 2012-06-16.

[26] FermiFred (2010-02-15). "C.R. Hagen discusses naming of Higgs Boson in 2010 Sakurai Prize Talk". Youtube.com. Retrieved 2012-06-16.

[27] Sample, Ian (2009-05-29). "Anything but the God particle by Ian Sample". Guardian. Retrieved 2012-06-16.

[28] G. 't Hooft and M. Veltman (1972). "Regularization and Renormalization of Gauge Fields". *Nuclear Physics B* **44** (1): 189–219. Bibcode:1972NuPhB..44..189T. doi:10.1016/0550-3213(72)90279-9.

[29] "Regularization and Renormalization of Gauge Fields by t'Hooft and Veltman (PDF)" (PDF). Retrieved 2012-06-16.

[30] Goldstone, J. (1961). "Field theories with " Superconductor " solutions".*Il Nuovo Cimento*19: 154–164.doi:10.1007/BF028122.

[31] Stueckelberg, E. C. G. (1938), "Die Wechselwirkungskräfte in der Elektrodynamik und in der Feldtheorie der Kräfte", *Helv. Phys. Acta.* **11:** 225

6.6 Further reading

- Schumm, Bruce A. (2004) *Deep Down Things.* Johns Hopkins Univ. Press. Chpt. 9.

6.7   External links

- Guralnik, G.S.; Hagen, C.R.; Kibble, T.W.B. (1964). "Global Conservation Laws and Massless Particles". *Physical Review Letters* **13** (20): 585–87. Bibcode:1964PhRvL..13..585G. doi:10.1103/PhysRevLett.13.585.

- Mark D. Roberts (1999) "A Generalized Higgs Model"

- 2010 Sakurai Prize - All Events - YouTube

- From BCS to the LHC - CERN Courier Jan 21, 2008, Steven Weinberg, University of Texas at Austin.

- Higgs, dark matter and supersymmetry: What the Large Hadron Collider will tell us (Steven Weinberg) - YouTube on YouTube 06-11-2009

- Gerry Guralnik speaks at Brown University about the 1964 PRL papers

- Guralnik, Gerald (2013). "Heretical Ideas that Provided the Cornerstone for the Standard Model of Particle Physics". SPG MITTEILUNGEN March 2013, No. 39, (p. 14)

- Steven Weinberg Praises Teams for Higgs Boson Theory

- Physical Review Letters – 50th Anniversary Milestone Papers

- Imperial College London on PRL 50th Anniversary Milestone Papers

- Englert–Brout–Higgs–Guralnik–Hagen–Kibble Mechanism on Scholarpedia

- History of Englert–Brout–Higgs–Guralnik–Hagen–Kibble Mechanism on Scholarpedia

- The Hunt for the Higgs at Tevatron

- The Mystery of Empty Space on YouTube. A lecture with UCSD physicist Kim Griest (43 minutes)

Chapter 7

Quantum field theory

"Relativistic quantum field theory" redirects here. For other uses, see Relativity.

In theoretical physics, **quantum field theory** (**QFT**) is a theoretical framework for constructing quantum mechanical models of subatomic particles in particle physics and quasiparticles in condensed matter physics. A QFT treats particles as excited states of an underlying physical field, so these are called field quanta.

In QFT, quantum mechanical interactions between particles are described by interaction terms between the corresponding underlying fields.

7.1 Definition

Quantum electrodynamics (QED) has one electron field and one photon field; quantum chromodynamics (QCD) has one field for each type of quark; and, in condensed matter, there is an atomic displacement field that gives rise to phonon particles. Edward Witten describes QFT as "by far" the most difficult theory in modern physics.[1]

7.1.1 Dynamics

See also: Relativistic dynamics

Ordinary quantum mechanical systems have a fixed number of particles, with each particle having a finite number of degrees of freedom. In contrast, the excited states of a QFT can represent any number of particles. This makes quantum field theories especially useful for describing systems where the particle count/number may change over time, a crucial feature of relativistic dynamics.

7.1.2 States

QFT interaction terms are similar in spirit to those between charges with electric and magnetic fields in Maxwell's equations. However, unlike the classical fields of Maxwell's theory, fields in QFT generally exist in quantum superpositions of states and are subject to the laws of quantum mechanics.

Because the fields are continuous quantities over space, there exist excited states with arbitrarily large numbers of particles in them, providing QFT systems with an effectively infinite number of degrees of freedom. Infinite degrees of freedom can easily lead to divergences of calculated quantities (i.e., the quantities become infinite). Techniques such as renormalization of QFT parameters or discretization of spacetime, as in lattice QCD, are often used to avoid such infinities so as to yield physically meaningful results.

7.1.3 Fields and radiation

The gravitational field and the electromagnetic field are the only two fundamental fields in nature that have infinite range and a corresponding classical low-energy limit, which greatly diminishes and hides their "particle-like" excitations. Albert Einstein in 1905, attributed "particle-like" and discrete exchanges of momenta and energy, characteristic of "field quanta", to the electromagnetic field. Originally, his principal motivation was to explain the thermodynamics of radiation. Although the photoelectric effect and Compton scattering strongly suggest the existence of the photon, it might alternately be explained by a mere quantization of emission; more definitive evidence of the quantum nature of radiation is now taken up into modern quantum optics as in the antibunching effect.[2]

7.2 Theories

There is currently no complete quantum theory of the remaining fundamental force, gravity. Many of the proposed theories to describe gravity as a QFT postulate the existence of a graviton particle that mediates the gravitational force. Presumably, the as yet unknown correct quantum field-theoretic treatment of the gravitational field will behave like Einstein's general theory of relativity in the low-energy limit. Quantum field theory of the fundamental forces itself has been postulated to be the low-energy effective field theory limit of a more fundamental theory such as superstring theory.

Most theories in standard particle physics are formulated as **relativistic quantum field theories**, such as QED, QCD, and the Standard Model. QED, the quantum field-theoretic description of the electromagnetic field, approximately reproduces Maxwell's theory of electrodynamics in the low-energy limit, with small non-linear corrections to the Maxwell equations required due to virtual electron–positron pairs.

In the perturbative approach to quantum field theory, the full field interaction terms are approximated as a perturbative expansion in the number of particles involved. Each term in the expansion can be thought of as forces between particles being mediated by other particles. In QED, the electromagnetic force between two electrons is caused by an exchange of photons. Similarly, intermediate vector bosons mediate the weak force and gluons mediate the strong force in QCD. The notion of a force-mediating particle comes from perturbation theory, and does not make sense in the context of non-perturbative approaches to QFT, such as with bound states.

7.3 History

Main article: History of quantum field theory

7.3.1 Foundations

The early development of the field involved Dirac, Fock, Pauli, Heisenberg and Bogolyubov. This phase of development culminated with the construction of the theory of quantum electrodynamics in the 1950s.

7.3.2 Gauge theory

Gauge theory was formulated and quantized, leading to the **unification of forces** embodied in the standard model of particle physics. This effort started in the 1950s with the work of Yang and Mills, was carried on by Martinus Veltman and a host of others during the 1960s and completed by the 1970s through the work of Gerard 't Hooft, Frank Wilczek, David Gross and David Politzer.

7.3.3   Grand synthesis

Parallel developments in the understanding of phase transitions in condensed matter physics led to the study of the renormalization group. This in turn led to the **grand synthesis** of theoretical physics, which unified theories of particle and condensed matter physics through quantum field theory. This involved the work of Michael Fisher and Leo Kadanoff in the 1970s, which led to the seminal reformulation of quantum field theory by Kenneth G. Wilson in 1975.

7.4   Principles

7.4.1   Classical and quantum fields

Main article: Classical field theory

A classical field is a function defined over some region of space and time.[3] Two physical phenomena which are described by classical fields are Newtonian gravitation, described by Newtonian gravitational field $\mathbf{g}(\mathbf{x}, t)$, and classical electromagnetism, described by the electric and magnetic fields $\mathbf{E}(\mathbf{x}, t)$ and $\mathbf{B}(\mathbf{x}, t)$. Because such fields can in principle take on distinct values at each point in space, they are said to have infinite degrees of freedom.[3]

Classical field theory does not, however, account for the quantum-mechanical aspects of such physical phenomena. For instance, it is known from quantum mechanics that certain aspects of electromagnetism involve discrete particles—photons—rather than continuous fields. The business of *quantum* field theory is to write down a field that is, like a classical field, a function defined over space and time, but which also accommodates the observations of quantum mechanics. This is a *quantum field*.

It is not immediately clear *how* to write down such a quantum field, since quantum mechanics has a structure very unlike a field theory. In its most general formulation, quantum mechanics is a theory of abstract operators (observables) acting on an abstract state space (Hilbert space), where the observables represent physically observable quantities and the state space represents the possible states of the system under study.[4] For instance, the fundamental observables associated with the motion of a single quantum mechanical particle are the position and momentum operators $\hat{x}$ and $\hat{p}$. Field theory, in contrast, treats x as a way to index the field rather than as an operator.[5]

There are two common ways of developing a quantum field: the path integral formalism and canonical quantization.[6] The latter of these is pursued in this article.

Lagrangian formalism

Quantum field theory frequently makes use of the Lagrangian formalism from classical field theory. This formalism is analogous to the Lagrangian formalism used in classical mechanics to solve for the motion of a particle under the influence of a field. In classical field theory, one writes down a Lagrangian density, $\mathcal{L}$, involving a field, $\varphi(\mathbf{x},t)$, and possibly its first derivatives ($\partial\varphi/\partial t$ and $\nabla\varphi$), and then applies a field-theoretic form of the Euler–Lagrange equation. Writing coordinates $(t, \mathbf{x}) = (x^0, x^1, x^2, x^3) = x^\mu$, this form of the Euler–Lagrange equation is[3]

$$\frac{\partial}{\partial x^\mu}\left[\frac{\partial\mathcal{L}}{\partial(\partial\phi/\partial x^\mu)}\right] - \frac{\partial\mathcal{L}}{\partial\phi} = 0,$$

where a sum over μ is performed according to the rules of Einstein notation.

By solving this equation, one arrives at the "equations of motion" of the field.[3] For example, if one begins with the Lagrangian density

$$\mathcal{L}(\phi, \nabla\phi) = -\rho(t, \mathbf{x})\,\phi(t, \mathbf{x}) - \frac{1}{8\pi G}|\nabla\phi|^2,$$

and then applies the Euler–Lagrange equation, one obtains the equation of motion

$$4\pi G\rho(t,\mathbf{x}) = \nabla^2\phi.$$

This equation is Newton's law of universal gravitation, expressed in differential form in terms of the gravitational potential $\varphi(t, \mathbf{x})$ and the mass density $\rho(t, \mathbf{x})$. Despite the nomenclature, the "field" under study is the gravitational potential, φ, rather than the gravitational field, $\mathbf{g}$. Similarly, when classical field theory is used to study electromagnetism, the "field" of interest is the electromagnetic four-potential $(V/c, \mathbf{A})$, rather than the electric and magnetic fields $\mathbf{E}$ and $\mathbf{B}$.

Quantum field theory uses this same Lagrangian procedure to determine the equations of motion for quantum fields. These equations of motion are then supplemented by commutation relations derived from the canonical quantization procedure described below, thereby incorporating quantum mechanical effects into the behavior of the field.

7.4.2 Single- and many-particle quantum mechanics

Main articles: Quantum mechanics and First quantization

In quantum mechanics, a particle (such as an electron or proton) is described by a complex wavefunction, $\psi(x, t)$, whose time-evolution is governed by the Schrödinger equation:

$$-\frac{\hbar^2}{2m}\frac{\partial^2}{\partial x^2}\psi(x,t) + V(x)\psi(x,t) = i\hbar\frac{\partial}{\partial t}\psi(x,t).$$

Here m is the particle's mass and $V(x)$ is the applied potential. Physical information about the behavior of the particle is extracted from the wavefunction by constructing expected values for various quantities; for example, the expected value of the particle's position is given by integrating $\psi^*(x)\, x\, \psi(x)$ over all space, and the expected value of the particle's momentum is found by integrating $-i\hbar\psi^*(x)\mathrm{d}\psi/\mathrm{d}x$. The quantity $\psi^*(x)\psi(x)$ is itself in the Copenhagen interpretation of quantum mechanics interpreted as a probability density function. This treatment of quantum mechanics, where a particle's wavefunction evolves against a classical background potential $V(x)$, is sometimes called *first quantization*.

This description of quantum mechanics can be extended to describe the behavior of multiple particles, so long as the number and the type of particles remain fixed. The particles are described by a wavefunction $\psi(x_1, x_2, \dots, xN, t)$, which is governed by an extended version of the Schrödinger equation.

Often one is interested in the case where N particles are all of the same type (for example, the 18 electrons orbiting a neutral argon nucleus). As described in the article on identical particles, this implies that the state of the entire system must be either symmetric (bosons) or antisymmetric (fermions) when the coordinates of its constituent particles are exchanged. This is achieved by using a Slater determinant as the wavefunction of a fermionic system (and a Slater permanent for a bosonic system), which is equivalent to an element of the symmetric or antisymmetric subspace of a tensor product.

For example, the general quantum state of a system of N bosons is written as

$$|\phi_1\cdots\phi_N\rangle = \sqrt{\frac{\prod_j N_j!}{N!}}\sum_{p\in S_N}|\phi_{p(1)}\rangle\otimes\cdots\otimes|\phi_{p(N)}\rangle,$$

where $|\psi_i\rangle$ are the single-particle states, Nj is the number of particles occupying state j, and the sum is taken over all possible permutations p acting on N elements. In general, this is a sum of $N!$ (N factorial) distinct terms. $\sqrt{\frac{\prod_j N_j!}{N!}}$ is a normalizing factor.

There are several shortcomings to the above description of quantum mechanics, which are addressed by quantum field theory. First, it is unclear how to extend quantum mechanics to include the effects of special relativity.[7] Attempted replacements for the Schrödinger equation, such as the Klein–Gordon equation or the Dirac equation, have many unsatisfactory qualities; for instance, they possess energy eigenvalues that extend to $-\infty$, so that there seems to be no easy definition of a ground state. It turns out that such inconsistencies arise from relativistic wavefunctions not having a well-defined probabilistic interpretation in position space, as probability conservation is not a relativistically covariant concept. The second shortcoming, related to the first, is that in quantum mechanics there is no mechanism to describe particle creation and annihilation;[8] this is crucial for describing phenomena such as pair production, which result from the conversion between mass and energy according to the relativistic relation $E = mc^2$.

7.4.3 Second quantization

Main article: Second quantization

In this section, we will describe a method for constructing a quantum field theory called **second quantization**. This basically involves choosing a way to index the quantum mechanical degrees of freedom in the space of multiple identical-particle states. It is based on the Hamiltonian formulation of quantum mechanics.

Several other approaches exist, such as the Feynman path integral,[9] which uses a Lagrangian formulation. For an overview of some of these approaches, see the article on quantization.

Bosons

For simplicity, we will first discuss second quantization for bosons, which form perfectly symmetric quantum states. Let us denote the mutually orthogonal single-particle states which are possible in the system by $|\phi_1\rangle, |\phi_2\rangle, |\phi_3\rangle$, and so on. For example, the 3-particle state with one particle in state $|\phi_1\rangle$ and two in state $|\phi_2\rangle$ is

$$\frac{1}{\sqrt{3}} \left[|\phi_1\rangle|\phi_2\rangle|\phi_2\rangle + |\phi_2\rangle|\phi_1\rangle|\phi_2\rangle + |\phi_2\rangle|\phi_2\rangle|\phi_1\rangle \right].$$

The first step in second quantization is to express such quantum states in terms of **occupation numbers**, by listing the number of particles occupying each of the single-particle states $|\phi_1\rangle, |\phi_2\rangle$, etc. This is simply another way of labelling the states. For instance, the above 3-particle state is denoted as

$$|1, 2, 0, 0, 0, \dots\rangle.$$

An N-particle state belongs to a space of states describing systems of N particles. The next step is to combine the individual N-particle state spaces into an extended state space, known as Fock space, which can describe systems of any number of particles. This is composed of the state space of a system with no particles (the so-called vacuum state, written as $|0\rangle$), plus the state space of a 1-particle system, plus the state space of a 2-particle system, and so forth. States describing a definite number of particles are known as Fock states: a general element of Fock space will be a linear combination of Fock states. There is a one-to-one correspondence between the occupation number representation and valid boson states in the Fock space.

At this point, the quantum mechanical system has become a quantum field in the sense we described above. The field's elementary degrees of freedom are the occupation numbers, and each occupation number is indexed by a number j indicating which of the single-particle states $|\phi_1\rangle, |\phi_2\rangle, \dots, |\phi_j\rangle, \dots$ it refers to:

$$|N_1, N_2, N_3, \dots, N_j, \dots\rangle.$$

The properties of this quantum field can be explored by defining creation and annihilation operators, which add and subtract particles. They are analogous to ladder operators in the quantum harmonic oscillator problem, which added and subtracted energy quanta. However, these operators literally create and annihilate particles of a given quantum state. The bosonic annihilation operator a_2 and creation operator $a_2^\dagger$ are easily defined in the occupation number representation as having the following effects:

$$a_2|N_1, N_2, N_3, \ldots\rangle = \sqrt{N_2}\,|\,N_1, (N_2 - 1), N_3, \ldots\rangle,$$

$$a_2^\dagger|N_1, N_2, N_3, \ldots\rangle = \sqrt{N_2 + 1}\,|\,N_1, (N_2 + 1), N_3, \ldots\rangle.$$

It can be shown that these are operators in the usual quantum mechanical sense, i.e. linear operators acting on the Fock space. Furthermore, they are indeed Hermitian conjugates, which justifies the way we have written them. They can be shown to obey the commutation relation

$$[a_i, a_j] = 0 \quad , \quad \left[a_i^\dagger, a_j^\dagger\right] = 0 \quad , \quad \left[a_i, a_j^\dagger\right] = \delta_{ij},$$

where δ stands for the Kronecker delta. These are precisely the relations obeyed by the ladder operators for an infinite set of independent quantum harmonic oscillators, one for each single-particle state. Adding or removing bosons from each state is therefore analogous to exciting or de-exciting a quantum of energy in a harmonic oscillator.

Applying an annihilation operator a_k followed by its corresponding creation operator $a_k^\dagger$ returns the number N_k of particles in the kth single-particle eigenstate:

$$a_k^\dagger a_k|\ldots, N_k, \ldots\rangle = N_k|\ldots, N_k, \ldots\rangle.$$

The combination of operators $a_k^\dagger a_k$ is known as the number operator for the kth eigenstate.

The Hamiltonian operator of the quantum field (which, through the Schrödinger equation, determines its dynamics) can be written in terms of creation and annihilation operators. For instance, for a field of free (non-interacting) bosons, the total energy of the field is found by summing the energies of the bosons in each energy eigenstate. If the kth single-particle energy eigenstate has energy E_k and there are N_k bosons in this state, then the total energy of these bosons is $E_k N_k$. The energy in the *entire* field is then a sum over k :

$$E_{\text{tot}} = \sum_k E_k N_k$$

This can be turned into the Hamiltonian operator of the field by replacing N_k with the corresponding number operator, $a_k^\dagger a_k$. This yields

$$H = \sum_k E_k\, a_k^\dagger a_k.$$

Fermions

It turns out that a different definition of creation and annihilation must be used for describing fermions. According to the Pauli exclusion principle, fermions cannot share quantum states, so their occupation numbers Ni can only take on the value 0 or 1. The fermionic annihilation operators c and creation operators $c^\dagger$ are defined by their actions on a Fock state thus

$$c_j|N_1, N_2, \ldots, N_j = 0, \ldots\rangle = 0$$

$$c_j|N_1, N_2, \ldots, N_j = 1, \ldots\rangle = (-1)^{(N_1 + \cdots + N_{j-1})}|N_1, N_2, \ldots, N_j = 0, \ldots\rangle$$

$$c_j^\dagger|N_1, N_2, \ldots, N_j = 0, \ldots\rangle = (-1)^{(N_1 + \cdots + N_{j-1})}|N_1, N_2, \ldots, N_j = 1, \ldots\rangle$$

$$c_j^\dagger |N_1, N_2, \ldots, N_j = 1, \ldots\rangle = 0.$$

These obey an anticommutation relation:

$$\{c_i, c_j\} = 0 \quad , \quad \left\{c_i^\dagger, c_j^\dagger\right\} = 0 \quad , \quad \left\{c_i, c_j^\dagger\right\} = \delta_{ij}.$$

One may notice from this that applying a fermionic creation operator twice gives zero, so it is impossible for the particles to share single-particle states, in accordance with the exclusion principle.

Field operators

We have previously mentioned that there can be more than one way of indexing the degrees of freedom in a quantum field. Second quantization indexes the field by enumerating the single-particle quantum states. However, as we have discussed, it is more natural to think about a "field", such as the electromagnetic field, as a set of degrees of freedom indexed by position.

To this end, we can define *field operators* that create or destroy a particle at a particular point in space. In particle physics, these operators turn out to be more convenient to work with, because they make it easier to formulate theories that satisfy the demands of relativity.

Single-particle states are usually enumerated in terms of their momenta (as in the particle in a box problem.) We can construct field operators by applying the Fourier transform to the creation and annihilation operators for these states. For example, the bosonic field annihilation operator $\phi(\mathbf{r})$ is

$$\phi(\mathbf{r}) \stackrel{\text{def}}{=} \sum_j e^{i\mathbf{k}_j \cdot \mathbf{r}} a_j.$$

The bosonic field operators obey the commutation relation

$$[\phi(\mathbf{r}), \phi(\mathbf{r}')] = 0 \quad , \quad [\phi^\dagger(\mathbf{r}), \phi^\dagger(\mathbf{r}')] = 0 \quad , \quad [\phi(\mathbf{r}), \phi^\dagger(\mathbf{r}')] = \delta^3(\mathbf{r} - \mathbf{r}')$$

where $\delta(x)$ stands for the Dirac delta function. As before, the fermionic relations are the same, with the commutators replaced by anticommutators.

The field operator is not the same thing as a single-particle wavefunction. The former is an operator acting on the Fock space, and the latter is a quantum-mechanical amplitude for finding a particle in some position. However, they are closely related, and are indeed commonly denoted with the same symbol. If we have a Hamiltonian with a space representation, say

$$H = -\frac{\hbar^2}{2m} \sum_i \nabla_i^2 + \sum_{i<j} U(|\mathbf{r}_i - \mathbf{r}_j|)$$

where the indices i and j run over all particles, then the field theory Hamiltonian (in the non-relativistic limit and for negligible self-interactions) is

$$H = -\frac{\hbar^2}{2m} \int d^3r \, \phi^\dagger(\mathbf{r}) \nabla^2 \phi(\mathbf{r}) + \frac{1}{2} \int d^3r \int d^3r' \, \phi^\dagger(\mathbf{r}) \phi^\dagger(\mathbf{r}') U(|\mathbf{r} - \mathbf{r}'|) \phi(\mathbf{r}') \phi(\mathbf{r}).$$

This looks remarkably like an expression for the expectation value of the energy, with ϕ playing the role of the wavefunction. This relationship between the field operators and wavefunctions makes it very easy to formulate field theories starting from space-projected Hamiltonians.

7.4.4   Dynamics

Once the Hamiltonian operator is obtained as part of the canonical quantization process, the time dependence of the state is described with the Schrödinger equation, just as with other quantum theories. Alternatively, the Heisenberg picture can be used where the time dependence is in the operators rather than in the states.

7.4.5   Implications

Unification of fields and particles

The "second quantization" procedure that we have outlined in the previous section takes a set of single-particle quantum states as a starting point. Sometimes, it is impossible to define such single-particle states, and one must proceed directly to quantum field theory. For example, a quantum theory of the electromagnetic field *must* be a quantum field theory, because it is impossible (for various reasons) to define a wavefunction for a single photon.[10] In such situations, the quantum field theory can be constructed by examining the mechanical properties of the classical field and guessing the corresponding quantum theory. For free (non-interacting) quantum fields, the quantum field theories obtained in this way have the same properties as those obtained using second quantization, such as well-defined creation and annihilation operators obeying commutation or anticommutation relations.

Quantum field theory thus provides a unified framework for describing "field-like" objects (such as the electromagnetic field, whose excitations are photons) and "particle-like" objects (such as electrons, which are treated as excitations of an underlying electron field), so long as one can treat interactions as "perturbations" of free fields. There are still unsolved problems relating to the more general case of interacting fields that may or may not be adequately described by perturbation theory. For more on this topic, see Haag's theorem.

Physical meaning of particle indistinguishability

The second quantization procedure relies crucially on the particles being identical. We would not have been able to construct a quantum field theory from a distinguishable many-particle system, because there would have been no way of separating and indexing the degrees of freedom.

Many physicists prefer to take the converse interpretation, which is that *quantum field theory explains what identical particles are*. In ordinary quantum mechanics, there is not much theoretical motivation for using symmetric (bosonic) or antisymmetric (fermionic) states, and the need for such states is simply regarded as an empirical fact. From the point of view of quantum field theory, particles are identical if and only if they are excitations of the same underlying quantum field. Thus, the question "why are all electrons identical?" arises from mistakenly regarding individual electrons as fundamental objects, when in fact it is only the electron field that is fundamental.

Particle conservation and non-conservation

During second quantization, we started with a Hamiltonian and state space describing a fixed number of particles (N), and ended with a Hamiltonian and state space for an arbitrary number of particles. Of course, in many common situations N is an important and perfectly well-defined quantity, e.g. if we are describing a gas of atoms sealed in a box. From the point of view of quantum field theory, such situations are described by quantum states that are eigenstates of the number operator $\hat{N}$, which measures the total number of particles present. As with any quantum mechanical observable, $\hat{N}$ is conserved if it commutes with the Hamiltonian. In that case, the quantum state is trapped in the N-particle subspace of the total Fock space, and the situation could equally well be described by ordinary N-particle quantum mechanics. (Strictly speaking, this is only true in the noninteracting case or in the low energy density limit of renormalized quantum field theories)

For example, we can see that the free-boson Hamiltonian described above conserves particle number. Whenever the Hamiltonian operates on a state, each particle destroyed by an annihilation operator a_k is immediately put back by the creation operator $a_k^\dagger$.

On the other hand, it is possible, and indeed common, to encounter quantum states that are *not* eigenstates of $\hat{N}$, which do not have well-defined particle numbers. Such states are difficult or impossible to handle using ordinary quantum mechanics, but they can be easily described in quantum field theory as quantum superpositions of states having different values of N. For example, suppose we have a bosonic field whose particles can be created or destroyed by interactions with a fermionic field. The Hamiltonian of the combined system would be given by the Hamiltonians of the free boson and free fermion fields, plus a "potential energy" term such as

$$H_I = \sum_{k,q} V_q (a_q + a_{-q}^\dagger) c_{k+q}^\dagger c_k,$$

where $a_k^\dagger$ and a_k denotes the bosonic creation and annihilation operators, $c_k^\dagger$ and c_k denotes the fermionic creation and annihilation operators, and V_q is a parameter that describes the strength of the interaction. This "interaction term" describes processes in which a fermion in state k either absorbs or emits a boson, thereby being kicked into a different eigenstate $k + q$. (In fact, this type of Hamiltonian is used to describe interaction between conduction electrons and phonons in metals. The interaction between electrons and photons is treated in a similar way, but is a little more complicated because the role of spin must be taken into account.) One thing to notice here is that even if we start out with a fixed number of bosons, we will typically end up with a superposition of states with different numbers of bosons at later times. The number of fermions, however, is conserved in this case.

In condensed matter physics, states with ill-defined particle numbers are particularly important for describing the various superfluids. Many of the defining characteristics of a superfluid arise from the notion that its quantum state is a superposition of states with different particle numbers. In addition, the concept of a coherent state (used to model the laser and the BCS ground state) refers to a state with an ill-defined particle number but a well-defined phase.

7.4.6 Axiomatic approaches

The preceding description of quantum field theory follows the spirit in which most physicists approach the subject. However, it is not mathematically rigorous. Over the past several decades, there have been many attempts to put quantum field theory on a firm mathematical footing by formulating a set of axioms for it. These attempts fall into two broad classes.

The first class of axioms, first proposed during the 1950s, include the Wightman, Osterwalder–Schrader, and Haag–Kastler systems. They attempted to formalize the physicists' notion of an "operator-valued field" within the context of functional analysis, and enjoyed limited success. It was possible to prove that any quantum field theory satisfying these axioms satisfied certain general theorems, such as the spin-statistics theorem and the CPT theorem. Unfortunately, it proved extraordinarily difficult to show that any realistic field theory, including the Standard Model, satisfied these axioms. Most of the theories that could be treated with these analytic axioms were physically trivial, being restricted to low-dimensions and lacking interesting dynamics. The construction of theories satisfying one of these sets of axioms falls in the field of constructive quantum field theory. Important work was done in this area in the 1970s by Segal, Glimm, Jaffe and others.

During the 1980s, a second set of axioms based on geometric ideas was proposed. This line of investigation, which restricts its attention to a particular class of quantum field theories known as topological quantum field theories, is associated most closely with Michael Atiyah and Graeme Segal, and was notably expanded upon by Edward Witten, Richard Borcherds, and Maxim Kontsevich. However, most of the physically relevant quantum field theories, such as the Standard Model, are not topological quantum field theories; the quantum field theory of the fractional quantum Hall effect is a notable exception. The main impact of axiomatic topological quantum field theory has been on mathematics, with important applications in representation theory, algebraic topology, and differential geometry.

Finding the proper axioms for quantum field theory is still an open and difficult problem in mathematics. One of the Millennium Prize Problems—proving the existence of a mass gap in Yang–Mills theory—is linked to this issue.

7.5   Associated phenomena

In the previous part of the article, we described the most general features of quantum field theories. Some of the quantum field theories studied in various fields of theoretical physics involve additional special ideas, such as renormalizability, gauge symmetry, and supersymmetry. These are described in the following sections.

7.5.1   Renormalization

Main article: Renormalization

Early in the history of quantum field theory, it was found that many seemingly innocuous calculations, such as the perturbative shift in the energy of an electron due to the presence of the electromagnetic field, give infinite results. The reason is that the perturbation theory for the shift in an energy involves a sum over all other energy levels, and there are infinitely many levels at short distances that each give a finite contribution which results in a divergent series.

Many of these problems are related to failures in classical electrodynamics that were identified but unsolved in the 19th century, and they basically stem from the fact that many of the supposedly "intrinsic" properties of an electron are tied to the electromagnetic field that it carries around with it. The energy carried by a single electron—its self energy— is not simply the bare value, but also includes the energy contained in its electromagnetic field, its attendant cloud of photons. The energy in a field of a spherical source diverges in both classical and quantum mechanics, but as discovered by Weisskopf with help from Furry, in quantum mechanics the divergence is much milder, going only as the logarithm of the radius of the sphere.

The solution to the problem, presciently suggested by Stueckelberg, independently by Bethe after the crucial experiment by Lamb, implemented at one loop by Schwinger, and systematically extended to all loops by Feynman and Dyson, with converging work by Tomonaga in isolated postwar Japan, comes from recognizing that all the infinities in the interactions of photons and electrons can be isolated into redefining a finite number of quantities in the equations by replacing them with the observed values: specifically the electron's mass and charge: this is called renormalization. The technique of renormalization recognizes that the problem is essentially purely mathematical, that extremely short distances are at fault. In order to define a theory on a continuum, first place a cutoff on the fields, by postulating that quanta cannot have energies above some extremely high value. This has the effect of replacing continuous space by a structure where very short wavelengths do not exist, as on a lattice. Lattices break rotational symmetry, and one of the crucial contributions made by Feynman, Pauli and Villars, and modernized by 't Hooft and Veltman, is a symmetry-preserving cutoff for perturbation theory (this process is called regularization). There is no known symmetrical cutoff outside of perturbation theory, so for rigorous or numerical work people often use an actual lattice.

On a lattice, every quantity is finite but depends on the spacing. When taking the limit of zero spacing, we make sure that the physically observable quantities like the observed electron mass stay fixed, which means that the constants in the Lagrangian defining the theory depend on the spacing. Hopefully, by allowing the constants to vary with the lattice spacing, all the results at long distances become insensitive to the lattice, defining a continuum limit.

The renormalization procedure only works for a certain class of quantum field theories, called **renormalizable quantum field theories**. A theory is **perturbatively renormalizable** when the constants in the Lagrangian only diverge at worst as logarithms of the lattice spacing for very short spacings. The continuum limit is then well defined in perturbation theory, and even if it is not fully well defined non-perturbatively, the problems only show up at distance scales that are exponentially small in the inverse coupling for weak couplings. The Standard Model of particle physics is perturbatively renormalizable, and so are its component theories (quantum electrodynamics/electroweak theory and quantum chromodynamics). Of the three components, quantum electrodynamics is believed to not have a continuum limit, while the asymptotically free $SU(2)$ and $SU(3)$ weak hypercharge and strong color interactions are nonperturbatively well defined.

The renormalization group describes how renormalizable theories emerge as the long distance low-energy effective field theory for any given high-energy theory. Because of this, renormalizable theories are insensitive to the precise nature of the underlying high-energy short-distance phenomena. This is a blessing because it allows physicists to formulate low energy theories without knowing the details of high energy phenomenon. It is also a curse, because once a renormalizable theory like the standard model is found to work, it gives very few clues to higher energy processes. The only way high energy processes can be seen in the standard model is when they allow otherwise forbidden events, or if they predict quantitative relations between the coupling constants.

7.5.2 Haag's theorem

See also: Haag's theorem

From a mathematically rigorous perspective, there exists no interaction picture in a Lorentz-covariant quantum field theory. This implies that the perturbative approach of Feynman diagrams in QFT is not strictly justified, despite producing vastly precise predictions validated by experiment. This is called Haag's theorem, but most particle physicists relying on QFT largely shrug it off.

7.5.3 Gauge freedom

A gauge theory is a theory that admits a symmetry with a local parameter. For example, in every quantum theory the global phase of the wave function is arbitrary and does not represent something physical. Consequently, the theory is invariant under a global change of phases (adding a constant to the phase of all wave functions, everywhere); this is a global symmetry. In quantum electrodynamics, the theory is also invariant under a *local* change of phase, that is – one may shift the phase of all wave functions so that the shift may be different at every point in space-time. This is a *local* symmetry. However, in order for a well-defined derivative operator to exist, one must introduce a new field, the gauge field, which also transforms in order for the local change of variables (the phase in our example) not to affect the derivative. In quantum electrodynamics this gauge field is the electromagnetic field. The change of local gauge of variables is termed gauge transformation. It is worth noting that by Noether's theorem, for every such symmetry there exists an associated conserved current. The aforementioned symmetry of the wavefunction under global phase changes implies the conservation of electric charge.

In quantum field theory the excitations of fields represent particles. The particle associated with excitations of the gauge field is the gauge boson, which is the photon in the case of quantum electrodynamics.

The degrees of freedom in quantum field theory are local fluctuations of the fields. The existence of a gauge symmetry reduces the number of degrees of freedom, simply because some fluctuations of the fields can be transformed to zero by gauge transformations, so they are equivalent to having no fluctuations at all, and they therefore have no physical meaning. Such fluctuations are usually called "non-physical degrees of freedom" or *gauge artifacts*; usually some of them have a negative norm, making them inadequate for a consistent theory. Therefore, if a classical field theory has a gauge symmetry, then its quantized version (i.e. the corresponding quantum field theory) will have this symmetry as well. In other words, a gauge symmetry cannot have a quantum anomaly. If a gauge symmetry is anomalous (i.e. not kept in the quantum theory) then the theory is non-consistent: for example, in quantum electrodynamics, had there been a gauge anomaly, this would require the appearance of photons with longitudinal polarization and polarization in the time direction, the latter having a negative norm, rendering the theory inconsistent; another possibility would be for these photons to appear only in intermediate processes but not in the final products of any interaction, making the theory non-unitary and again inconsistent (see optical theorem).

In general, the gauge transformations of a theory consist of several different transformations, which may not be commutative. These transformations are together described by a mathematical object known as a gauge group. Infinitesimal gauge transformations are the gauge group generators. Therefore the number of gauge bosons is the group dimension (i.e. number of generators forming a basis).

All the fundamental interactions in nature are described by gauge theories. These are:

- Quantum chromodynamics, whose gauge group is $\mathbf{SU}(3)$. The gauge bosons are eight gluons.

- The electroweak theory, whose gauge group is $\mathbf{U}(1) \times \mathbf{SU}(2)$, (a direct product of $\mathbf{U}(1)$ and $\mathbf{SU}(2)$).

- Gravity, whose classical theory is general relativity, admits the equivalence principle, which is a form of gauge symmetry. However, it is explicitly non-renormalizable.

7.5.4 Multivalued gauge transformations

The gauge transformations which leave the theory invariant involve, by definition, only single-valued gauge functions $\Lambda(x_i)$ which satisfy the Schwarz integrability criterion

$$\partial_{x_i x_j}\Lambda = \partial_{x_j x_i}\Lambda.$$

An interesting extension of gauge transformations arises if the gauge functions $\Lambda(x_i)$ are allowed to be multivalued functions which violate the integrability criterion. These are capable of changing the physical field strengths and are therefore not proper symmetry transformations. Nevertheless, the transformed field equations describe correctly the physical laws in the presence of the newly generated field strengths. See the textbook by H. Kleinert cited below for the applications to phenomena in physics.

7.5.5 Supersymmetry

Main article: Supersymmetry

Supersymmetry assumes that every fundamental fermion has a superpartner that is a boson and vice versa. It was introduced in order to solve the so-called Hierarchy Problem, that is, to explain why particles not protected by any symmetry (like the Higgs boson) do not receive radiative corrections to its mass driving it to the larger scales (GUT, Planck...). It was soon realized that supersymmetry has other interesting properties: its gauged version is an extension of general relativity (Supergravity), and it is a key ingredient for the consistency of string theory.

The way supersymmetry protects the hierarchies is the following: since for every particle there is a superpartner with the same mass, any loop in a radiative correction is cancelled by the loop corresponding to its superpartner, rendering the theory UV finite.

Since no superpartners have yet been observed, if supersymmetry exists it must be broken (through a so-called soft term, which breaks supersymmetry without ruining its helpful features). The simplest models of this breaking require that the energy of the superpartners not be too high; in these cases, supersymmetry is expected to be observed by experiments at the Large Hadron Collider. The Higgs particle has been detected at the LHC, and no such superparticles have been discovered.

7.6 See also

- Abraham–Lorentz force

- Basic concepts of quantum mechanics

- Common integrals in quantum field theory

- Constructive quantum field theory

- Einstein–Maxwell–Dirac equations

- Feynman path integral

- Form factor (quantum field theory)

- Fundamental equation of unified field theory

- Green–Kubo relations

- Green's function (many-body theory)

- Invariance mechanics

- List of quantum field theories

- Pauli exclusion principle

- Photon polarization

- Pseudoscalar Field

- Quantum field theory in curved spacetime

- Quantum flavordynamics

- Quantum geometrodynamics

- Quantum hydrodynamics

- Quantum magnetodynamics

- Quantum triviality

- Relation between Schrödinger's equation and the path integral formulation of quantum mechanics

- Relationship between string theory and quantum field theory

- Schwinger–Dyson equation

- Static forces and virtual-particle exchange

- Symmetry in quantum mechanics

- Theoretical and experimental justification for the Schrödinger equation

- Ward–Takahashi identity

- Wheeler–Feynman absorber theory

- Wigner's classification

- Wigner's theorem

7.7 Notes

7.8 References

[1] "Beautiful Minds, Vol. 20: Ed Witten". la Repubblica. 2010. Retrieved 22 June 2012. See here.

[2] J. J. Thorn et al. (2004). Observing the quantum behavior of light in an undergraduate laboratory. . J. J. Thorn, M. S. Neel, V. W. Donato, G. S. Bergreen, R. E. Davies, and M. Beck. American Association of Physics Teachers, 2004.DOI: 10.1119/1.1737397.

[3] David Tong, *Lectures on Quantum Field Theory*, chapter 1.

[4] Srednicki, Mark. *Quantum Field Theory* (1st ed.). p. 19.

[5] Srednicki, Mark. *Quantum Field Theory* (1st ed.). pp. 25–6.

[6] Zee, Anthony. *Quantum Field Theory in a Nutshell* (2nd ed.). p. 61.

[7] David Tong, *Lectures on Quantum Field Theory*, Introduction.

[8] Zee, Anthony. *Quantum Field Theory in a Nutshell* (2nd ed.). p. 3.

[9] Abraham Pais, *Inward Bound: Of Matter and Forces in the Physical World* ISBN 0-19-851997-4. Pais recounts how his astonishment at the rapidity with which Feynman could calculate using his method. Feynman's method is now part of the standard methods for physicists.

[10] Newton, T.D.; Wigner, E.P. (1949). "Localized states for elementary systems". *Reviews of Modern Physics* **21** (3): 400–406. Bibcode:1949RvMP...21..400N. doi:10.1103/RevModPhys.21.400.

7.9 Further reading

General readers

- Feynman, R.P. (2001) [1964]. *The Character of Physical Law*. MIT Press. ISBN 0-262-56003-8.

- Feynman, R.P. (2006) [1985]. *QED: The Strange Theory of Light and Matter*. Princeton University Press. ISBN 0-691-12575-9.

- Gribbin, J. (1998). *Q is for Quantum: Particle Physics from A to Z*. Weidenfeld & Nicolson. ISBN 0-297-81752-3.

- Schumm, Bruce A. (2004) *Deep Down Things*. Johns Hopkins Univ. Press. Chpt. 4.

Introductory texts

- McMahon, D. (2008). *Quantum Field Theory*. McGraw-Hill. ISBN 978-0-07-154382-8.

- Bogoliubov, N.; Shirkov, D. (1982). *Quantum Fields*. Benjamin-Cummings. ISBN 0-8053-0983-7.

- Frampton, P.H. (2000). *Gauge Field Theories. Frontiers in Physics (2nd ed.). Wiley.*

- Greiner, W; Müller, B. (2000). *Gauge Theory of Weak Interactions*. Springer. ISBN 3-540-67672-4.

- Itzykson, C.; Zuber, J.-B. (1980). *Quantum Field Theory*. McGraw-Hill. ISBN 0-07-032071-3.

- Kane, G.L. (1987). *Modern Elementary Particle Physics*. Perseus Books. ISBN 0-201-11749-5.

- Kleinert, H.; Schulte-Frohlinde, Verena (2001). *Critical Properties of φ^4-Theories*. World Scientific. ISBN 981-02-4658-7.

- Kleinert, H. (2008). *Multivalued Fields in Condensed Matter, Electrodynamics, and Gravitation* (PDF). World Scientific. ISBN 978-981-279-170-2.

- Loudon, R (1983). *The Quantum Theory of Light*. Oxford University Press. ISBN 0-19-851155-8.

- Mandl, F.; Shaw, G. (1993). *Quantum Field Theory*. John Wiley & Sons. ISBN 978-0-471-94186-6.

- Peskin, M.; Schroeder, D. (1995). *An Introduction to Quantum Field Theory*. Westview Press. ISBN 0-201-50397-2.

- Ryder, L.H. (1985). *Quantum Field Theory*. Cambridge University Press. ISBN 0-521-33859-X.

- Schwartz, M.D. (2014). *Quantum Field Theory and the Standard Model*. Cambridge University Press. ISBN 978-1107034730.

- Srednicki, Mark (2007) *Quantum Field Theory*. Cambridge Univ. Press.

- Yndauráin, F.J. (1996). *Relativistic Quantum Mechanics and Introduction to Field Theory* (1st ed.). Springer. ISBN 978-3-540-60453-2.

- Zee, A. (2003). *Quantum Field Theory in a Nutshell*. Princeton University Press. ISBN 0-691-01019-6.

Advanced texts

- Brown, Lowell S. (1994). *Quantum Field Theory*. Cambridge University Press. ISBN 978-0-521-46946-3.

- Bogoliubov, N.; Logunov, A.A.; Oksak, A.I.; Todorov, I.T. (1990). *General Principles of Quantum Field Theory*. Kluwer Academic Publishers. ISBN 978-0-7923-0540-8.

- Weinberg, S. (1995). *The Quantum Theory of Fields* **1–3**. Cambridge University Press.

Articles:

- Gerard 't Hooft (2007) "The Conceptual Basis of Quantum Field Theory" in Butterfield, J., and John Earman, eds., *Philosophy of Physics, Part A*. Elsevier: 661–730.

- Frank Wilczek (1999) "Quantum field theory", *Reviews of Modern Physics* 71: S83–S95. Also doi=10.1103/Rev. Mod. Phys. 71.

7.10 External links

- Hazewinkel, Michiel, ed. (2001), "Quantum field theory", *Encyclopedia of Mathematics*, Springer, ISBN 978-1-55608-010-4

- Stanford Encyclopedia of Philosophy: "Quantum Field Theory", by Meinard Kuhlmann.

- Siegel, Warren, 2005. *Fields*. A free text, also available from arXiv:hep-th/9912205.

- Quantum Field Theory by P. J. Mulders

Chapter 8

Supersymmetry algebra

In theoretical physics, a **supersymmetry algebra** (or **SUSY algebra**) is a mathematical formalism for describing the relation between bosons and fermions. The supersymmetry algebra contains not only the Poincaré algebra and a compact subalgebra of internal symmetries, but also contains some fermionic supercharges, transforming as a sum of N real spinor representations of the Poincaré group. Such symmetries are allowed by the Haag–Lopuszanski–Sohnius theorem. When $N>1$ the algebra is said to have **extended supersymmetry**. The supersymmetry algebra is a semidirect product of a central extension of the super-Poincaré algebra by a compact Lie algebra B of internal symmetries.

Bosonic fields commute while fermionic fields anticommute. In order to have a transformation that relates the two kinds of fields, the introduction of a $\mathbf{Z}_2$-grading under which the even elements are bosonic and the odd elements are fermionic is required. Such an algebra is called a Lie superalgebra.

Just as one can have representations of a Lie algebra, one can also have representations of a Lie superalgebra, called supermultiplets. For each Lie algebra, there exists an associated Lie group which is connected and simply connected, unique up to isomorphism, and the representations of the algebra can be extended to create group representations. In the same way, representations of a Lie superalgebra can sometimes be extended into representations of a Lie supergroup.

8.1   Structure of a supersymmetry algebra

The general supersymmetry algebra for spacetime dimension d, and with the fermionic piece consisting of a sum of N irreducible real spinor representations, has a structure of the form

$(P{\times}Z).Q.(L{\times}B)$

where

- P is a bosonic abelian vector normal subalgebra of dimension d, normally identified with translations of spacetime. It is a vector representation of L.

- Z is a scalar bosonic algebra in the center whose elements are called central charges.

- Q is an abelian fermionic spinor subquotient algebra, and is a sum of N real spinor representations of L. (When the signature of spacetime is divisible by 4 there are two different spinor representations of L, so there is some ambiguity about the structure of Q as a representation of L.) The elements of Q, or rather their inverse images in the supersymmetry algebra, are called supercharges. The subalgebra $(P{\times}Z).Q$ is sometimes also called the super-symmetry algebra and is nilpotent of length at most 2, with the Lie bracket of two supercharges lying in $P{\times}Z$.

- L is a bosonic subalgebra, isomorphic to the Lorentz algebra in d dimensions, of dimension $d(d-1)/2$

- *B* is a scalar bosonic subalgebra, given by the Lie algebra of some compact group, called the group of internal symmetries. It commutes with *P,Z*, and *L*, but may act non-trivially on the supercharges *Q*.

The terms "bosonic" and "fermionic" refer to even and odd subspaces of the superalgebra.

The terms "scalar", "spinor", "vector", refer to the behavior of subalgebras under the action of the Lorentz algebra *L*.

The number N is the number of irreducible real spin representations. When the signature of spacetime is divisible by 4 this is ambiguous as in this case there are two different irreducible real spinor representations, and the number N is sometimes replaced by a pair of integers (N_1, N_2).

The supersymmetry algebra is sometimes regarded as a real super algebra, and sometimes as a complex algebra with a hermitian conjugation. These two views are essentially equivalent, as the real algebra can be constructed from the complex algebra by taking the skew-Hermitian elements, and the complex algebra can be constructed from the real one by taking tensor product with the complex numbers.

The bosonic part of the superalgebra is isomorphic to the product of the Poincaré algebra *P.L* with the algebra *Z×B* of internal symmetries.

When $N>1$ the algebra is said to have **extended supersymmetry**.

When *Z* is trivial, the subalgebra *P.Q.L* is the Super-Poincaré algebra.

8.2 See also

- adinkra symbols

- super-Poincaré algebra

- superconformal algebra

- $N = 1$ supersymmetry algebra in $1 + 1$ dimensions

- $N = 2$ superconformal algebra

8.3 References

- Bagger, Jonathan; Wess, Julius (1992), *Supersymmetry and supergravity*, Princeton Series in Physics (2nd ed.), Princeton University Press, ISBN 0-691-02530-4, MR 1152804

- Haag, Rudolf; Sohnius, Martin; Łopuszański, Jan T. (1975), "All possible generators of supersymmetries of the S-matrix", *Nuclear Physics B* **88**: 257–274, Bibcode:1975NuPhB..88..257H, doi:10.1016/0550-3213(75)90279-5, MR 0411396

Chapter 9

Supercommutative algebra

In mathematics, a **supercommutative algebra** is a superalgebra (i.e. a $\mathbf{Z}_2$-graded algebra) such that for any two homogeneous elements x, y we have[1]

$$yx = (-1)^{|x||y|}xy.$$

Equivalently, it is a superalgebra where the supercommutator

$$[x, y] = xy - (-1)^{|x||y|}yx$$

always vanishes. Algebraic structures which supercommute in the above sense are sometimes referred to as **skew-commutative associative algebras** to emphasize the anti-commutation, or, to emphasize the grading, **graded-commutative** or, if the supercommutativity is understood, simply **commutative**.

Any commutative algebra is a supercommutative algebra if given the trivial gradation (i.e. all elements are even). Grassmann algebras (also known as exterior algebras) are the most common examples of nontrivial supercommutative algebras. The **supercenter** of any superalgebra is the set of elements that supercommute with all elements, and is a supercommutative algebra.

The even subalgebra of a supercommutative algebra is always a commutative algebra. That is, even elements always commute. Odd elements, on the other hand, always anticommute. That is,

$$xy + yx = 0$$

for odd x and y. In particular, the square of any odd element x vanishes whenever 2 is invertible:

$$x^2 = 0.$$

Thus a commutative superalgebra (with 2 invertible and nonzero degree one component) always contains nilpotent elements.

9.1   See also

- Commutative algebra
- Lie superalgebra

9.2 References

[1] Varadarajan, V. S. *Supersymmetry for Mathematicians: An Introduction*. American Mathematical Society. p. 76. ISBN 9780821883518.

Chapter 10

Supersymmetric quantum mechanics

In theoretical physics, **supersymmetric quantum mechanics** is an area of research where mathematical concepts from high-energy physics are applied to the field of quantum mechanics.

10.1   Introduction

Understanding the consequences of supersymmetry has proven mathematically daunting, and it has likewise been difficult to develop theories that could account for symmetry breaking, *i.e.*, the lack of observed partner particles of equal mass. To make progress on these problems, physicists developed *supersymmetric quantum mechanics*, an application of the supersymmetry (SUSY) superalgebra to quantum mechanics as opposed to quantum field theory. It was hoped that studying SUSY's consequences in this simpler setting would lead to new understanding; remarkably, the effort created new areas of research in quantum mechanics itself.

For example, as of 2004 students are typically taught to "solve" the hydrogen atom by a laborious process which begins by inserting the Coulomb potential into the Schrödinger equation. After a considerable amount of work using many differential equations, the analysis produces a recursion relation for the Laguerre polynomials. The final outcome is the spectrum of hydrogen-atom energy states (labeled by quantum numbers n and l). Using ideas drawn from SUSY, the final result can be derived with significantly greater ease, in much the same way that operator methods are used to solve the harmonic oscillator.[1] A similar supersymmetric approach can also be used to more accurately find the hydrogen spectrum using the Dirac equation.[2] Oddly enough, this approach is analogous to the way Erwin Schrödinger first solved the hydrogen atom.[3][4] Of course, he did not *call* his solution supersymmetric, as SUSY was thirty years in the future.

The SUSY solution of the hydrogen atom is only one example of the very general class of solutions which SUSY provides to *shape-invariant potentials*, a category which includes most potentials taught in introductory quantum mechanics courses.

SUSY quantum mechanics involves pairs of Hamiltonians which share a particular mathematical relationship, which are called *partner Hamiltonians*. (The potential energy terms which occur in the Hamiltonians are then called *partner potentials*.) An introductory theorem shows that for every eigenstate of one Hamiltonian, its partner Hamiltonian has a corresponding eigenstate with the same energy (except possibly for zero energy eigenstates). This fact can be exploited to deduce many properties of the eigenstate spectrum. It is analogous to the original description of SUSY, which referred to bosons and fermions. We can imagine a "bosonic Hamiltonian", whose eigenstates are the various bosons of our theory. The SUSY partner of this Hamiltonian would be "fermionic", and its eigenstates would be the theory's fermions. Each boson would have a fermionic partner of equal energy—but, in the relativistic world, energy and mass are interchangeable, so we can just as easily say that the partner particles have equal mass.

SUSY concepts have provided useful extensions to the WKB approximation in the form of a modified version of the Bohr-Sommerfeld quantization condition. In addition, SUSY has been applied to non-quantum statistical mechanics through the Fokker-Planck equation, showing that even if the original inspiration in high-energy particle physics turns out to be a blind alley, its investigation has brought about many useful benefits.

10.2 Example: the harmonic oscillator

The Schrödinger equation for the harmonic oscillator takes the form

$$H^{HO}\psi_n(x) = \left(\frac{-\hbar^2}{2m}\frac{d^2}{dx^2} + \frac{m\omega^2}{2}x^2\right)\psi_n(x) = E_n^{HO}\psi_n(x),$$

where $\psi_n(x)$ is the n th energy eigenstate of H^{HO} with energy E_n^{HO}. We want to find an expression for E_n^{HO} in terms of n. We define the operators

$$A = \frac{\hbar}{\sqrt{2m}}\frac{d}{dx} + W(x)$$

and

$$A^\dagger = -\frac{\hbar}{\sqrt{2m}}\frac{d}{dx} + W(x),$$

where $W(x)$, which we need to choose, is called the superpotential of H^{HO}. We also define the aforementioned partner Hamiltonians $H^{(1)}$ and $H^{(2)}$ as

$$H^{(1)} = A^\dagger A = \frac{-\hbar^2}{2m}\frac{d^2}{dx^2} - \frac{\hbar}{\sqrt{2m}}W'(x) + W^2(x)$$

$$H^{(2)} = AA^\dagger = \frac{-\hbar^2}{2m}\frac{d^2}{dx^2} + \frac{\hbar}{\sqrt{2m}}W'(x) + W^2(x).$$

A zero energy ground state $\psi_0^{(1)}(x)$ of $H^{(1)}$ would satisfy the equation

$$H^{(1)}\psi_0^{(1)}(x) = A^\dagger A\psi_0^{(1)}(x) = A^\dagger\left(\frac{\hbar}{\sqrt{2m}}\frac{d}{dx} + W(x)\right)\psi_0^{(1)}(x) = 0.$$

Assuming that we know the ground state of the harmonic oscillator $\psi_0(x)$, we can solve for $W(x)$ as

$$W(x) = \frac{-\hbar}{\sqrt{2m}}\left(\frac{\psi_0'(x)}{\psi_0(x)}\right) == x\sqrt{m\omega^2/2}$$

We then find that

$$H^{(1)} = \frac{-\hbar^2}{2m}\frac{d^2}{dx^2} - \frac{m\omega^2}{2}x^2 - \frac{\hbar\omega}{2}$$

$$H^{(2)} = \frac{-\hbar^2}{2m}\frac{d^2}{dx^2} - \frac{m\omega^2}{2}x^2 + \frac{\hbar\omega}{2}.$$

We can now see that

$$H^{(1)} = H^{(2)} - \hbar\omega = H^{HO} - \frac{\hbar\omega}{2}.$$

This is a special case of shape invariance, discussed below. Taking without proof the introductory theorem mentioned above, it is apparent that the spectrum of $H^{(1)}$ will start with $E_0 = 0$ and continue upwards in steps of $\hbar\omega$. The spectra of $H^{(2)}$ and H^{HO} will have the same even spacing, but will be shifted up by amounts $\hbar\omega$ and $\hbar\omega/2$, respectively. It follows that the spectrum of H^{HO} is therefore the familiar $E_n^{HO} = \hbar\omega(n + 1/2)$.

10.3   The SUSY QM superalgebra

In fundamental quantum mechanics, we learn that an algebra of operators is defined by commutation relations among those operators. For example, the canonical operators of position and momentum have the commutator [x,p]=i. (Here, we use "natural units" where Planck's constant is set equal to 1.) A more intricate case is the algebra of angular momentum operators; these quantities are closely connected to the rotational symmetries of three-dimensional space. To generalize this concept, we define an *anticommutator,* which relates operators the same way as an ordinary commutator, but with the opposite sign:

$$\{A, B\} = AB + BA.$$

If operators are related by anticommutators as well as commutators, we say they are part of a *Lie superalgebra.* Let's say we have a quantum system described by a Hamiltonian $\mathcal{H}$ and a set of N self-adjoint operators Qi. We shall call this system *supersymmetric* if the following anticommutation relation is valid for all $i, j = 1, \ldots, N$:

$$\{Q_i, Q_j^\dagger\} = \mathcal{H}\delta_{ij}.$$

If this is the case, then we call Q_i the system's *supercharges.*

10.4   Example

Let's look at the example of a one-dimensional nonrelativistic particle with a 2D (*i.e.,* two states) internal degree of freedom called "spin" (it's not really spin because "real" spin is a property of 3D particles). Let b be an operator which transforms a "spin up" particle into a "spin down" particle. Its adjoint $b^\dagger$ then transforms a spin down particle into a spin up particle; the operators are normalized such that the anticommutator $\{b, b^\dagger\} = 1$. And of course, $b^2 = 0$. Let p be the momentum of the particle and x be its position with $[x, p] = i$. Let W (the "superpotential") be an arbitrary complex analytic function of x and define the supersymmetric operators

$$Q_1 = \frac{1}{2}\left[(p - iW)b + (p + iW^\dagger)b^\dagger\right]$$

$$Q_2 = \frac{i}{2}\left[(p - iW)b - (p + iW^\dagger)b^\dagger\right]$$

Note that Q_1 and Q_2 are self-adjoint. Let the Hamiltonian

$$H = \{Q_1, Q_1\} = \{Q_2, Q_2\} = \frac{(p + \Im\{W\})^2}{2} + \frac{\Re\{W\}^2}{2} + \frac{\Re\{W\}'}{2}(bb^\dagger - b^\dagger b)$$

where W' is the derivative of W. Also note that $\{Q_1,Q_2\}$=0. This is nothing other than $N = 2$ supersymmetry. Note that $\Im\{W\}$ acts like an electromagnetic vector potential.

Let's also call the spin down state "bosonic" and the spin up state "fermionic". This is only in analogy to quantum field theory and should not be taken literally. Then, Q_1 and Q_2 maps "bosonic" states into "fermionic" states and vice versa.

Let's reformulate this a bit:

Define

$$Q = (p - iW)b$$

and of course,

$$Q^\dagger = (p + iW^\dagger)b^\dagger$$

$$\{Q, Q\} = \{Q^\dagger, Q^\dagger\} = 0$$

and

$$\{Q^\dagger, Q\} = 2H$$

An operator is "bosonic" if it maps "bosonic" states to "bosonic" states and "fermionic" states to "fermionic" states. An operator is "fermionic" if it maps "bosonic" states to "fermionic" states and vice versa. Any operator can be expressed uniquely as the sum of a bosonic operator and a fermionic operator. Define the supercommutator [,} as follows: Between two bosonic operators or a bosonic and a fermionic operator, it is none other than the commutator but between two fermionic operators, it is an anticommutator.

Then, x and p are bosonic operators and b, $b^\dagger$, Q and $Q^\dagger$ are fermionic operators.

Let's work in the Heisenberg picture where x, b and $b^\dagger$ are functions of time.

Then,

$$[Q, x\} = -ib$$

$$[Q, b\} = 0$$

$$[Q, b^\dagger\} = \frac{dx}{dt} - i\Re\{W\}$$

$$[Q^\dagger, x\} = ib^\dagger$$

$$[Q^\dagger, b\} = \frac{dx}{dt} + i\Re\{W\}$$

$$[Q^\dagger, b^\dagger\} = 0$$

This is nonlinear in general: *i.e.*, x(t), b(t) and $b^\dagger(t)$ do not form a linear SUSY representation because $\Re\{W\}$ isn't necessarily linear in *x*. To avoid this problem, define the self-adjoint operator $F = \Re\{W\}$. Then,

$$[Q, x\} = -ib$$

$$[Q, b\} = 0$$

$$[Q, b^\dagger\} = \frac{dx}{dt} - iF$$

$$[Q, F\} = -\frac{db}{dt}$$

$$[Q^\dagger, x\} = ib^\dagger$$

$$[Q^\dagger, b\} = \frac{dx}{dt} + iF$$

$$[Q^\dagger, b^\dagger\} = 0$$

$$[Q^\dagger, F\} = \frac{db^\dagger}{dt}$$

and we see that we have a linear SUSY representation.

Now let's introduce two "formal" quantities, θ ; and $\bar{\theta}$ with the latter being the adjoint of the former such that

$$\{\theta,\theta\} = \{\bar{\theta},\bar{\theta}\} = \{\bar{\theta},\theta\} = 0$$

and both of them commute with bosonic operators but anticommute with fermionic ones.

Next, we define a construct called a superfield:

$$f(t,\bar{\theta},\theta) = x(t) - i\theta b(t) - i\bar{\theta}b^\dagger(t) + \bar{\theta}\theta F(t)$$

f is self-adjoint, of course. Then,

$$[Q,f] = \frac{\partial}{\partial\theta}f - i\bar{\theta}\frac{\partial}{\partial t}f,$$

$$[Q^\dagger,f] = \frac{\partial}{\partial\bar{\theta}}f - i\theta\frac{\partial}{\partial t}f.$$

Incidentally, there's also a U(1)R symmetry, with p and x and W having zero R-charges and $b^\dagger$ having an R-charge of 1 and b having an R-charge of -1.

10.5   Shape invariance

Suppose W is real for all real x . Then we can simplify the expression for the Hamiltonian to

$$H = \frac{(p)^2}{2} + \frac{W^2}{2} + \frac{W'}{2}(bb^\dagger - b^\dagger b)$$

There are certain classes of superpotentials such that both the bosonic and fermionic Hamiltonians have similar forms. Specifically

$$V_+(x,a_1) = V_-(x,a_2) + R(a_1)$$

where the a 's are parameters. For example, the hydrogen atom potential with angular momentum l can be written this way.

$$\frac{-e^2}{4\pi\epsilon_0}\frac{1}{r} + \frac{h^2 l(l+1)}{2m}\frac{1}{r^2} - E_0$$

This corresponds to V_- for the superpotential

$$W = \frac{\sqrt{2m}}{h}\frac{e^2}{24\pi\epsilon_0(l+1)} - \frac{h(l+1)}{r\sqrt{2m}}$$

$$V_+ = \frac{-e^2}{4\pi\epsilon_0}\frac{1}{r} + \frac{h^2(l+1)(l+2)}{2m}\frac{1}{r^2} + \frac{e^4 m}{32\pi^2 h^2\epsilon_0^2(l+1)^2}$$

This is the potential for $l+1$ angular momentum shifted by a constant. After solving the $l=0$ ground state, the supersymmetric operators can be used to construct the rest of the bound state spectrum.

In general, since V_- and V_+ are partner potentials, they share the same energy spectrum except the one extra ground energy. We can continue this process of finding partner potentials with the shape invariance condition, giving the following formula for the energy levels in terms of the parameters of the potential

$$E_n = \sum_{i=1}^{n} R(a_i)$$

where a_i are the parameters for the multiple partnered potentials.

10.6　See also

- Supersymmetry algebra
- Superalgebra

10.7　References

[1] Valance, A.; Morgan, T. J.; Bergeron, H. (1990), "Eigensolution of the Coulomb Hamiltonian via supersymmetry", *American Journal of Physics* (AAPT) **58** (5): 487–491, Bibcode:1990AmJPh..58..487V, doi:10.1119/1.16452

[2] Thaller, B. (1992). The Dirac Equation. Texts and Monographs in Physics. Springer.

[3] Schrödinger, Erwin (1940), "A Method of Determining Quantum-Mechanical Eigenvalues and Eigenfunctions", *Proceedings of the Royal Irish Academy* (Royal Irish Academy) **46**: 9–16

[4] Schrödinger, Erwin (1941), "Further Studies on Solving Eigenvalue Problems by Factorization", *Proceedings of the Royal Irish Academy* (Royal Irish Academy) **46**: 183–206

10.8　Sources

- F. Cooper, A. Khare and U. Sukhatme, "Supersymmetry and Quantum Mechanics", Phys.Rept.251:267-385, 1995.

10.9　External links

- References from INSPIRE-HEP

Chapter 11

Supersymmetry as a quantum group

The concept in theoretical physics of supersymmetry can be reinterpreted in the language of noncommutative geometry and quantum groups. In particular, it involves a mild form of noncommutativity, namely supercommutativity.

11.1 Unitary $(-1)^F$ operator

Following is the essence of supersymmetry, which is encapsulated within the following minimal quantum group. We have the two dimensional Hopf algebra generated by (-1)F subject to

$$(-1)^{F^2} = 1$$

with the counit

$$\epsilon((-1)^F) = 1$$

and the coproduct

$$\Delta(-1)^F = (-1)^F \otimes (-1)^F$$

and the antipode

$$S(-1)^F = (-1)^F$$

Thus far, there is nothing supersymmetric about this Hopf algebra at all; it is isomorphic to the Hopf algebra of the two element group $\mathbb{Z}_2$. Supersymmetry comes in when introducing the nontrivial quasitriangular structure

$$\mathcal{R} = \frac{1}{2} \left[1 \otimes 1 + (-1)^F \otimes 1 + 1 \otimes (-1)^F - (-1)^F \otimes (-1)^F \right]$$

where +1 eigenstates of $(-1)^F$ are called bosons and -1 eigenstates are called fermions.

This describes a fermionic braiding; don't pick up a phase factor when interchanging two bosons or a boson and a fermion, but multiply by -1 when interchanging two fermions. This provides the essence of the boson/fermion distinction.

11.2   Fermionic operators

The previous analysis only introduced the concept of fermions, and is not actual supersymmetry. The Hopf algebra is $\mathbb{Z}_2$ graded and contains even and odd elements. Even elements commute with $(-1)^F$; odd ones anticommute. The subalgebra not containing (-1)^F is supercommutative.

Let's say we are dealing with a super Lie algebra with even generators x and odd generators y.

Then,

$$\Delta x = x \otimes 1 + 1 \otimes x$$

$$\Delta y = y \otimes 1 + (-1)^F \otimes y$$

This is compatible with $\mathcal{R}$.

Supersymmetry is the symmetry over systems where interchanging two fermions attain a minus sign.

11.3   See also

- Introduction to quantum mechanics
- Group theory
- Symmetry group

Chapter 12

Superspace

"**Superspace**" is the coordinate space of a theory exhibiting supersymmetry. In such a formulation, along with ordinary space dimensions x, y, z, ..., there are also "anticommuting" dimensions whose coordinates are labeled in Grassmann numbers rather than real numbers. The ordinary space dimensions correspond to bosonic degrees of freedom, the anticommuting dimensions to fermionic degrees of freedom.

See also supermanifold (although the definition of a superspace as a supermanifold here does not agree with the definition used in that article).

$\mathbf{R}^{m|n}$ is the $\mathbf{Z}_2$-graded vector space with $\mathbf{R}^m$ as the even subspace and $\mathbf{R}^n$ as the odd subspace. The same definition applies to $\mathbf{C}^{m|n}$.

The word "superspace" was first used by John Wheeler in an unrelated sense to describe the configuration space of general relativity; for example, this usage may be seen in his 1973 textbook *Gravitation*.

12.1 Examples

12.1.1 Trivial examples

The smallest superspace is a point which contains neither bosonic nor fermionic directions. Other trivial examples include the n-dimensional real plane $\mathbf{R}^n$, which is a vector space extending in n real, bosonic directions and no fermionic directions. The vector space $\mathbf{R}^{0|n}$, which is the n-dimensional real Grassmann algebra. The space $\mathbf{R}^{1|1}$ of one even and one odd direction is known as the space of dual numbers, introduced by William Clifford in 1873.

12.1.2 The superspace of supersymmetric quantum mechanics

Supersymmetric quantum mechanics with N supercharges is often formulated in the superspace $\mathbf{R}^{1|2N}$, which contains one real direction t identified with time and N complex Grassmann directions which are spanned by Θi and $\Theta^* i$, where i runs for 1 to N.

Consider the special case $N = 1$. The superspace $\mathbf{R}^{1|2}$ is a 3-dimensional vector space. A given coordinate therefore may be written as a triple (t, Θ, Θ^*). The coordinates form a Lie superalgebra, in which the gradation degree of t is even and that of Θ and Θ^* is odd. This means that a bracket may be defined between any two elements of this vector space, and that this bracket reduces to the commutator on two even coordinates and on one even and one odd coordinate while it is an anticommutator on two odd coordinates. This superspace is an abelian Lie superalgebra, which means that all of the forementioned brackets vanish

$$[t, t] = [t, \theta] = [t, \theta^*] = \{\theta, \theta\} = \{\theta, \theta^*\} = \{\theta^*, \theta^*\} = 0$$

where $[a, b]$ is the commutator of a and b and $\{a, b\}$ is the anticommutator of a and b.

One may define functions from this vectorspace to itself, which are called superfields. The above algebraic relations imply that, if we expand our superfield as a power series in Θ and Θ^* then we will only find terms at the zeroeth and first orders, because $\Theta^2 = \Theta^{*2} = 0$. Therefore superfields may be written as arbitrary functions of t multiplied by the zeroeth and first order terms in the two Grassmann coordinates

$$\Phi(t, \Theta, \Theta^*) = \phi(t) + \Theta\Psi(t) - \Theta^*\Phi^*(t) + \Theta\Theta^*F(t)$$

Superfields, which are representations of the supersymmetry of superspace, generalize the notion of tensors, which are representations of the rotation group of a bosonic space.

One may then define derivatives in the Grassmann directions, which take the first order term in the expansion of a superfield to the zeroeth order term and annihilate the zeroeth order term. One can choose sign conventions such that the derivatives satisfy the anticommutation relations

$$\left\{\frac{\partial}{\partial\theta}, \Theta\right\} = \left\{\frac{\partial}{\partial\theta^*}, \Theta^*\right\} = 1$$

These derivatives may be assembled into supercharges

$$Q = \frac{\partial}{\partial\theta} + i\Theta^*\frac{\partial}{\partial t} \quad \text{and} \quad Q^\dagger = \frac{\partial}{\partial\theta^*} + i\Theta\frac{\partial}{\partial t}$$

whose anticommutators identify them as the fermionic generators of a supersymmetry algebra

$$\{Q, Q^\dagger\} = 2i\frac{\partial}{\partial t}$$

where i times the time derivative is the Hamiltonian operator in quantum mechanics. Both Q and its adjoint anticommute with themselves. The supersymmetry variation with supersymmetry parameter ε of a superfield Φ is defined to be

$$\delta_\epsilon\Phi = (\epsilon^*Q + \epsilon Q^\dagger)\Phi.$$

We can evaluate this variation using the action of Q on the superfields

$$[Q, \Phi] = \left(\frac{\partial}{\partial\theta} + i\theta^*\frac{\partial}{\partial t}\right)\Phi = \psi + \theta^*\left(F + i\dot\phi\right) - i\theta\theta^*\dot\psi.$$

Similarly one may define covariant derivatives on superspace

$$D = \frac{\partial}{\partial\theta} - i\theta\frac{\partial}{\partial t} \quad \text{and} \quad D^\dagger = \frac{\partial}{\partial\theta^*} - i\theta\frac{\partial}{\partial t}$$

which anticommute with the supercharges and satisfy a wrong sign supersymmetry algebra

$$\{D, D^\dagger\} = -2i\frac{\partial}{\partial t}$$

The fact that the covariant derivatives anticommute with the supercharges means the supersymmetry transformation of a covariant derivative of a superfield is equal to the covariant derivative of the same supersymmetry transformation of the same superfield. Thus, generalizing the covariant derivative in bosonic geometry which constructs tensors from tensors, the superspace covariant derivative constructs superfields from superfields.

12.1.3   Four-dimensional $N = 1$ superspace

Perhaps the most popular superspace in physics is d=4 N=1 super Minkowski space $\mathbf{R}^{4|4}$, which is the direct sum of four real bosonic dimensions and four real Grassmann dimensions. In supersymmetric quantum field theories one is interested in superspaces which furnish representations of a Lie superalgebra called a supersymmetry algebra. The bosonic part of the supersymmetry algebra is the Poincaré algebra, while the fermionic part is constructed using spinors of Grassmann numbers.

For this reason, in physical applications one considers an action of the supersymmetry algebra on the four fermionic directions of $\mathbf{R}^{4|4}$ such that they transform as a spinor under the Poincaré subalgebra. In four dimensions there are three distinct irreducible 4-component spinors. There is the Majorana spinor, the left-handed Weyl spinor and the right-handed Weyl spinor. The CPT theorem implies that in a unitary, Poincaré invariant theory, which is a theory in which the S-matrix is a unitary matrix and the same Poincaré generators act on the asymptotic in-states as on the asymptotic out-states, the supersymmetry algebra must contain an equal number of left-handed and right-handed Weyl spinors. However, since each Weyl spinor has four components, this means that if one includes any Weyl spinors one must have 8 fermionic directions. Such a theory is said to have extended supersymmetry, and such models have received a lot of attention. For example, supersymmetric gauge theories with eight supercharges and fundamental matter have been solved by Nathan Seiberg and Edward Witten, see Seiberg–Witten gauge theory. However in this subsection we are considering the superspace with four fermionic components and so no Weyl spinors are consistent with the CPT theorem.

Note: There are many sign conventions in use and this is only one of them.

This leaves us with one possibility, the four fermionic directions transform as a Majorana spinor $\theta\alpha$. We can also form a conjugate spinor

$$\overline{\theta} \overset{\text{def}}{=} i\theta^\dagger \gamma^0 = -\theta^\perp C$$

where C is the charge conjugation matrix, which is defined by the property that when it conjugates a gamma matrix, the gamma matrix is negated and transposed. The first equality is the definition of θ while the second is a consequence of the Majorana spinor condition $\theta^* = i\gamma_0 C\theta$. The conjugate spinor plays a role similar to that of θ^* in the superspace $\mathbf{R}^{1|2}$, except that the Majorana condition, as manifested in the above equation, imposes that θ and θ^* are not independent.

In particular we may construct the supercharges

$$Q = -\frac{\partial}{\partial\overline{\theta}} + \gamma^\mu \theta \partial_\mu$$

which satisfy the supersymmetry algebra

$$\{Q, Q\} = \{\overline{Q}, Q\} C = 2\gamma^\mu \partial_\mu C = -2i\gamma^\mu P_\mu C$$

where $P = i\partial_\mu$ is the 4-momentum operator. Again the covariant derivative is defined like the supercharge but with the second term negated and it anticommutes with the supercharges. Thus the covariant derivative of a supermultiplet is another supermultiplet.

12.2 See also

12.2.1 Spaces

- Chiral superspace

- Harmonic superspace

- Projective superspace

12.2.2 Formalisms

- ADM formalism

- Hamilton–Jacobi–Einstein equation

- Wheeler–DeWitt equation

12.3 References

- Duplij, Steven; Siegel, Warren; Bagger, Jonathan, eds. (2005), *Concise Encyclopedia of Supersymmetry And Non-commutative Structures in Mathematics and Physics*, Berlin, New York: Springer, ISBN 978-1-4020-1338-6 (Second printing)

Chapter 13

Supergeometry

Supergeometry is differential geometry of modules over graded commutative algebras, supermanifolds and graded manifolds. Supergeometry is part and parcel of many classical and quantum field theories involving odd fields, e.g., SUSY field theory, BRST theory, or supergravity.

Supergeometry is formulated in terms of $\mathbb{Z}_2$-graded modules and sheaves over $\mathbb{Z}_2$-graded commutative algebras (supercommutative algebras). In particular, superconnections are defined as Koszul connections on these modules and sheaves. However, supergeometry is not particular noncommutative geometry because of a different definition of a graded derivation.

Graded manifolds and supermanifolds also are phrased in terms of sheaves of graded commutative algebras. Graded manifolds are characterized by sheaves on smooth manifolds, while supermanifolds are constructed by gluing of sheaves of supervector spaces. Note that there are different types of supermanifolds. These are smooth supermanifolds (H^∞-, G^∞-, GH^∞-supermanifolds), G-supermanifolds, and DeWitt supermanifolds. In particular, supervector bundles and principal superbundles are considered in the category of G-supermanifolds. Note that definitions of principal superbundles and principal superconnections straightforwardly follow that of smooth principal bundles and principal connections. Principal graded bundles also are considered in the category of graded manifolds.

There is a different class of Quillen–Ne'eman superbundles and superconnections. These superconnections have been applied to computing the Chern character in K-theory, noncommutative geometry, and BRST formalism.

13.1   See also

- Supermanifold

- Graded manifold

- Supersymmetry

- Connection (algebraic framework)

- Supermetric

13.2   References

- Bartocci, C.; Bruzzo, U.; Hernandez Ruiperez, D. (1991), *The Geometry of Supermanifolds*, Kluwer, ISBN 0-7923-1440-9.

- Rogers, A. (2007), *Supermanifolds: Theory and Applications*, World Scientific, ISBN 981-02-1228-3.

- Mangiarotti, L.; Sardanashvily, G. (2000), *Connections in Classical and Quantum Field Theory*, World Scientific, ISBN 981-02-2013-8.

13.3 External links

- G. Sardanashvily, Lectures on supergeometry, arXiv: 0910.0092.

Chapter 14

Supergroup (physics)

The concept of "**supergroup**" is a generalization of that of group. In other words, every supergroup carries a natural group structure, and vice versa, but there may be more than one way to structure a given group as a supergroup. A supergroup is like a Lie group in that there is a well defined notion of smooth function defined on them. However the functions may have even and odd parts. Moreover a supergroup has a super Lie algebra which plays a role similar to that of a Lie algebra for Lie groups in that they determine most of the representation theory and which is the starting point for classification.

More formally, a **Lie supergroup** is a supermanifold G together with a multiplication morphism $\mu : G \times G \to G$, an inversion morphism $i : G \to G$ and a unit morphism $e : 1 \to G$ which makes G a group object in the category of supermanifolds. This means that, formulated as commutative diagrams, the usual associativity and inversion axioms of a group continue to hold. Since every manifold is a super manifold, a Lie supergroup generalises the notion of a Lie group.

There are many possible supergroups. The ones of most interest in theoretical physics are the ones which extend the Poincaré group or the conformal group. Of particular interest are the **orthosymplectic groups** *Osp(N/M)* and the **superunitary groups** *SU(N/M)*.

An equivalent algebraic approach starts from the observation that a super manifold is determined by its ring of supercommutative smooth functions, and that a morphism of super manifolds corresponds one to one with an algebra homomorphism between their functions in the opposite direction, i.e. that the category of supermanifolds is opposite to the category of algebras of smooth graded commutative functions. Reversing all the arrows in the commutative diagrams that define a Lie supergroup then shows that functions over the supergroup have the structure of a $\mathbf{Z}_2$-graded Hopf algebra. Likewise the representations of this Hopf algebra turn out to be $\mathbf{Z}_2$-graded comodules. This Hopf algebra gives the global properties of the supergroup.

There is another related Hopf algebra which is the dual of the previous Hopf algebra. It can be identified with the Hopf algebra of graded differential operators at the origin. It only gives the local properties of the symmetries i.e., it only gives information about infinitesimal supersymmetry transformations. The representations of this Hopf algebra are modules. Like in the non graded case, this Hopf algebra can be described purely algebraically as the universal enveloping algebra of the Lie superalgebra.

In a similar way one can define an affine algebraic supergroup as a group object in the category of superalgebraic affine varieties. An affine algebraic supergroup has a similar one to one relation to its Hopf algebra of superpolynomials. Using the language of schemes, which combines the geometric and algebraic point of view, algebraic supergroup schemes can be defined including super Abelian varieties.

Chapter 15

Supergravity

In theoretical physics, **supergravity** (**supergravity theory**; **SUGRA** for short) is a field theory that combines the principles of supersymmetry and general relativity. Together, these imply that, in supergravity, the supersymmetry is a local symmetry (in contrast to non-gravitational supersymmetric theories, such as the Minimal Supersymmetric Standard Model). Since the generators of supersymmetry (SUSY) are convoluted with the Poincaré group to form a super-Poincaré algebra, it can be seen that supergravity follows naturally from supersymmetry.[1]

15.1   Gravitons

Like any field theory of gravity, a supergravity theory contains a spin-2 field whose quantum is the graviton. Supersymmetry requires the graviton field to have a superpartner. This field has spin 3/2 and its quantum is the gravitino. The number of gravitino fields is equal to the number of supersymmetries.

15.2   History

15.2.1   Gauge supersymmetry

The first theory[1] of local supersymmetry was proposed in 1975 by Dick Arnowitt and Pran Nath and was called **gauge supersymmetry**.

15.2.2   SUGRA

SUGRA, or supergravity, was discovered in 1976 by Dan Freedman, Sergio Ferrara and Peter van Nieuwenhuizen,[1] but was quickly generalized to many different theories in various numbers of dimensions and additional (N) supersymmetry charges. Supergravity theories with N>1 are usually referred to as extended supergravity (SUEGRA). Some supergravity theories were shown to be equivalent to certain higher-dimensional supergravity theories via dimensional reduction (e.g. $N = 1$ **11-dimensional** supergravity is dimensionally reduced on S^7 to $N = 8$, $d = 4$ SUGRA). The resulting theories were sometimes referred to as Kaluza–Klein theories as Kaluza and Klein constructed in 1919 a 5-dimensional gravitational theory, that when dimensionally reduced on circle, its 4-dimensional non-massive modes describe electromagnetism coupled to gravity.

15.2.3   mSUGRA

mSUGRA means minimal SUper GRAvity. The construction of a realistic model of particle interactions within the $N = 1$ supergravity framework where supersymmetry (SUSY) is broken by a super Higgs mechanism was carried out by Ali Chamseddine, Richard Arnowitt and Pran Nath in 1982. In these classes of models collectively now known as minimal supergravity Grand Unification Theories (mSUGRA GUT), gravity mediates the breaking of SUSY through the existence of a hidden sector. mSUGRA naturally generates the Soft SUSY breaking terms which are a consequence of the Super Higgs effect. Radiative breaking of electroweak symmetry through Renormalization Group Equations (RGEs) follows as an immediate consequence. mSUGRA is one of the most widely investigated models of particle physics due to its predictive power—requiring only four input parameters and a sign to determine the low energy phenomenology from the scale of Grand Unification.

See also: Gravity-Mediated Supersymmetry Breaking in the MSSM

15.2.4   11d: the maximal SUGRA

One of these supergravities, the 11-dimensional theory, generated considerable excitement as the first potential candidate for the theory of everything. This excitement was built on four pillars, two of which have now been largely discredited:

- Werner Nahm showed[2] that 11 dimensions was the largest number of dimensions consistent with a single graviton, and that a theory with more dimensions would also have particles with spins greater than 2. These problems are avoided in 12 dimensions if two of these dimensions are timelike, as has been often emphasized by Itzhak Bars.

- In 1981, Ed Witten showed[3] that 11 was the smallest number of dimensions that was big enough to contain the gauge groups of the Standard Model, namely SU(3) for the strong interactions and SU(2) times U(1) for the electroweak interactions. Today many techniques exist to embed the standard model gauge group in supergravity in any number of dimensions. For example, in the mid and late 1980s, the obligatory gauge symmetry in type I and heterotic string theories was often used. In type II string theory they could also be obtained by compactifying on certain Calabi–Yau manifolds. Today one may also use D-branes to engineer gauge symmetries.

- In 1978, Eugène Cremmer, Bernard Julia and Joël Scherk (CJS) found[4] the classical action for an 11-dimensional supergravity theory. This remains today the only known classical 11-dimensional theory with local supersymmetry and no fields of spin higher than two. Other 11-dimensional theories are known that are quantum-mechanically inequivalent to the CJS theory, but classically equivalent (that is, they reduce to the CJS theory when one imposes the classical equations of motion). For example, in the mid 1980s Bernard de Wit and Hermann Nicolai found an alternate theory in D=11 Supergravity with Local SU(8) Invariance. This theory, while not manifestly Lorentz-invariant, is in many ways superior to the CJS theory in that, for example, it dimensionally-reduces to the 4-dimensional theory without recourse to the classical equations of motion.

- In 1980, Peter Freund and M. A. Rubin showed that compactification from 11 dimensions preserving all the SUSY generators could occur in two ways, leaving only 4 or 7 macroscopic dimensions (the other 7 or 4 being compact).[5] Unfortunately, the noncompact dimensions have to form an anti-de Sitter space. Today it is understood that there are many possible compactifications, but that the Freund-Rubin compactifications are invariant under all of the supersymmetry transformations that preserve the action.

Thus, the first two results appeared to establish 11 dimensions uniquely, the third result appeared to specify the theory, and the last result explained why the observed universe appears to be four-dimensional.

Many of the details of the theory were fleshed out by Peter van Nieuwenhuizen, Sergio Ferrara and Daniel Z. Freedman.

15.2.5 The end of the SUGRA era

The initial excitement over 11-dimensional supergravity soon waned, as various failings were discovered, and attempts to repair the model failed as well. Problems included:

- The compact manifolds which were known at the time and which contained the standard model were not compatible with supersymmetry, and could not hold quarks or leptons. One suggestion was to replace the compact dimensions with the 7-sphere, with the symmetry group SO(8), or the squashed 7-sphere, with symmetry group SO(5) times SU(2).

- Until recently, the physical neutrinos seen in experiments were believed to be massless, and appeared to be left-handed, a phenomenon referred to as the chirality of the Standard Model. It was very difficult to construct a chiral fermion from a compactification — the compactified manifold needed to have singularities, but physics near singularities did not begin to be understood until the advent of orbifold conformal field theories in the late 1980s.

- Supergravity models generically result in an unrealistically large cosmological constant in four dimensions, and that constant is difficult to remove, and so require fine-tuning. This is still a problem today.

- Quantization of the theory led to quantum field theory gauge anomalies rendering the theory inconsistent. In the intervening years physicists have learned how to cancel these anomalies.

Some of these difficulties could be avoided by moving to a 10-dimensional theory involving superstrings. However, by moving to 10 dimensions one loses the sense of uniqueness of the 11-dimensional theory.

The core breakthrough for the 10-dimensional theory, known as the first superstring revolution, was a demonstration by Michael B. Green, John H. Schwarz and David Gross that there are only three supergravity models in 10 dimensions which have gauge symmetries and in which all of the gauge and gravitational anomalies cancel. These were theories built on the groups SO(32) and $E_8 \times E_8$, the direct product of two copies of E_8. Today we know that, using D-branes for example, gauge symmetries can be introduced in other 10-dimensional theories as well.[6]

15.2.6 The second superstring revolution

Initial excitement about the 10-dimensional theories, and the string theories that provide their quantum completion, died by the end of the 1980s. There were too many Calabi–Yaus to compactify on, many more than Yau had estimated, as he admitted in December 2005 at the 23rd International Solvay Conference in Physics. None quite gave the standard model, but it seemed as though one could get close with enough effort in many distinct ways. Plus no one understood the theory beyond the regime of applicability of string perturbation theory.

There was a comparatively quiet period at the beginning of the 1990s; however, several important tools were developed. For example, it became apparent that the various superstring theories were related by "string dualities", some of which relate weak string-coupling (i.e. perturbative) physics in one model with strong string-coupling (i.e. non-perturbative) in another.

Then it all changed, in what is known as the second superstring revolution. Joseph Polchinski realized that obscure string theory objects, called D-branes, which he had discovered six years earlier, are stringy versions of the p-branes that were known in supergravity theories. The treatment of these p-branes was not restricted by string perturbation theory; in fact, thanks to supersymmetry, p-branes in supergravity were understood well beyond the limits in which string theory was understood.

Armed with this new nonperturbative tool, Edward Witten and many others were able to show that all of the perturbative string theories were descriptions of different states in a single theory which Witten named M-theory. Furthermore he argued that M-theory's long wavelength limit (i.e. when the quantum wavelength associated to objects in the theory are much larger than the size of the 11th dimension) should be described by the 11-dimensional supergravity that had fallen out of favor with the first superstring revolution 10 years earlier, accompanied by the 2- and 5-branes.

Historically, then, supergravity has come "full circle". It is a commonly used framework in understanding features of string theories, M-theory and their compactifications to lower spacetime dimensions.

15.3   Relation to superstrings

Particular 10-dimensional supergravity theories are considered "low energy limits" of the 10-dimensional superstring theories; more precisely, these arise as the massless, tree-level approximation of string theories. True effective field theories of string theories, rather than truncations, are rarely available. Due to string dualities, the conjectured 11-dimensional M-theory is required to have 11-dimensional supergravity as a "low energy limit". However, this doesn't necessarily mean that string theory/M-theory is the only possible UV completion of supergravity; supergravity research is useful independent of those relations.

15.4   4D $N = 1$ SUGRA

Before we move on to SUGRA proper, let's recapitulate some important details about general relativity. We have a 4D differentiable manifold M with a Spin(3,1) principal bundle over it. This principal bundle represents the local Lorentz symmetry. In addition, we have a vector bundle T over the manifold with the fiber having four real dimensions and transforming as a vector under Spin(3,1). We have an invertible linear map from the tangent bundle TM to T. This map is the vierbein. The local Lorentz symmetry has a gauge connection associated with it, the spin connection.

The following discussion will be in superspace notation, as opposed to the component notation, which isn't manifestly covariant under SUSY. There are actually *many* different versions of SUGRA out there which are inequivalent in the sense that their actions and constraints upon the torsion tensor are different, but ultimately equivalent in that we can always perform a field redefinition of the supervierbeins and spin connection to get from one version to another.

In 4D N=1 SUGRA, we have a 4|4 real differentiable supermanifold M, i.e. we have 4 real bosonic dimensions and 4 real fermionic dimensions. As in the nonsupersymmetric case, we have a Spin(3,1) principal bundle over M. We have an $\mathbf{R}^{4|4}$ vector bundle T over M. The fiber of T transforms under the local Lorentz group as follows; the four real bosonic dimensions transform as a vector and the four real fermionic dimensions transform as a Majorana spinor. This Majorana spinor can be reexpressed as a complex left-handed Weyl spinor and its complex conjugate right-handed Weyl spinor (they're not independent of each other). We also have a spin connection as before.

We will use the following conventions; the spatial (both bosonic and fermionic) indices will be indicated by M, N, ... The bosonic spatial indices will be indicated by μ, ν, ..., the left-handed Weyl spatial indices by α, β,..., and the right-handed Weyl spatial indices by $\dot{\alpha}$, $\dot{\beta}$, ... The indices for the fiber of T will follow a similar notation, except that they will be hatted like this: $\hat{M}, \hat{\alpha}$. See van der Waerden notation for more details. $M = (\mu, \alpha, \dot{\alpha})$. The supervierbein is denoted by $e_N^{\hat{M}}$, and the spin connection by $\omega_{\hat{M}\hat{N}P}$. The *inverse* supervierbein is denoted by $E_{\hat{M}}^N$.

The supervierbein and spin connection are real in the sense that they satisfy the reality conditions

$$e_N^{\hat{M}}(x,\overline{\theta},\theta)^* = e_{N*}^{\hat{M}*}(x,\theta,\overline{\theta}) \text{ where } \mu^* = \mu, \, \alpha^* = \dot{\alpha}, \text{ and } \dot{\alpha}^* = \alpha \text{ and } \omega(x,\overline{\theta},\theta)^* = \omega(x,\theta,\overline{\theta}) .$$

The covariant derivative is defined as

$$D_{\hat{M}}f = E_{\hat{M}}^N \left(\partial_N f + \omega_N[f]\right)$$

The covariant exterior derivative as defined over supermanifolds needs to be super graded. This means that every time we interchange two fermionic indices, we pick up a +1 sign factor, instead of -1.

The presence or absence of R symmetries is optional, but if R-symmetry exists, the integrand over the full superspace has to have an R-charge of 0 and the integrand over chiral superspace has to have an R-charge of 2.

A chiral superfield X is a superfield which satisfies $\overline{D}_{\hat{\alpha}}X = 0$. In order for this constraint to be consistent, we require the integrability conditions that $\left\{\overline{D}_{\hat{\alpha}}, \overline{D}_{\hat{\beta}}\right\} = c_{\hat{\alpha}\hat{\beta}}^{\hat{\gamma}}\overline{D}_{\hat{\gamma}}$ for some coefficients c.

Unlike nonSUSY GR, the torsion has to be nonzero, at least with respect to the fermionic directions. Already, even in flat superspace, $D_{\hat{\alpha}}e_{\hat{\beta}} + \overline{D}_{\hat{\beta}}e_{\hat{\alpha}} \neq 0$. In one version of SUGRA (but certainly not the only one), we have the following constraints upon the torsion tensor:

$$T^{\hat{\gamma}}_{\underline{\hat{\alpha}}\underline{\hat{\beta}}} = 0$$

$$T^{\hat{\mu}}_{\hat{\alpha}\hat{\beta}} = 0$$

$$T^{\hat{\mu}}_{\dot{\hat{\alpha}}\dot{\hat{\beta}}} = 0$$

$$T^{\hat{\mu}}_{\hat{\alpha}\dot{\hat{\beta}}} = 2i\sigma^{\hat{\mu}}_{\hat{\alpha}\dot{\hat{\beta}}}$$

$$T^{\hat{\nu}}_{\hat{\mu}\underline{\hat{\alpha}}} = 0$$

$$T^{\hat{\rho}}_{\hat{\mu}\hat{\nu}} = 0$$

Here, $\underline{\alpha}$ is a shorthand notation to mean the index runs over either the left or right Weyl spinors.

The superdeterminant of the supervierbein, $|e|$, gives us the volume factor for M. Equivalently, we have the volume 4|4-superform $e^{\hat{\mu}=0} \wedge \cdots \wedge e^{\hat{\mu}=3} \wedge e^{\hat{\alpha}=1} \wedge e^{\hat{\alpha}=2} \wedge e^{\dot{\hat{\alpha}}=1} \wedge e^{\dot{\hat{\alpha}}=2}$.

If we complexify the superdiffeomorphisms, there is a gauge where $E^{\mu}_{\hat{\alpha}} = 0$, $E^{\beta}_{\dot{\hat{\alpha}}} = 0$ and $E^{\dot{\beta}}_{\dot{\hat{\alpha}}} = \delta^{\dot{\beta}}_{\dot{\alpha}}$. The resulting chiral superspace has the coordinates x and Θ.

R is a scalar valued chiral superfield derivable from the supervielbeins and spin connection. If f is any superfield, $(\bar{D}^2 - 8R) f$ is always a chiral superfield.

The action for a SUGRA theory with chiral superfields X, is given by

$$S = \int d^4x d^2\Theta 2\mathcal{E} \left[\frac{3}{8} \left(\bar{D}^2 - 8R \right) e^{-K(\bar{X},X)/3} + W(X) \right] + c.c.$$

where K is the Kähler potential and W is the superpotential, and $\mathcal{E}$ is the chiral volume factor.

Unlike the case for flat superspace, adding a constant to either the Kähler or superpotential is now physical. A constant shift to the Kähler potential changes the effective Planck constant, while a constant shift to the superpotential changes the effective cosmological constant. As the effective Planck constant now depends upon the value of the chiral superfield X, we need to rescale the supervierbeins (a field redefinition) to get a constant Planck constant. This is called the **Einstein frame**.

15.5   N = 8 supergravity in 4 dimensions

N=8 Supergravity is the most symmetric quantum field theory which involves gravity and a finite number of fields. It can be found from a dimensional reduction of 11D supergravity by making the size of 7 of the dimensions go to zero. It has 8 supersymmetries which is the most any gravitational theory can have since there are 8 half-steps between spin 2 and spin −2. (A graviton has the highest spin in this theory which is a spin 2 particle). More supersymmetries would mean the particles would have superpartners with spins higher than 2. The only theories with spins higher than 2 which are consistent involve an infinite number of particles (such as String Theory). Stephen Hawking in his *A Brief History of Time* speculated that this theory could be the Theory of Everything. However in later years this was abandoned in favour of String Theory. There has been renewed interest in the 21st century with the possibility that this theory may be finite.

15.6   Higher-dimensional SUGRA

Main article: Higher-dimensional supergravity

Higher-dimensional SUGRA is the higher-dimensional, supersymmetric generalization of general relativity. Supergravity can be formulated in any number of dimensions up to eleven. Higher-dimensional SUGRA focuses upon supergravity in greater than four dimensions.

The number of supercharges in a spinor depends on the dimension and the signature of spacetime. The supercharges occur in spinors. Thus the limit on the number of supercharges cannot be satisfied in a spacetime of arbitrary dimension. Some theoretical examples in which this is satisfied are:

- 12-dimensional two-time theory

- 11-dimensional maximal SUGRA

- 10-dimensional SUGRA theories

  - Type IIA SUGRA: N = (1, 1)

  - IIA SUGRA from 11d SUGRA

  - Type IIB SUGRA: N = (2, 0)

  - Type I gauged SUGRA: N = (1, 0)

- 9d SUGRA theories

  - Maximal 9d SUGRA from 10d

  - T-duality

  - N = 1 Gauged SUGRA

The supergravity theories that have attracted the most interest contain no spins higher than two. This means, in particular, that they do not contain any fields that transform as symmetric tensors of rank higher than two under Lorentz transformations. The consistency of interacting higher spin field theories is, however, presently a field of very active interest.

15.7   See also

15.8   Notes

[1] P. van Nieuwenhuizen, Phys. Rep. 68, 189 (1981)

[2] Werner Nahm, "Supersymmetries and their representations". *Nuclear Physics B* **135** no 1 (1978) pp 149-166, doi:10.1016/0550-3213(78)90218-3

[3] Ed Witten, "Search for a realistic Kaluza-Klein theory". *Nuclear Physics B* **186** no 3 (1981) pp 412-428, doi:10.1016/0550-3213(81)90021-3

[4] E. Cremmer, B. Julia and J. Scherk, "Supergravity theory in eleven dimensions", *Physics Letters* **B76** (1978) pp 409-412,

[5] Peter G.O. Freund; Mark A. Rubin (1980). "Dynamics of dimensional reduction". *Physics Letters B* **97** (2): 233–235. Bibcode:1980PhLB...97..233F. doi:10.1016/0370-2693(80)90590-0.

[6] Blumenhagen, R.; Cvetic, M.; Langacker, P.; Shiu, G. (2005). "Toward Realistic Intersecting D-Brane Models". arXiv:hep-th/0502005 [hep-th].

15.9 References

15.9.1 Historical

- P. Nath and R. Arnowitt, "Generalized Super-Gauge Symmetry as a New Framework for Unified Gauge Theories", *Physics Letters B '56* (1975) 177.

- D.Z. Freedman, P. van Nieuwenhuizen and S. Ferrara, "Progress Toward A Theory Of Supergravity", *Physical Review* **D13** (1976) pp 3214–3218.

- E. Cremmer, B. Julia and J. Scherk, "Supergravity theory in eleven dimensions", *Physics Letters* **B76** (1978) pp 409–412. scanned version

- P. Freund and M. Rubin, "Dynamics of dimensional reduction", *Physics Letters* **B97** (1980) pp 233–235.

- Ali H. Chamseddine, R. Arnowitt, Pran Nath, "Locally Supersymmetric Grand Unification", " Phys. Rev.Lett.49982"

- Michael B. Green, John H. Schwarz, "Anomaly Cancellation in Supersymmetric D=10 Gauge Theory and Superstring Theory", *Physics Letters* **B149** (1984) pp117–122.

15.9.2 General

- Bernard de Wit(2002) Supergravity

- A Supersymmetry Primer (1998); updated in (2006).

- Adel Bilal, Introduction to supersymmetry (2001) ArXiv hep-th/0101055, (*a comprehensive introduction to supersymmetry*).

- Friedemann Brandt, Lectures on supergravity (2002) ArXiv hep-th/0204035, (*an introduction to 4-dimensional N = 1 supergravity*).

- Wess, Julius; Bagger, Jonathan (1992). *Supersymmetry and Supergravity*. Princeton University Press. p. 260. ISBN 0-691-02530-4.

Chapter 16

Supercharge

For other uses, see Supercharge (disambiguation).

In theoretical physics, a **supercharge** is a generator of supersymmetry transformations.

Supercharge, denoted by the symbol Q, is an operator which transforms bosons into fermions, and vice versa. Since the supercharge operator changes a particle with spin one-half to a particle with spin one or zero, the supercharge itself is a spinor that carries one half unit of spin.[1]

16.1 Commutation

Supercharge is described by the Super-Poincaré algebra

Supercharge commutes with the Hamiltonian operator:

$$[\,Q\,,\,H\,] = 0$$

16.2 References

[1] Supersymmetry to the rescue?

Chapter 17

Superfield

In theoretical physics, one often analyzes theories with supersymmetry in which **superfields** play a very important role. A superfield is a function defined in superspace which properly packages the various fields of a supermultiplet, namely, the array of fermion and boson fields related among themselves by supersymmetry. Superfields were introduced by Abdus Salam and John Strathdee in 1979.[1]

17.1 Overview

In four dimensions, the simplest example—namely, the minimal $N = 1$ supersymmetry—may be written as a vector in a superspace with four extra fermionic coordinates $\theta^1, \theta^2, \bar{\theta}^1, \bar{\theta}^2$, transforming as a two-component spinor and its conjugate. More generally, there are 4N extra fermionic (Grassmann number) coordinates .

A more coordinate-free description of the superspace is that it's the quotient space of the super-Poincaré group divided by the Lorentz group.

Every superfield, i.e. a field that depends on all coordinates of the superspace (or in other words, an element of a module of the algebra of functions over superspace), may be expanded with respect to the new fermionic coordinates. There exists a special kind of superfields, the so-called *chiral superfields*, that, in the chiral representation of supersymmetry, depend only on the variables θ but not their conjugates. The last term in the corresponding expansion, namely $F\theta^1\theta^2$, is called the *F-term*. Other superfields include vector superfields.

There also exist superfields in theories with larger supersymmetry.

Manifestly supersymmetric actions may also be written as integrals over the whole superspace. Some special terms, such as the superpotential, may be written as integrals over θs only. They are also referred to as F-terms, much like the terms in the ordinary potential that arise from these terms of the supersymmetric Lagrangian.

17.2 References

[1]Salam, A.; Strathdee, J. (1974). "Super-gauge transformations".*Nuclear Physics B***76**(3): 477–201.Bibcode:1974NuPhB..76..4S. doi:10.1016/0550-3213(74)90537-9.

Chapter 18

Superpartner

In particle physics, a **Superpartner** (also **Sparticle**) is a hypothetical elementary particle. Supersymmetry is one of the synergistic theories in current high-energy physics that predicts the existence of these "shadow" particles.[1][2]

The word *superpartner* is a portmanteau of *supersymmetry* and *partner*. The word *sparticle* is a portmanteau of *supersymmetry* and *particle*.

18.1 Theoretical predictions

According to the supersymmetry theory, each fermion should have a partner boson, the fermion's superpartner, and each boson should have a partner fermion. Exact *unbroken* supersymmetry would predict that a particle and its superpartners would have the same mass. No superpartners of the Standard Model particles have yet been found. This may indicate that supersymmetry is incorrect, or it may also be the result of the fact that supersymmetry is not an exact, *unbroken* symmetry of nature. If superpartners are found, their masses would indicate the scale at which supersymmetry is broken.[1][3]

For particles that are real scalars (such as an axion), there is a fermion superpartner as well as a second, real scalar field. For axions, these particles are often referred to as axinos and saxions.

In extended supersymmetry there may be more than one superparticle for a given particle. For instance, with two copies of supersymmetry in four dimensions, a photon would have two fermion superpartners and a scalar superpartner.

In zero dimensions it is possible to have supersymmetry, but no superpartners. However, this is the only situation where supersymmetry does not imply the existence of superpartners.

18.2 Recreating superpartners

If the supersymmetry theory is correct, it should be possible to recreate these particles in high-energy particle accelerators. Doing so will not be an easy task; these particles may have masses up to a thousand times greater than their corresponding "real" particles.[1]

Some researchers have hoped the Large Hadron Collider at CERN might produce evidence for the existence of superpartner particles.[1] However, as of 2013, no such evidence has been found.[4]

18.3 See also

- Chargino

- Gluino

- Gravitino as a superpartner of the hypothetical graviton

- Neutralino

- Sfermion

- Higgsino

18.4 References

[1] Langacker, Paul (November 22, 2010). Sprouse, Gene D., ed. "Meet a superpartner at the LHC". *Physics* (New York: American Physical Society) **3** (98). Bibcode:2010PhyOJ...3...98L. doi:10.1103/Physics.3.98. ISSN 1943-2879. OCLC 233971234. Archived from the original on 2011-02-22. Retrieved 21 February 2011.

[2] Overbye, Dennis (May 15, 2007). "A Giant Takes On Physics' Biggest Questions". *The New York Times* (Manhattan, New York: Arthur Ochs Sulzberger, Jr.). p. F1. ISSN 0362-4331. OCLC 1645522. Retrieved 21 February 2011.

[3] Quigg, Chris (January 17, 2008). "Sidebar: Solving the Higgs Puzzle". *Scientific American* (Nature Publishing Group). ISSN 0036-8733. OCLC 1775222. Archived from the original on 2011-02-22. Retrieved 21 February 2011.

[4] Jamieson, Valerie (13 December 2013). "Higgs Nobel bash: I was at the party of the universe". *New Scientist*. Retrieved 20 December 2013. So far the Higgs hasn't given many supersymmetric clues.

18.5 External links

- Argonne National Laboratory

- Large Hadron Collider

- CERN homepage

Chapter 19

Graviton

This article is about the hypothetical particle. For other uses, see Graviton (disambiguation).

In physics, the **graviton** is a hypothetical elementary particle that mediates the force of gravitation in the framework of quantum field theory. If it exists, the graviton is expected to be massless (because the gravitational force appears to have unlimited range) and must be a spin-2 boson. The spin follows from the fact that the source of gravitation is the stress–energy tensor, a second-rank tensor (compared to electromagnetism's spin-1 photon, the source of which is the four-current, a first-rank tensor). Additionally, it can be shown that any massless spin-2 field would give rise to a force indistinguishable from gravitation, because a massless spin-2 field must couple to (interact with) the stress–energy tensor in the same way that the gravitational field does. Seeing as the graviton is hypothetical, its discovery would unite quantum theory with gravity.[4] This result suggests that, if a massless spin-2 particle is discovered, it must be the graviton, so that the only experimental verification needed for the graviton may simply be the discovery of a massless spin-2 particle.[5]

19.1 Theory

The four other known forces of nature are mediated by elementary particles: electromagnetism by the photon, the strong interaction by the gluons, the Higgs field by the Higgs Boson, and the weak interaction by the W and Z bosons. The hypothesis is that the gravitational interaction is likewise mediated by an – as yet undiscovered – elementary particle, dubbed as *the graviton*. In the classical limit, the theory would reduce to general relativity and conform to Newton's law of gravitation in the weak-field limit.[6][7][8]

19.1.1 Gravitons and renormalization

When describing graviton interactions, the classical theory (i.e., the tree diagrams) and semiclassical corrections (one-loop diagrams) behave normally, but Feynman diagrams with two (or more) loops lead to ultraviolet divergences; that is, infinite results that cannot be removed because the quantized general relativity is not renormalizable, unlike quantum electrodynamics. That is, the usual ways physicists calculate the probability that a particle will emit or absorb a graviton give nonsensical answers and the theory loses its predictive power. These problems, together with some conceptual puzzles, led many physicists to believe that a theory more complete than quantized general relativity must describe the behavior near the Planck scale.

19.1.2 Comparison with other forces

Unlike the force carriers of the other forces, gravitation plays a special role in general relativity in defining the spacetime in which events take place. In some descriptions, matter modifies the 'shape' of spacetime itself, and gravity is a result of this shape, an idea which at first glance may appear hard to match with the idea of a force acting between particles.[9]

Because the diffeomorphism invariance of the theory does not allow any particular space-time background to be singled out as the "true" space-time background, general relativity is said to be background independent. In contrast, the Standard Model is *not* background independent, with Minkowski space enjoying a special status as the fixed background space-time.[10] A theory of quantum gravity is needed in order to reconcile these differences.[11] Whether this theory should be background independent is an open question. The answer to this question will determine our understanding of what specific role gravitation plays in the fate of the universe.[12]

19.1.3 Gravitons in speculative theories

String theory predicts the existence of gravitons and their well-defined interactions. A graviton in perturbative string theory is a closed string in a very particular low-energy vibrational state. The scattering of gravitons in string theory can also be computed from the correlation functions in conformal field theory, as dictated by the AdS/CFT correspondence, or from matrix theory.

A feature of gravitons in string theory is that, as closed strings without endpoints, they would not be bound to branes and could move freely between them. If we live on a brane (as hypothesized by brane theories) this "leakage" of gravitons from the brane into higher-dimensional space could explain why gravitation is such a weak force, and gravitons from other branes adjacent to our own could provide a potential explanation for dark matter. However if gravitons were to move completely freely between branes this would dilute gravity too much, causing a violation of Newton's inverse square law. To combat this, Lisa Randall found that a three-brane (such as ours) would have a gravitational pull of its own, preventing gravitons from drifting freely, possibly resulting in the diluted gravity we observe while roughly maintaining Newton's inverse square law.[13] See brane cosmology.

A theory by Ahmed Farag Ali and Saurya Das adds quantum mechanical corrections (using Bohm trajectories) to general relativistic geodesics. If gravitons are given a small but non-zero mass, it could explain the cosmological constant without need for dark energy and solve the smallness problem.[14]

19.2 Experimental observation

Unambiguous detection of individual gravitons, though not prohibited by any fundamental law, is impossible with any physically reasonable detector.[15] The reason is the extremely low cross section for the interaction of gravitons with matter. For example, a detector with the mass of Jupiter and 100% efficiency, placed in close orbit around a neutron star, would only be expected to observe one graviton every 10 years, even under the most favorable conditions. It would be impossible to discriminate these events from the background of neutrinos, since the dimensions of the required neutrino shield would ensure collapse into a black hole.[15]

However, experiments to detect gravitational waves, which may be viewed as coherent states of many gravitons, are underway (e.g., LIGO and VIRGO). Although these experiments cannot detect individual gravitons, they might provide information about certain properties of the graviton.[16] For example, if gravitational waves were observed to propagate slower than *c* (the speed of light in a vacuum), that would imply that the graviton has mass (however, gravitational waves must propagate slower than "c" in a region with non-zero mass density if they are to be detectable).[17] Astronomical observations of the kinematics of galaxies, especially the galaxy rotation problem and modified Newtonian dynamics, might point toward gravitons having non-zero mass.[18]

19.3 Difficulties and outstanding issues

Most theories containing gravitons suffer from severe problems. Attempts to extend the Standard Model or other quantum field theories by adding gravitons run into serious theoretical difficulties at high energies (processes involving energies close to or above the Planck scale) because of infinities arising due to quantum effects (in technical terms, gravitation is nonrenormalizable). Since classical general relativity and quantum mechanics seem to be incompatible at such energies, from a theoretical point of view, this situation is not tenable. One possible solution is to replace particles with strings.

String theories are quantum theories of gravity in the sense that they reduce to classical general relativity plus field theory at low energies, but are fully quantum mechanical, contain a graviton, and are believed to be mathematically consistent.[19]

19.4   See also

- Gravitomagnetism

- Gravitational wave

- Planck mass

- Gravitation

- Static forces and virtual-particle exchange

- Multiverse

- Gravitino

19.5   References

[1] G is used to avoid confusion with gluons (symbol g)

[2] Rovelli, C. (2001). "Notes for a brief history of quantum gravity". arXiv:gr-qc/0006061 [gr-qc].

[3] Blokhintsev, D. I.; Gal'perin, F. M. (1934). "Gipoteza neitrino i zakon sokhraneniya energii" [Neutrino hypothesis and conservation of energy]. *Pod Znamenem Marxisma* (in Russian) **6**: 147–157.

[4] Lightman, A. P.; Press, W. H.; Price, R. H.; Teukolsky, S. A. (1975). "Problem 12.16". *Problem book in Relativity and Gravitation*. Princeton University Press. ISBN 0-691-08162-X.

[5] For a comparison of the geometric derivation and the (non-geometric) spin-2 field derivation of general relativity, refer to box 18.1 (and also 17.2.5) of Misner, C. W.; Thorne, K. S.; Wheeler, J. A. (1973). *Gravitation*. W. H. Freeman. ISBN 0-7167-0344-0.

[6] Feynman, R. P.; Morinigo, F. B.; Wagner, W. G.; Hatfield, B. (1995). *Feynman Lectures on Gravitation*. Addison-Wesley. ISBN 0-201-62734-5.

[7] Zee, A. (2003). *Quantum Field Theory in a Nutshell*. Princeton University Press. ISBN 0-691-01019-6.

[8] Randall, L. (2005). *Warped Passages: Unraveling the Universe's Hidden Dimensions*. Ecco Press. ISBN 0-06-053108-8.

[9] See the other articles on General relativity, Gravitational field, Gravitational wave, etc

[10] Colosi, D. et al. (2005). "Background independence in a nutshell: The dynamics of a tetrahedron". *Classical and Quantum Gravity* **22** (14): 2971. arXiv:gr-qc/0408079. Bibcode:2005CQGra..22.2971C. doi:10.1088/0264-9381/22/14/008.

[11] Witten, E. (1993). "Quantum Background Independence In String Theory". arXiv:hep-th/9306122 [hep-th].

[12] Smolin, L. (2005). "The case for background independence". arXiv:hep-th/0507235 [hep-th].

[13] Kaku, Michio (2006). *Parallel Worlds - The science of alternative universes and our future in the Cosmos*. pp. 218–221.

[14] Ali, Ahmed Farang (2014). "Cosmology from quantum potential". *Physical Letters B* **741**: 276–279. arXiv:1404.3093v3. doi:10.1016/j.physletb.2014.12.057.

[15] Rothman, T.; Boughn, S. (2006). "Can Gravitons be Detected?". *Foundations of Physics* **36** (12): 1801–1825. arXiv:gr-qc/0601043. Bibcode:2006FoPh...36.1801R. doi:10.1007/s10701-006-9081-9.

[16] Freeman Dyson (8 October 2013). "Is a graviton detectable?". *International Journal of Modern Physics A* **28** (25): 1330041-1–1330035–14. Bibcode:2013IJMPA..2830041D. doi:10.1142/S0217751X1330041X.

[17] Will, C. M. (1998). "Bounding the mass of the graviton using gravitational-wave observations of inspiralling compact binaries". *Physical Review D* **57** (4): 2061–2068. arXiv:gr-qc/9709011. Bibcode:1998PhRvD..57.2061W. doi:10.1103/PhysRevD.57.2061.

[18] Trippe, S. (2013), "A Simplified Treatment of Gravitational Interaction on Galactic Scales", J. Kor. Astron. Soc. **46**, 41. arXiv:1211.4692

[19] Sokal, A. (July 22, 1996). "Don't Pull the String Yet on Superstring Theory". *The New York Times*. Retrieved March 26, 2010.

19.6 External links

-

- Graviton on *In Our Time* at the BBC. (listen now)

Chapter 20

Neutralino

In supersymmetry, the **neutralino**[1] is a hypothetical particle. There are four neutralinos that are fermions and are electrically neutral, the lightest of which is typically stable. They are typically labeled N0
1 (the lightest), N0
2, N0
3 and N0
4 (the heaviest) although sometimes $\tilde{\chi}_1^0, \ldots, \tilde{\chi}_4^0$ is also used when $\tilde{\chi}_i^\pm$ is used to refer to charginos. These four states are mixtures of the bino and the neutral wino (which are the neutral electroweak gauginos), and the neutral higgsinos. As the neutralinos are Majorana fermions, each of them is identical to its antiparticle. Because these particles only interact with the weak vector bosons, they are not directly produced at hadron colliders in copious numbers. They would primarily appear as particles in cascade decays of heavier particles (decays that happen in multiple steps) usually originating from colored supersymmetric particles such as squarks or gluinos.

In R-parity conserving models, the lightest neutralino is stable and all supersymmetric cascade-decays end up decaying into this particle which leaves the detector unseen and its existence can only be inferred by looking for unbalanced momentum in a detector.

The heavier neutralinos typically decay through a neutral Z boson to a lighter neutralino or through a charged W boson to a light chargino:[2]

The mass splittings between the different neutralinos will dictate which patterns of decays are allowed.

Up to present, neutralinos have never been observed or detected in an experiment.

20.1 Origins in supersymmetric theories

In supersymmetry models, all Standard Model particles have partner particles with the same quantum numbers except for the quantum number spin, which differs by 1/2 from its partner particle. Since the superpartners of the Z boson (zino), the photon (photino) and the neutral higgs (higgsino) have the same quantum numbers, they can mix to form four eigenstates of the mass operator called "neutralinos". In many models the lightest of the four neutralinos turns out to be the lightest supersymmetric particle (LSP), though other particles may also take on this role.

20.2 Phenomenology

The exact properties of each neutralino will depend on the details of the mixing[1] (e.g. whether they are more higgsino-like or gaugino-like), but they tend to have masses at the weak scale (100 GeV – 1 TeV) and couple to other particles with strengths characteristic of the weak interaction. In this way they are phenomenologically similar to neutrinos, and so are not directly observable in particle detectors at accelerators.

In models in which R-parity is conserved and the lightest of the four neutralinos is the LSP, the lightest neutralino is stable and is eventually produced in the decay chain of all other superpartners.[3] In such cases supersymmetric processes at accelerators are characterized by a large discrepancy in energy and momentum between the visible initial and final state particles, with this energy being carried off by a neutralino which departs the detector unnoticed.[4][5] This is an important signature to discriminate supersymmetry from Standard Model backgrounds.

20.3   Relationship to dark matter

As a heavy, stable particle, the lightest neutralino is an excellent candidate to form the universe's cold dark matter.[6][7][8] In many models the lightest neutralino can be produced thermally in the hot early universe and leave approximately the right relic abundance to account for the observed dark matter. A lightest neutralino of roughly 10–10000 GeV is the leading weakly interacting massive particle (WIMP) dark matter candidate.[9]

Neutralino dark matter could be observed experimentally in nature either indirectly or directly. For indirect observation, gamma ray and neutrino telescopes look for evidence of neutralino annihilation in regions of high dark matter density such as the galactic or solar centre.[4] For direct observation, special purpose experiments such as the Cryogenic Dark Matter Search (CDMS) seek to detect the rare impacts of WIMPs in terrestrial detectors. These experiments have begun to probe interesting supersymmetric parameter space, excluding some models for neutralino dark matter, and upgraded experiments with greater sensitivity are under development.

20.4   See also

- Lightest Supersymmetric Particle

- Real neutral particle

20.5   Notes

[1] Martin, pp. 71–74

[2] J.-F. Grivaz & the Particle Data Group (2010). "Supersymmetry, Part II (Experiment)" (PDF). *Journal of Physics G* **37** (7): 1309–1319.

[3] Martin, p. 83

[4] Feng, Jonathan L (2010). "Dark Matter Candidates from Particle Physics and Methods of Detection". *Annual Review of Astronomy and Astrophysics* **48**: 495–545. arXiv:1003.0904. Bibcode:2010ARA&A..48..495F. doi:10.1146/annurev-astro-082708-101659. |chapter= ignored (help)

[5] Ellis, John; Olive, Keith A. (2010). "Supersymmetric Dark Matter Candidates". arXiv:1001.3651 [astro-ph]. Also published as Chapter 8 in Bertone

[6] M. Drees; G. Gerbier & the Particle Data Group (2010). "Dark Matter" (PDF). *Journal of Physics G* **37** (7A): 255–260.

[7] Martin, p. 99

[8] Bertone, p. 8

[9] Martin, p. 124

20.6   References

- Martin, Stephen P. (2008). "A Supersymmetry Primer". v5. arXiv:hep-ph/9709356 [hep-ph]. Also published as Chapter 1 in Kane, Gordon L, ed. (2010). *Perspectives on Supersymmetry II*. World Scientific. p. 604. ISBN 978-981-4307-48-2.

- Bertone, Gianfranco, ed. (2010). *Particle Dark Matter: Observations, Models and Searches*. Cambridge University Press. p. 762. ISBN 978-0-521-76368-4.

Chapter 21

Theory of everything

 A theory of everything (ToE) or final theory, ultimate theory, or master theory is a hypothetical single, all-encompassing, coherent theoretical framework of physics that fully explains and links together all physical aspects of the universe. Finding a ToE is one of the major unsolved problems in physics. Over the past few centuries, two theoretical frameworks have been developed that, as a whole, most closely resemble a ToE.

 The two theories upon which all modern physics rests are general relativity (GR) and quantum field theory (QFT). GR is a theoretical framework that only focuses on the force of gravity for understanding the universe in regions of both large-scale and high-mass: stars, galaxies & clusters of galaxies. On the other hand, QFT is a theoretical framework that only focuses on three non-gravitational forces for understanding the universe in regions of both small scale and low mass: sub-atomic particles, atoms, molecules, etc. QFT successfully implemented the Standard Model and unified the interactions (the so-called Grand Unification Theory) between the three non-gravitational forces: weak, strong and the electro-magnetic force.

Through years of research, physicists have experimentally confirmed with tremendous accuracy virtually every prediction made by these two theories when in their appropriate domains of applicability. In accordance with their findings, scientists also learned that GR and QFT, as they are currently formulated, are mutually incompatible - they cannot both be right. Since the usual domains of applicability of GR and QFT are so different, most situations require that only one of the two theories be used.[3][4]:842–844 As it turns out, this incompatibility between GR and QFT is only an apparent issue in regions of extremely small-scale and high-mass, such as those that exist within a black hole or during the beginning stages of the universe (i.e., the moment immediately following the Big Bang). To resolve this conflict, a theoretical framework revealing a deeper underlying reality, unifying gravity with the other three interactions, must be discovered to harmoniously integrate the realms of GR and QFT into a seamless whole: a single theory that, in principle, is capable of describing all phenomena. In pursuit of this goal, quantum gravity has recently become an area of active research.

Over the past few decades, a single explanatory framework, called "string theory", has emerged that may turn out to be the ultimate theory of the universe. Many physicists believe that, at the beginning of the universe (up to 10^{-43} seconds after the Big Bang), the four fundamental forces were once a single fundamental force. Unlike most (if not all) other theories, string theory may be on its way to successfully incorporating each of the four fundamental forces into a unified whole. According to string theory, every particle in the universe, at its most microscopic level (Planck length), consists of varying combinations of vibrating strings (or strands) with preferred patterns of vibration. String theory claims that it is through these specific oscillatory patterns of strings that a particle of unique mass and force charge is created (that is to say, the electron is a type of string that vibrates one way, while the up-quark is a type of string vibrating another way, and so forth).

Initially, the term *theory of everything* was used with an ironic connotation to refer to various overgeneralized theories. For example, a grandfather of Ijon Tichy — a character from a cycle of Stanisław Lem's science fiction stories of the 1960s — was known to work on the "General Theory of Everything". Physicist John Ellis[5] claims to have introduced the term into the technical literature in an article in *Nature* in 1986.[6] Over time, the term stuck in popularizations of theoretical physics research.

21.1   Historical antecedents

21.1.1   From ancient Greece to Einstein

Archimedes was possibly the first scientist known to have described nature with axioms (or principles) and then deduce new results from them.[7] He thus tried to describe "everything" starting from a few axioms. Any "theory of everything" is similarly expected to be based on axioms and to deduce all observable phenomena from them.[8]:340

The concept of 'atom', introduced by Democritus, unified all phenomena observed in nature as the motion of atoms. In ancient Greek times philosophers speculated that the apparent diversity of observed phenomena was due to a single type of interaction, namely the collisions of atoms. Following atomism, the mechanical philosophy of the 17th century posited that all forces could be ultimately reduced to contact forces between the atoms, then imagined as tiny solid particles.[9]:184[10]

In the late 17th century, Isaac Newton's description of the long-distance force of gravity implied that not all forces in nature result from things coming into contact. Newton's work in his *Principia* dealt with this in a further example of unification, in this case unifying Galileo's work on terrestrial gravity, Kepler's laws of planetary motion and the phenomenon of tides by explaining these apparent actions at a distance under one single law: the law of universal gravitation.[11]

In 1814, building on these results, Laplace famously suggested that a sufficiently powerful intellect could, if it knew the position and velocity of every particle at a given time, along with the laws of nature, calculate the position of any particle at any other time:[12]:ch 7

> An intellect which at a certain moment would know all forces that set nature in motion, and all positions of all items of which nature is composed, if this intellect were also vast enough to submit these data to analysis, it would embrace in a single formula the movements of the greatest bodies of the universe and those of the tiniest atom; for such an intellect nothing would be uncertain and the future just like the past would be present before its eyes.
> —*Essai philosophique sur les probabilités*, Introduction. 1814

Laplace thus envisaged a combination of gravitation and mechanics as a theory of everything. Modern quantum mechanics implies that uncertainty is inescapable, and thus that Laplace's vision has to be amended: a theory of everything must include gravitation and quantum mechanics.

In 1820, Hans Christian Ørsted discovered a connection between electricity and magnetism, triggering decades of work that culminated in 1865, in James Clerk Maxwell's theory of electromagnetism. During the 19th and early 20th centuries, it gradually became apparent that many common examples of forces – contact forces, elasticity, viscosity, friction, and pressure – result from electrical interactions between the smallest particles of matter.

In his experiments of 1849–50, Michael Faraday was the first to search for a unification of gravity with electricity and magnetism.[13] However, he found no connection.

In 1900, David Hilbert published a famous list of mathematical problems. In Hilbert's sixth problem, he challenged researchers to find an axiomatic basis to all of physics. In this problem he thus asked for what today would be called a theory of everything.[14]

In the late 1920s, the new quantum mechanics showed that the chemical bonds between atoms were examples of (quantum) electrical forces, justifying Dirac's boast that "the underlying physical laws necessary for the mathematical theory of a large part of physics and the whole of chemistry are thus completely known".[15]

After 1915, when Albert Einstein published the theory of gravity (general relativity), the search for a unified field theory combining gravity with electromagnetism began with a renewed interest. In Einstein's day, the strong and the weak forces had not yet been discovered, yet, he found the potential existence of two other distinct forces -gravity and electromagnetism- far more alluring. This launched his thirty-year voyage in search of the so-called "unified field theory" that he hoped would show that these two forces are really manifestations of one grand underlying principle. During these last few decades of his life, this quixotic quest isolated Einstein from the mainstream of physics. Understandably, the mainstream was instead far more excited about the newly emerging framework of quantum mechanics. Einstein wrote to a friend in the early 1940s, "I have become a lonely old chap who is mainly known because he doesn't wear socks and who is exhibited as a curiosity on special occasions." Prominent contributors were Gunnar Nordström, Hermann Weyl, Arthur Eddington, Theodor Kaluza, Oskar Klein, and most notably, Albert Einstein and his collaborators. Einstein intensely searched for, but ultimately failed to find, a unifying theory.[16]:ch 17 (But see:Einstein–Maxwell–Dirac equations.) More than a half a century later, Einstein's dream of discovering a unified theory has become the Holy Grail of modern physics.

21.1.2 Twentieth century and the nuclear interactions

In the twentieth century, the search for a unifying theory was interrupted by the discovery of the strong and weak nuclear forces (or interactions), which differ both from gravity and from electromagnetism. A further hurdle was the acceptance that in a ToE, quantum mechanics had to be incorporated from the start, rather than emerging as a consequence of a deterministic unified theory, as Einstein had hoped.

Gravity and electromagnetism could always peacefully coexist as entries in a list of classical forces, but for many years it seemed that gravity could not even be incorporated into the quantum framework, let alone unified with the other fundamental forces. For this reason, work on unification, for much of the twentieth century, focused on understanding the three "quantum" forces: electromagnetism and the weak and strong forces. The first two were combined in 1967–68 by Sheldon Glashow, Steven Weinberg, and Abdus Salam into the "electroweak" force.[17] Electroweak unification is a broken symmetry: the electromagnetic and weak forces appear distinct at low energies because the particles carrying the weak force, the W and Z bosons, have non-zero masses of 80.4 GeV/c^2 and 91.2 GeV/c^2, whereas the photon, which carries the electromagnetic force, is massless. At higher energies Ws and Zs can be created easily and the unified nature of the force becomes apparent.

While the strong and electroweak forces peacefully coexist in the Standard Model of particle physics, they remain distinct. So far, the quest for a theory of everything is thus unsuccessful on two points: neither a unification of the strong and electroweak forces – which Laplace would have called 'contact forces' – has been achieved, nor has a unification of these forces with gravitation been achieved.

21.2 Modern physics

21.2.1 Conventional sequence of theories

A Theory of Everything would unify all the fundamental interactions of nature: gravitation, strong interaction, weak interaction, and electromagnetism. Because the weak interaction can transform elementary particles from one kind into another, the ToE should also yield a deep understanding of the various different kinds of possible particles. The usual assumed path of theories is given in the following graph, where each unification step leads one level up:

In this graph, electroweak unification occurs at around 100 GeV, grand unification is predicted to occur at 10^{16} GeV, and unification of the GUT force with gravity is expected at the Planck energy, roughly 10^{19} GeV.

Several Grand Unified Theories (GUTs) have been proposed to unify electromagnetism and the weak and strong forces. Grand unification would imply the existence of an electronuclear force; it is expected to set in at energies of the order of 10^{16} GeV, far greater than could be reached by any possible Earth-based particle accelerator. Although the simplest GUTs have been experimentally ruled out, the general idea, especially when linked with supersymmetry, remains a favorite candidate in the theoretical physics community. Supersymmetric GUTs seem plausible not only for their theoretical "beauty", but because they naturally produce large quantities of dark matter, and because the inflationary force may be related to GUT physics (although it does not seem to form an inevitable part of the theory). Yet GUTs are clearly not the final answer; both the current standard model and all proposed GUTs are quantum field theories which require the problematic technique of renormalization to yield sensible answers. This is usually regarded as a sign that these are only effective field theories, omitting crucial phenomena relevant only at very high energies.[3]

The final step in the graph requires resolving the separation between quantum mechanics and gravitation, often equated with general relativity. Numerous researchers concentrate their efforts on this specific step; nevertheless, no accepted theory of quantum gravity – and thus no accepted theory of everything – has emerged yet. It is usually assumed that the ToE will also solve the remaining problems of GUTs.

In addition to explaining the forces listed in the graph, a ToE may also explain the status of at least two candidate forces suggested by modern cosmology: an inflationary force and dark energy. Furthermore, cosmological experiments also suggest the existence of dark matter, supposedly composed of fundamental particles outside the scheme of the standard model. However, the existence of these forces and particles has not been proven yet.

21.2.2 String theory and M-theory

Since the 1990s, many physicists believe that 11-dimensional M-theory, which is described in some limits by one of the five perturbative superstring theories, and in another by the maximally-supersymmetric 11-dimensional supergravity, is the theory of everything. However, there is no widespread consensus on this issue.

A surprising property of string/M-theory is that extra dimensions are required for the theory's consistency. In this regard, string theory can be seen as building on the insights of the Kaluza–Klein theory, in which it was realized that applying general relativity to a five-dimensional universe (with one of them small and curled up) looks from the four-dimensional perspective like the usual general relativity together with Maxwell's electrodynamics. This lent credence to the idea of unifying gauge and gravity interactions, and to extra dimensions, but didn't address the detailed experimental requirements. Another important property of string theory is its supersymmetry, which together with extra dimensions are the two main proposals for resolving the hierarchy problem of the standard model, which is (roughly) the question of why gravity is so much weaker than any other force. The extra-dimensional solution involves allowing gravity to propagate into the other dimensions while keeping other forces confined to a four-dimensional spacetime, an idea that has been realized with explicit stringy mechanisms.[18]

Research into string theory has been encouraged by a variety of theoretical and experimental factors. On the experimental side, the particle content of the standard model supplemented with neutrino masses fits into a spinor representation of $SO(10)$, a subgroup of E8 that routinely emerges in string theory, such as in heterotic string theory[19] or (sometimes equivalently) in F-theory.[20][21] String theory has mechanisms that may explain why fermions come in three hierarchical generations, and explain the mixing rates between quark generations.[22] On the theoretical side, it has begun to address some of the key questions in quantum gravity, such as resolving the black hole information paradox, counting the correct entropy of black holes[23][24] and allowing for topology-changing processes.[25][26][27] It has also led to many insights in pure mathematics and in ordinary, strongly-coupled gauge theory due to the Gauge/String duality.

In the late 1990s, it was noted that one major hurdle in this endeavor is that the number of possible four-dimensional universes is incredibly large. The small, "curled up" extra dimensions can be compactified in an enormous number of different ways (one estimate is 10^{500}) each of which leads to different properties for the low-energy particles and forces. This array of models is known as the string theory landscape.[8]:347

One proposed solution is that many or all of these possibilities are realised in one or another of a huge number of universes, but that only a small number of them are habitable, and hence the fundamental constants of the universe are ultimately the result of the anthropic principle rather than dictated by theory. This has led to criticism of string theory,[28] arguing that it cannot make useful (i.e., original, falsifiable, and verifiable) predictions and regarding it as a pseudoscience. Others disagree,[29] and string theory remains an extremely active topic of investigation in theoretical physics.

21.2.3 Loop quantum gravity

Current research on loop quantum gravity may eventually play a fundamental role in a ToE, but that is not its primary aim.[30] Also loop quantum gravity introduces a lower bound on the possible length scales.

There have been recent claims that loop quantum gravity may be able to reproduce features resembling the Standard Model. So far only the first generation of fermions (leptons and quarks) with correct parity properties have been modelled by Sundance Bilson-Thompson using preons constituted of braids of spacetime as the building blocks.[31] However, there is no derivation of the Lagrangian that would describe the interactions of such particles, nor is it possible to show that such particles are fermions, nor that the gauge groups or interactions of the Standard Model are realised. Utilization of quantum computing concepts made it possible to demonstrate that the particles are able to survive quantum fluctuations.[32]

This model leads to an interpretation of electric and colour charge as topological quantities (electric as number and chirality of twists carried on the individual ribbons and colour as variants of such twisting for fixed electric charge).

Bilson-Thompson's original paper suggested that the higher-generation fermions could be represented by more complicated braidings, although explicit constructions of these structures were not given. The electric charge, colour, and parity properties of such fermions would arise in the same way as for the first generation. The model was expressly generalized for an infinite number of generations and for the weak force bosons (but not for photons or gluons) in a 2008 paper by Bilson-Thompson, Hackett, Kauffman and Smolin.[33]

21.2.4   Other attempts

A recent development is the theory of causal fermion systems,[34] giving all three current physical theories (quantum mechanics, general relativity and quantum field theory) as limiting cases.

A recent and very prolific attempt is called Causal Sets. As some of the approaches mentioned above, its direct goal isn't necessarily to achieve a ToE but primarily a working theory of quantum gravity, which might eventually include the standard model and become a candidate for a ToE. Its founding principle is that spacetime is fundamentally discrete and that the spacetime events are related by a partial order. This partial order has the physical meaning of the causality relations between relative past and future distinguishing spacetime events.

Outside the previously mentioned attempts there is Garrett Lisi's E8 proposal. This theory provides an attempt of identifying general relativity and the standard model within the Lie group E8. The theory doesn't provide a novel quantization procedure and the author suggests its quantization might follow the Loop Quantum Gravity approach above mentioned.[35]

Christoph Schiller's Strand Model attempts to account for the gauge symmetry of the Standard Model of particle physics, U(1)×SU(2)×SU(3), with the three Reidemeister moves of knot theory by equating each elementary particle to a different tangle of one, two, or three strands (selectively a long prime knot or unknotted curve, a rational tangle, or a braided tangle respectively).

21.2.5   Present status

At present, there is no candidate theory of everything that includes the standard model of particle physics and general relativity. For example, no candidate theory is able to calculate the fine structure constant or the mass of the electron. Most particle physicists expect that the outcome of the ongoing experiments – the search for new particles at the large particle accelerators and for dark matter – are needed in order to provide further input for a ToE.

21.3   Theory of everything and philosophy

Main article: Theory of everything (philosophy)

The philosophical implications of a physical ToE are frequently debated. For example, if philosophical physicalism is true, a physical ToE will coincide with a philosophical theory of everything.

The "system building" style of metaphysics attempts to answer *all* the important questions in a coherent way, providing a complete picture of the world. Plato and Aristotle could be said to have created early examples of comprehensive systems. In the early modern period (17th and 18th centuries), the system-building *scope* of philosophy is often linked to the rationalist *method* of philosophy, which is the technique of deducing the nature of the world by pure *a priori* reason. Examples from the early modern period include the Leibniz's Monadology, Descarte's Dualism, and Spinoza's Monism. Hegel's Absolute idealism and Whitehead's Process philosophy were later systems.

21.4   Arguments against a theory of everything

In parallel to the intense search for a ToE, various scholars have seriously debated the possibility of its discovery.

21.4.1 Gödel's incompleteness theorem

A number of scholars claim that Gödel's incompleteness theorem suggests that any attempt to construct a ToE is bound to fail. Gödel's theorem, informally stated, asserts that any formal theory expressive enough for elementary arithmetical facts to be expressed and strong enough for them to be proved is either inconsistent (both a statement and its denial can be derived from its axioms) or incomplete, in the sense that there is a true statement that can't be derived in the formal theory.

Stanley Jaki, in his 1966 book *The Relevance of Physics*, pointed out that, because any "theory of everything" will certainly be a consistent non-trivial mathematical theory, it must be incomplete. He claims that this dooms searches for a deterministic theory of everything.[36] In a later reflection, Jaki states that it is wrong to say that a final theory is impossible, but rather that "when it is on hand one cannot know rigorously that it is a final theory."[37]

Freeman Dyson has stated that

Stephen Hawking was originally a believer in the Theory of Everything but, after considering Gödel's Theorem, concluded that one was not obtainable.

Jürgen Schmidhuber(1997) has argued against this view; he points out that Gödel's theorems are irrelevant for computable physics.[38] In 2000, Schmidhuber explicitly constructed limit-computable, deterministic universes whose pseudo-randomness based on undecidable, Gödel-like halting problems is extremely hard to detect but does not at all prevent formal ToEs describable by very few bits of information.[39]

Related critique was offered by Solomon Feferman,[40] among others. Douglas S. Robertson offers Conway's game of life as an example:[41] The underlying rules are simple and complete, but there are formally undecidable questions about the game's behaviors. Analogously, it may (or may not) be possible to completely state the underlying rules of physics with a finite number of well-defined laws, but there is little doubt that there are questions about the behavior of physical systems which are formally undecidable on the basis of those underlying laws.

Since most physicists would consider the statement of the underlying rules to suffice as the definition of a "theory of everything", most physicists argue that Gödel's Theorem does *not* mean that a ToE cannot exist. On the other hand, the scholars invoking Gödel's Theorem appear, at least in some cases, to be referring not to the underlying rules, but to the understandability of the behavior of all physical systems, as when Hawking mentions arranging blocks into rectangles, turning the computation of prime numbers into a physical question.[42] This definitional discrepancy may explain some of the disagreement among researchers.

21.4.2 Fundamental limits in accuracy

No physical theory to date is believed to be precisely accurate. Instead, physics has proceeded by a series of "successive approximations" allowing more and more accurate predictions over a wider and wider range of phenomena. Some physicists believe that it is therefore a mistake to confuse theoretical models with the true nature of reality, and hold that the series of approximations will never terminate in the "truth". Einstein himself expressed this view on occasions.[43] Following this view, we may reasonably hope for *a* theory of everything which self-consistently incorporates all currently known forces, but we should not expect it to be the final answer.

On the other hand it is often claimed that, despite the apparently ever-increasing complexity of the mathematics of each new theory, in a deep sense associated with their underlying gauge symmetry and the number of fundamental physical constants, the theories are becoming simpler. If this is the case, the process of simplification cannot continue indefinitely.

21.4.3 Lack of fundamental laws

There is a philosophical debate within the physics community as to whether a theory of everything deserves to be called *the* fundamental law of the universe.[44] One view is the hard reductionist position that the ToE is the fundamental law and that all other theories that apply within the universe are a consequence of the ToE. Another view is that emergent laws, which govern the behavior of complex systems, should be seen as equally fundamental. Examples of emergent laws are the second law of thermodynamics and the theory of natural selection. The advocates of emergence argue that emergent laws, especially those describing complex or living systems are independent of the low-level, microscopic laws. In this view, emergent laws are as fundamental as a ToE.

The debates do not make the point at issue clear. Possibly the only issue at stake is the right to apply the high-status term "fundamental" to the respective subjects of research. A well-known one took place between Steven Weinberg and Philip Anderson

21.4.4   Impossibility of being "of everything"

Although the name "theory of everything" suggests the determinism of Laplace's quotation, this gives a very misleading impression. Determinism is frustrated by the probabilistic nature of quantum mechanical predictions, by the extreme sensitivity to initial conditions that leads to mathematical chaos, by the limitations due to event horizons, and by the extreme mathematical difficulty of applying the theory. Thus, although the current standard model of particle physics "in principle" predicts almost all known non-gravitational phenomena, in practice only a few quantitative results have been derived from the full theory (e.g., the masses of some of the simplest hadrons), and these results (especially the particle masses which are most relevant for low-energy physics) are less accurate than existing experimental measurements. The ToE would almost certainly be even harder to apply for the prediction of experimental results, and thus might be of limited use.

A motive for seeking a ToE, apart from the pure intellectual satisfaction of completing a centuries-long quest, is that prior examples of unification have predicted new phenomena, some of which (e.g., electrical generators) have proved of great practical importance. And like in these prior examples of unification, the ToE would probably allow us to confidently define the domain of validity and residual error of low-energy approximations to the full theory.

21.4.5   Infinite number of onion layers

Lee Smolin regularly argues that the layers of nature may be like the layers of an onion, and that the number of layers might be infinite. This would imply an infinite sequence of physical theories.

The argument is not universally accepted, because it is not obvious that infinity is a concept that applies to the foundations of nature.

21.4.6   Impossibility of calculation

Weinberg[45] points out that calculating the precise motion of an actual projectile in the Earth's atmosphere is impossible. So how can we know we have an adequate theory for describing the motion of projectiles? Weinberg suggests that we know *principles* (Newton's laws of motion and gravitation) that work "well enough" for simple examples, like the motion of planets in empty space. These principles have worked so well on simple examples that we can be reasonably confident they will work for more complex examples. For example, although general relativity includes equations that do not have exact solutions, it is widely accepted as a valid theory because all of its equations with exact solutions have been experimentally verified. Likewise, a ToE must work for a wide range of simple examples in such a way that we can be reasonably confident it will work for every situation in physics.

21.5   See also

- Absolute (philosophy)

- An Exceptionally Simple Theory of Everything

- Argument from beauty

- Attractor

- Beyond the standard model

- Big Bang

- Brownian motion

- Chaos theory

- Chronology of the universe

- Electroweak interaction

- Holographic principle

- Mathematical beauty

- Mathematical universe hypothesis

- Multiverse

- Standard Model (mathematical formulation)

- *The Theory of Everything (2014 film)* - a feature film about Prof.Stephen Hawking and his first wife Jane Hawking.

- Timeline of the Big Bang

- Zero-energy universe

21.6 References

21.6.1 Footnotes

[1] Steven Weinberg. *Dreams of a Final Theory: The Scientist's Search for the Ultimate Laws of Nature*. Knopf Doubleday Publishing Group. ISBN 978-0-307-78786-6.

[2] Stephen W. Hawking (28 February 2006). *The Theory of Everything: The Origin and Fate of the Universe*. Phoenix Books; Special Anniv. ISBN 978-1-59777-508-3.

[3] Carlip, Steven (2001). "Quantum Gravity: a Progress Report". *Reports on Progress in Physics* **64** (8). arXiv:gr-qc/0108040. Bibcode:2001RPPh...64..885C. doi:10.1088/0034-4885/64/8/301.

[4] Susanna Hornig Priest (14 July 2010). *Encyclopedia of Science and Technology Communication*. SAGE Publications. ISBN 978-1-4522-6578-0.

[5]Ellis, John (2002). "Physics gets physical (correspondence)".*Nature***415**(6875): 957.Bibcode:2002Natur.415..957E.doi:15957b.

[6]Ellis, John (1986). "The Superstring: Theory of Everything, or of Nothing?".*Nature***323**(6089): 595–598.Bibcode:1986Nat5E. doi:10.1038/323595a0.

[7] Rorres, Chris (2009). "ARCHIMEDES AND THE QUEST FOR THE THEORY OF EVERYTHING".

[8] Chris Impey (26 March 2012). *How It Began: A Time-Traveler's Guide to the Universe*. W. W. Norton. ISBN 978-0-393-08002-5.

[9] William E. Burns (1 January 2001). *The Scientific Revolution: An Encyclopedia*. ABC-CLIO. ISBN 978-0-87436-875-8.

[10] Shapin, Steven (1996). *The Scientific Revolution*. University of Chicago Press. ISBN 0-226-75021-3.

[11] Newton, Sir Isaac (1729). *The Mathematical Principles of Natural Philosophy* **II**. p. 255.

[12] Sean Carroll (7 January 2010). *From Eternity to Here: The Quest for the Ultimate Theory of Time*. Penguin Group US. ISBN 978-1-101-15215-7.

[13] Faraday, M. (1850). "Experimental Researches in Electricity. Twenty-Fourth Series. On the Possible Relation of Gravity to Electricity". *Abstracts of the Papers Communicated to the Royal Society of London* **5**: 994–995. doi:10.1098/rspl.1843.0267.

[14] A.N. Gorban, I. Karlin, Hilbert's 6th Problem: exact and approximate hydrodynamic manifolds for kinetic equations, Bull. Amer. Math. Soc., 51 (2014), no. 2, 186-246, doi:10.1090/S0273-0979-2013-01439-3

[15] Dirac, P.A.M. (1929). "Quantum mechanics of many-electron systems". *Proceedings of the Royal Society of London A* **123** (792): 714. Bibcode:1929RSPSA.123..714D. doi:10.1098/rspa.1929.0094.

[16] Abraham Pais (23 September 1982). *Subtle is the Lord : The Science and the Life of Albert Einstein: The Science and the Life of Albert Einstein.* Oxford University Press. ISBN 978-0-19-152402-8.

[17] Weinberg (1993), Ch. 5

[18] Holloway, M (2005). "The Beauty of Branes"(PDF).*Scientific American*(Scientific American)**293**(4): 38.Bibcode:2005SciAmH. doi:10.1038/scientificamerican1005-38. PMID 16196251. Retrieved August 13, 2012.

[19] Nilles, Hans Peter; Ramos-Sánchez, Saúl; Ratz, Michael; Vaudrevange, Patrick K. S. (2008). "From strings to the MSSM". *The European Physical Journal C* **59** (2): 249. arXiv:0806.3905. Bibcode:2009EPJC...59..249N. doi:10.1140/epjc/s10052-008-0740-1.

[20] Beasley, Chris; Heckman, Jonathan J; Vafa, Cumrun (2009). "GUTs and exceptional branes in F-theory — I". *Journal of High Energy Physics* **2009**: 058. arXiv:0802.3391. Bibcode:2009JHEP...01..058B. doi:10.1088/1126-6708/2009/01/058.

[21] Donagi, Ron and Wijnholt, Martijn (2008) Model Building with F-Theory

[22] Heckman, Jonathan J. and Vafa, Cumrun (2008) Flavor Hierarchy From F-theory

[23] Strominger, Andrew; Vafa, Cumrun (1996). "Microscopic origin of the Bekenstein-Hawking entropy". *Physics Letters B* **379**: 99. arXiv:hep-th/9601029. Bibcode:1996PhLB..379...99S. doi:10.1016/0370-2693(96)00345-0.

[24] Horowitz, Gary (1996) The Origin of Black Hole Entropy in String Theory

[25] Greene, Brian R.; Morrison, David R.; Strominger, Andrew (1995). "Black hole condensation and the unification of string vacua". *Nuclear Physics B* **451**: 109. arXiv:hep-th/9504145. Bibcode:1995NuPhB.451..109G. doi:10.1016/0550-3213(95)0-X.

[26] Aspinwall, Paul S.; Greene, Brian R.; Morrison, David R. (1994). "Calabi-Yau moduli space, mirror manifolds and spacetime topology change in string theory". *Nuclear Physics B* **416** (2): 414. arXiv:hep-th/9309097. Bibcode:1994NuPhB.416..414A. doi:10.1016/0550-3213(94)90321-2.

[27] Adams, Allan; Liu, Xiao; McGreevy, John; Saltman, Alex; Silverstein, Eva (2005). "Things fall apart: Topology change from winding tachyons". *Journal of High Energy Physics* **2005** (10): 033. arXiv:hep-th/0502021. Bibcode:2005JHEP...10..033A. doi:10.1088/1126-6708/2005/10/033.

[28] Smolin, Lee (2006). *The Trouble With Physics: The Rise of String Theory, the Fall of a Science, and What Comes Next.* Houghton Mifflin. ISBN 978-0-618-55105-7.

[29] Duff, M. J. (2011). "String and M-Theory: Answering the Critics". *Foundations of Physics* **43**: 182. arXiv:1112.0788. Bibcode:2013FoPh...43..182D. doi:10.1007/s10701-011-9618-4.

[30] Potter, Franklin (15 February 2005). "Leptons And Quarks In A Discrete Spacetime" (PDF). *Frank Potter's Science Gems.* Retrieved 2009-12-01.

[31] Bilson-Thompson, Sundance O.; Markopoulou, Fotini; Smolin, Lee (2007). "Quantum gravity and the standard model". *Classical and Quantum Gravity* **24** (16): 3975–3994. arXiv:hep-th/0603022. Bibcode:2007CQGra..24.3975B. doi:10.1088/0264-9381/24/16/002.

[32] Castelvecchi, Davide; Valerie Jamieson (August 12, 2006). "You are made of space-time". *New Scientist* (2564).

[33] Sundance Bilson-Thompson; Jonathan Hackett; Lou Kauffman; Lee Smolin (2008). "Particle Identifications from Symmetries of Braided Ribbon Network Invariants". arXiv:0804.0037 [hep-th].

[34] F. Finster; J. Kleiner (2015). "Causal fermion systems as a candidate for a unified physical theory". arXiv:1502.03587 [math-ph].

[35] A. G. Lisi (2007). "An Exceptionally Simple Theory of Everything". arXiv:0711.0770 [hep-th].

[36] Jaki, S.L. (1966). *The Relevance of Physics*. Chicago Press. pp. 127–130.

[37] Stanley L. Jaki (2004) "A Late Awakening to Gödel in Physics", pp. 8–9.

[38] Schmidhuber, Jürgen (1997). *A Computer Scientist's View of Life, the Universe, and Everything*. *Lecture Notes in Computer Science*. Springer. pp. 201–208. doi:10.1007/BFb0052071. ISBN 978-3-540-63746-2.

[39] Schmidhuber, Jürgen (2002). "Hierarchies of generalized Kolmogorov complexities and nonenumerable universal measures computable in the limit". arXiv:quant-ph/0011122.

[40] Feferman, Solomon (17 November 2006). "The nature and significance of Gödel's incompleteness theorems" (PDF). Institute for Advanced Study. Retrieved 2009-01-12.

[41] Robertson, Douglas S. (2007). "Goedel's Theorem, the Theory of Everything, and the Future of Science and Mathematics". *Complexity* **5** (5): 22–27. doi:10.1002/1099-0526(200005/06)5:5<22::AID-CPLX4>3.0.CO;2-0.

[42] Hawking, Stephen (20 July 2002). "Gödel and the end of physics". Retrieved 2009-12-01.

[43] Einstein, letter to Felix Klein, 1917. (On determinism and approximations.) Quoted in Pais (1982), Ch. 17.

[44] Weinberg (1993), Ch 2.

[45] Weinberg (1993) p. 5

21.6.2   Bibliography

- Pais, Abraham (1982) *Subtle is the Lord...: The Science and the Life of Albert Einstein* (Oxford University Press, Oxford, . Ch. 17, ISBN 0-19-853907-X

- Weinberg, Steven (1993) *Dreams of a Final Theory: The Search for the Fundamental Laws of Nature*, Hutchinson Radius, London, ISBN 0-09-177395-4

21.7   External links

- The Elegant Universe, *Nova* episode about the search for the theory of everything and string theory.

- Theory of Everything, freeview video by the Vega Science Trust, BBC and Open University.

- The Theory of Everything: Are we getting closer, or is a final theory of matter and the universe impossible? Debate between John Ellis (physicist), Frank Close and Nicholas Maxwell.

- Why The World Exists, a discussion between physicist Laura Mersini-Houghton, cosmologist George Francis Rayner Ellis and philosopher David Wallace about dark matter, parallel universes and explaining why these and the present Universe exist.

Chapter 22

Superstring theory

"Superstring" redirects here. For the converse relation of "substring", see Superstring (formal languages). For the bundle of firecrackers, see Superstring (fireworks).

Superstring theory is an attempt to explain all of the particles and fundamental forces of nature in one theory by modelling them as vibrations of tiny supersymmetric strings.

'Superstring theory' is a shorthand for **supersymmetric string theory** because unlike bosonic string theory, it is the version of string theory that incorporates fermions and supersymmetry.

Since the second superstring revolution the five superstring theories are regarded as different limits of a single theory tentatively called M-theory, or simply string theory.

22.1 Background

The deepest problem in theoretical physics is harmonizing the theory of general relativity, which describes gravitation and applies to large-scale structures (stars, galaxies, super clusters), with quantum mechanics, which describes the other three fundamental forces acting on the atomic scale.

The development of a quantum field theory of a force invariably results in infinite (and therefore useless) probabilities. Physicists have developed mathematical techniques (renormalization) to eliminate these infinities which work for three of the four fundamental forces—electromagnetic, strong nuclear and weak nuclear forces—but not for gravity. The development of a quantum theory of gravity must therefore come about by different means than those used for the other forces.[1]

According to the theory, the fundamental constituents of reality are strings of the Planck length (about 10^{-33} cm) which vibrate at resonant frequencies. Every string, in theory, has a unique resonance, or harmonic. Different harmonics determine different fundamental particles. The tension in a string is on the order of the Planck force (10^{44} newtons). The graviton (the proposed messenger particle of the gravitational force), for example, is predicted by the theory to be a string with wave amplitude zero.

22.1.1 Lack of experimental evidence

Superstring theory is based on supersymmetry. No supersymmetric particles have been discovered and recent research at LHC and Tevatron has excluded some of the ranges.[2][3][4][5] For instance, the mass constraint of the Minimal Supersymmetric Standard Model squarks has been up to 1.1 TeV, and gluinos up to 500 GeV.[6] No report on suggesting large extra dimensions has been delivered from LHC. There have been no principles so far to limit the number of vacua in the concept of a landscape of vacua.[7]

Some particle physicists became disappointed[8] by the lack of experimental verification of supersymmetry, and some have already discarded it; Jon Butterworth at the University College London said that we had no sign of supersymmetry, even in higher energy region, excluding the superpartners of the top quark up to a few TeV. Ben Allanach at the University of Cambridge states that if we do not discover any new particles in the next trial at the LHC, then we can say it is unlikely to discover supersymmetry at CERN in the foreseeable future.[8]

22.2   Extra dimensions

See also: Why does consistency require 10 dimensions?

Our physical space is observed to have only three large dimensions and—taken together with duration as the fourth dimension—a physical theory must take this into account. However, nothing prevents a theory from including more than 4 dimensions. In the case of string theory, consistency requires spacetime to have 10 (3+1+6) dimensions. The fact that we see only 3 dimensions of space can be explained by one of two mechanisms: either the extra dimensions are compactified on a very small scale, or else our world may live on a 3-dimensional submanifold corresponding to a brane, on which all known particles besides gravity would be restricted.

If the extra dimensions are compactified, then the extra six dimensions must be in the form of a Calabi–Yau manifold. Within the more complete framework of M-theory, they would have to take form of a G2 manifold. Calabi-Yaus are interesting mathematical spaces in their own right. A particular exact symmetry of string/M-theory called T-duality (which exchanges momentum modes for winding number and sends compact dimensions of radius R to radius 1/R),[9] has led to the discovery of equivalences between different Calabi-Yaus called Mirror Symmetry.

Superstring theory is not the first theory to propose extra spatial dimensions. It can be seen as building upon the Kaluza–Klein theory which proposed a 4+1-dimensional theory of gravity. When compactified on a circle, the gravity in the extra dimension precisely describes electromagnetism from the perspective of the 3 remaining large space dimensions. Thus the original Kaluza–Klein theory is a prototype for the unification of gauge and gravity interactions, at least at the classical level, however it is known to be insufficient to describe nature for a variety of reasons (missing weak and strong forces, lack of parity violation, etc.) A more complex compact geometry is needed to reproduce the known gauge forces. This is not all: In order to obtain a consistent, fundamental, quantum theory the upgrade to string theory is also necessary, not just the extra dimensions.

22.3   Number of superstring theories

Theoretical physicists were troubled by the existence of five separate string theories. A possible solution for this dilemma was suggested at the beginning of what is called the second superstring revolution in the 1990s, which suggests that the five string theories might be different limits of a single underlying theory, called M-theory. This remains a conjecture.[10]

The five consistent superstring theories are:

- The type I string has one supersymmetry in the ten-dimensional sense (16 supercharges). This theory is special in the sense that it is based on unoriented open and closed strings, while the rest are based on oriented closed strings.

- The type II string theories have two supersymmetries in the ten-dimensional sense (32 supercharges). There are actually two kinds of type II strings called type IIA and type IIB. They differ mainly in the fact that the IIA theory is non-chiral (parity conserving) while the IIB theory is chiral (parity violating).

- The heterotic string theories are based on a peculiar hybrid of a type I superstring and a bosonic string. There are two kinds of heterotic strings differing in their ten-dimensional gauge groups: the heterotic $E_8 \times E_8$ string and the heterotic SO(32) string. (The name heterotic SO(32) is slightly inaccurate since among the SO(32) Lie groups, string theory singles out a quotient Spin(32)/Z_2 that is not equivalent to SO(32).)

Chiral gauge theories can be inconsistent due to anomalies. This happens when certain one-loop Feynman diagrams cause a quantum mechanical breakdown of the gauge symmetry. The anomalies were canceled out via the Green–Schwarz mechanism.

Even though there are only five superstring theories, in order to make detailed predictions for real experiments, information is needed about exactly what physical configuration the theory is in. This considerably complicates efforts to test string theory because there is an astronomically high number – 10^{500} or more – of configurations that meet some of the basic requirements to be consistent with our world. Along with the extreme remoteness of the Planck scale, this is the other major reason it is hard to test superstring theory.

Another approach to the number of superstring theories refers to the mathematical structure called composition algebra. In the findings of abstract algebra there are just seven composition algebras over the field of real numbers. In 1990 physicists R. Foot and G.C. Joshi in Australia stated that "the seven classical superstring theories are in one-to-one correspondence to the seven composition algebras."[11]

22.4 Integrating general relativity and quantum mechanics

General relativity typically deals with situations involving large mass objects in fairly large regions of spacetime whereas quantum mechanics is generally reserved for scenarios at the atomic scale (small spacetime regions). The two are very rarely used together, and the most common case in which they are combined is in the study of black holes. Having "peak density", or the maximum amount of matter possible in a space, and very small area, the two must be used in synchrony in order to predict conditions in such places; yet, when used together, the equations fall apart, spitting out impossible answers, such as imaginary distances and less than one dimension.

The major problem with their congruence is that, at Planck scale (a fundamental small unit of length) lengths, general relativity predicts a smooth, flowing surface, while quantum mechanics predicts a random, warped surface, neither of which are anywhere near compatible. Superstring theory resolves this issue, replacing the classical idea of point particles with loops. These loops have an average diameter of the Planck length, with extremely small variances, which completely ignores the quantum mechanical predictions of Planck-scale length dimensional warping.

Singularities are avoided because the observed consequences of "Big Crunches" never reach zero size. In fact, should the universe begin a "big crunch" sort of process, string theory dictates that the universe could never be smaller than the size of a string, at which point it would actually begin expanding.

22.5 Mathematics

22.5.1 D-branes

D-branes are membrane-like objects in 10D string theory. They can be thought of as occurring as a result of a Kaluza–Klein compactification of 11D M-theory which contains membranes. Because compactification of a geometric theory produces extra vector fields the D-branes can be included in the action by adding an extra U(1) vector field to the string action.

$$\partial_z \to \partial_z + iA_z(z, \overline{z})$$

In **type I** open string theory, the ends of open strings are always attached to D-brane surfaces. A string theory with more gauge fields such as SU(2) gauge fields would then correspond to the compactification of some higher-dimensional theory above 11 dimensions which is not thought to be possible to date. Furthermore, the tachyons attached to the D-branes, show, the instability of those d-branes with respect to the annihilation.We will consider that tachyon total energy is (or reflects) the total energy of the D-branes.

22.5.2   Why five superstring theories?

For a 10 dimensional supersymmetric theory we are allowed a 32-component Majorana spinor. This can be decomposed into a pair of 16-component Majorana-Weyl (chiral) spinors. There are then various ways to construct an invariant depending on whether these two spinors have the same or opposite chiralities:

The heterotic superstrings come in two types SO(32) and $E_8 \times E_8$ as indicated above and the type I superstrings include open strings.

22.6   Beyond superstring theory

It is conceivable that the five superstring theories are approximated to a theory in higher dimensions possibly involving membranes. Because the action for this involves quartic terms and higher so is not Gaussian, the functional integrals are very difficult to solve and so this has confounded the top theoretical physicists. Edward Witten has popularised the concept of a theory in 11 dimensions M-theory involving membranes interpolating from the known symmetries of superstring theory. It may turn out that there exist membrane models or other non-membrane models in higher dimensions which may become acceptable when new unknown symmetries of nature are found, such as noncommutative geometry for example. It is thought, however, that 16 is probably the maximum since O(16) is a maximal subgroup of E8 the largest exceptional lie group and also is more than large enough to contain the Standard Model. Quartic integrals of the non-functional kind are easier to solve so there is hope for the future. This is the series solution which is always convergent when a is non-zero and negative:

$$\int_{-\infty}^{\infty} \exp(ax^4 + bx^3 + cx^2 + dx + f)\, dx = e^f \sum_{n,m,p=0}^{\infty} \frac{b^{4n}}{(4n)!} \frac{c^{2m}}{(2m)!} \frac{d^{4p}}{(4p)!} \frac{\Gamma(3n + m + p + \frac{1}{4})}{a^{3n+m+p+\frac{1}{4}}}$$

In the case of membranes the series would correspond to sums of various membrane interactions that are not seen in string theory.

22.6.1   Compactification

Investigating theories of higher dimensions often involves looking at the 10 dimensional superstring theory and interpreting some of the more obscure results in terms of compactified dimensions. For example D-branes are seen as compactified membranes from 11D M-theory. Theories of higher dimensions such as 12D F-theory and beyond will produce other effects such as gauge terms higher than $U(1)$. The components of the extra vector fields (A) in the D-brane actions can be thought of as extra coordinates (X) in disguise. However, the *known* symmetries including supersymmetry currently restrict the spinors to have 32-components which limits the number of dimensions to 11 (or 12 if you include two time dimensions.) Some commentators (e.g. John Baez et al.) have speculated that the exceptional lie groups E_6, E_7 and E_8 having maximum orthogonal subgroups O(10), O(12) and O(16) may be related to theories in 10, 12 and 16 dimensions; 10 dimensions corresponding to string theory and the 12 and 16 dimensional theories being yet undiscovered but would be theories based on 3-branes and 7-branes respectively. However this is a minority view within the string community. Since E_7 is in some sense F_4 quaternified and E_8 is F_4 octonified, then the 12 and 16 dimensional theories, if they did exist, may involve the noncommutative geometry based on the quaternions and octonions respectively. From the above discussion, it can be seen that physicists have many ideas for extending superstring theory beyond the current 10 dimensional theory, but so far none have been successful.

22.6.2   Kac–Moody algebras

Since strings can have an infinite number of modes, the symmetry used to describe string theory is based on infinite dimensional Lie algebras. Some Kac–Moody algebras that have been considered as symmetries for M-theory have been E_{10} and E_{11} and their supersymmetric extensions.

22.7 See also

- AdS/CFT

- dS/CFT correspondence

- Grand unification theory

- Large Hadron Collider

- List of string theory topics

- Quantum gravity

- String field theory

22.8 Notes

[1] Polchinski, Joseph. *String Theory: Volume I*. Cambridge University Press, p. 4.

[2] Woit, Peter (February 22, 2011). "Implications of Initial LHC Searches for Supersymmetry".

[3] Cassel, S.; Ghilencea, D. M.; Kraml, S.; Lessa, A.; Ross, G. G. (2011). "Fine-tuning implications for complementary dark matter and LHC SUSY searches". *Journal of High Energy Physics* **2011** (5): 120. arXiv:1101.4664. Bibcode:2011JHEP...05..120C. doi:10.1007/JHEP05(2011)120.

[4] Falkowski, Adam (Jester) (February 16, 2011). "What LHC tells about SUSY". *resonaances.blogspot.com*. Archived from the original on March 22, 2014. Retrieved March 22, 2014.

[5] Tapper, Alex (24 March 2010). "Early SUSY searches at the LHC" (PDF). Imperial College London.

[6] CMS Collaboration (2011). "Search for Supersymmetry at the LHC in Events with Jets and Missing Transverse Energy". *Physical Review Letters* **107** (22): 221804. arXiv:1109.2352. Bibcode:2011PhRvL.107v1804C. doi:10.1103/PhysRevLett.107.221. PMID 22182023.

[7] Shifman, M. (2012). "Frontiers Beyond the Standard Model: Reflections and Impressionistic Portrait of the Conference". *Modern Physics Letters A* **27** (40): 1230043. Bibcode:2012MPLA...2730043S. doi:10.1142/S0217732312300431.

[8] Jha, Alok (August 6, 2013). "One year on from the Higgs boson find, has physics hit the buffers?". *The Guardian*. photograph: Harold Cunningham/Getty Images (London: GMG). ISSN 0261-3077. OCLC 60623878. Archived from the original on March 22, 2014. Retrieved March 22, 2014.

[9] Polchinski, Joseph. *String Theory: Volume I*. Cambridge University Press, p. 247.

[10] Polchinski, Joseph. *String Theory: Volume II*. Cambridge University Press, p. 198.

[11] Foot, R.; Joshi, G. C. (1990). "Nonstandard signature of spacetime, superstrings, and the split composition algebras". *Letters in Mathematical Physics* **19**: 65–71. Bibcode:1990LMaPh..19...65F. doi:10.1007/BF00402262.

22.9 References

- Kaku, Michio (1999). *Introduction to Superstring and M-Theory* (2nd ed.). New York, USA: Springer-Verlag.

- Shen, Sinyan (1982). *Introduction to Superfluidity* (2nd ed.). Beijing, China: Science Press.

- Greene, Brian (2000). *The Elegant Universe: Superstrings, Hidden Dimensions, and the Quest for the Ultimate Theory*. Random House Inc.

22.10 External links

- Wellcome Collection video on superstring theory

- The Official Superstring theory website: http://superstringtheory.com/index.html

Chapter 23

Coleman–Mandula theorem

The **Coleman–Mandula theorem**, named after Sidney Coleman and Jeffrey Mandula, is a no-go theorem in theoretical physics. It states that "space-time and internal symmetries cannot be combined in any but a trivial way".[1] The only conserved quantities in a "realistic" theory with a mass gap, apart from the generators of the Poincaré group, must be Lorentz scalars.

23.1 Description

Every quantum field theory satisfying certain technical assumptions about its S-matrix that has non-trivial interactions can only have a symmetry Lie algebra which is always a direct product of the Poincaré group and an internal group if there is a mass gap: no mixing between these two is possible. As the authors say in the introduction to the 1967 publication, "We prove a new theorem on the impossibility of combining space-time and internal symmetries in any but a trivial way."[2][3]

Note that this theorem only constrains the symmetries of the S-matrix itself. As such, it places no constraints on spontaneously broken symmetries which do not show up directly on the S-matrix level. In fact, it is easy to construct spontaneously broken symmetries (in interacting theories) which unify spatial and internal symmetries.

This theorem also only applies to Lie algebras and not Lie groups. As such, it does not apply to discrete symmetries or globally for Lie groups. As an example of the latter, we might have a model where a rotation by 2π (a spacetime symmetry) is identified with an involutive internal symmetry which commutes with all the other internal symmetries.

If there is no mass gap, it could be a tensor product of the conformal algebra with an internal Lie algebra. But in the absence of a mass gap, there are also other possibilities. For example, quantum electrodynamics has vector and tensor conserved charges. See infraparticle for more details.

Supersymmetry may be considered a possible "loophole" of the theorem because it contains additional generators (supercharges) that are not scalars but rather spinors. This loophole is possible because supersymmetry is a Lie superalgebra, not a Lie algebra. The corresponding theorem for supersymmetric theories with a mass gap is the Haag–Lopuszanski–Sohnius theorem.

Quantum group symmetry, present in some two-dimensional integrable quantum field theories like the sine-Gordon model, exploits a similar loophole.

23.2 Notes

[1] Pelc, Oskar; Horwitz, L. P. (1997). "Generalization of the Coleman–Mandula theorem to higher dimension". *Journal of Mathematical Physics* **38** (1): 139. arXiv:hep-th/9605147. Bibcode:1997JMP....38..139P. doi:10.1063/1.531846.

[2] Valuing Negativity | Cosmic Variance

[3] *Physical Review* **159** (5): 1251. 1967. Bibcode:1967PhRv..159.1251C. doi:10.1103/PhysRev.159.1251. Missing or empty |title= (help)

23.3 References

• Sidney Coleman and Jeffrey Mandula (1967). "All Possible Symmetries of the S Matrix". *Physical Review* **159** (5): 1251–1256. Bibcode:1967PhRv..159.1251C. doi:10.1103/PhysRev.159.1251.

Chapter 24

Haag–Lopuszanski–Sohnius theorem

In theoretical physics, the **Haag–Lopuszanski–Sohnius theorem** shows that the possible symmetries of a consistent 4-dimensional quantum field theory do not only consist of internal symmetries and Poincaré symmetry, but can also include supersymmetry as a nontrivial extension of the Poincaré algebra. This significantly generalized the Coleman–Mandula theorem.

One of the important results is that the fermionic part of the Lie superalgebra has to have spin-1/2 (spin 3/2 or higher are ruled out).

24.1   History

Prior to the Haag–Lopuszanski–Sohnius theorem, the Coleman–Mandula theorem was the strongest of a series of no-go theorems, stating that the symmetry group of a consistent 4-dimensional quantum field theory is the direct product of the internal symmetry group and the Poincaré group.

In 1975, Rudolf Haag, Jan Łopuszański, and Martin Sohnius published their proof that weakening the assumptions of the Coleman–Mandula theorem by allowing both commuting and anticommuting symmetry generators, there is a nontrivial extension of the Poincaré algebra, namely the supersymmetry algebra.

24.2   Importance

What is most fundamental in this result (and thus in supersymmetry), is that there can be an interplay of spacetime symmetry with internal symmetry (in the sense of "mixing particles"): the supersymmetry generators transform bosonic particles into fermionic ones and vice versa, but the anticommutator of two such transformations yields a translation in spacetime. Precisely such an interplay seemed excluded by the Coleman–Mandula theorem, which stated that (bosonic) internal symmetries cannot interact non-trivially with spacetime symmetry.

This theorem was also an important justification of the previously found Wess–Zumino model, an interacting four-dimensional quantum field theory with supersymmetry, leading to a renormalizable theory.

24.3   See also

- Supergravity

- S-matrix

24.4   References

- Haag, Rudolf; Sohnius, Martin; Łopuszański, Jan T. (1975), "All possible generators of supersymmetries of the S-matrix", *Nuclear Physics B* **88**: 257–274, Bibcode:1975NuPhB..88..257H, doi:10.1016/0550-3213(75)90279-5, MR 0411396

Chapter 25

Magnetic monopole

A **magnetic monopole** is a hypothetical elementary particle in particle physics that is an isolated magnet with only one magnetic pole (a north pole without a south pole or vice versa).[1][2] In more technical terms, a magnetic monopole would have a net "magnetic charge". Modern interest in the concept stems from particle theories, notably the grand unified and superstring theories, which predict their existence.[3][4]

Magnetism in bar magnets and electromagnets does not arise from magnetic monopoles. There is no conclusive experimental evidence that magnetic monopoles exist at all in our universe.

Some condensed matter systems contain effective (non-isolated) magnetic monopole *quasi*-particles,[5] or contain phenomena that are mathematically analogous to magnetic monopoles.[6]

25.1 Historical background

25.1.1 Pre-twentieth century

Many early scientists attributed the magnetism of lodestones to two different "magnetic fluids" ("effluvia"), a north-pole fluid at one end and a south-pole fluid at the other, which attracted and repelled each other in analogy to positive and negative electric charge.[7][8] However, an improved understanding of electromagnetism in the nineteenth century showed that the magnetism of lodestones was properly explained by Ampère's circuital law, not magnetic monopole fluids. Gauss's law for magnetism, one of Maxwell's equations, is the mathematical statement that magnetic monopoles do not exist. Nevertheless, it was pointed out by Pierre Curie in 1894[9] that magnetic monopoles *could* conceivably exist, despite not having been seen so far.

25.1.2 Twentieth century

The *quantum* theory of magnetic charge started with a paper by the physicist Paul A.M. Dirac in 1931.[10] In this paper, Dirac showed that if *any* magnetic monopoles exist in the universe, then all electric charge in the universe must be quantized.[11] The electric charge *is*, in fact, quantized, which is consistent with (but does not prove) the existence of monopoles.[11]

Since Dirac's paper, several systematic monopole searches have been performed. Experiments in 1975[12] and 1982[13] produced candidate events that were initially interpreted as monopoles, but are now regarded as inconclusive.[14] Therefore, it remains an open question whether monopoles exist. Further advances in theoretical particle physics, particularly developments in grand unified theories and quantum gravity, have led to more compelling arguments (detailed below) that monopoles do exist. Joseph Polchinski, a string-theorist, described the existence of monopoles as "one of the safest bets that one can make about physics not yet seen".[15] These theories are not necessarily inconsistent with the experimental evidence. In some theoretical models, magnetic monopoles are unlikely to be observed, because they are too massive to

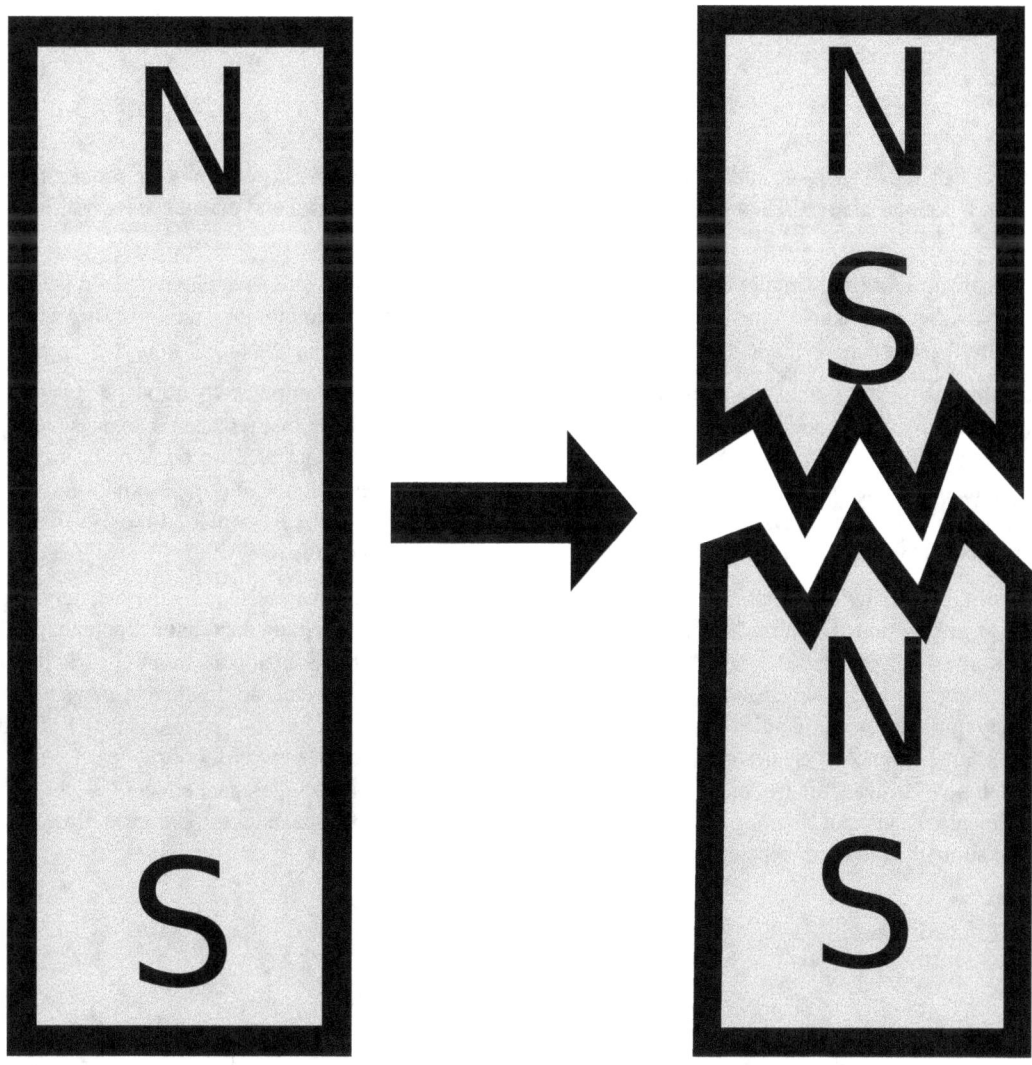

*It is impossible to make **magnetic monopoles** from a bar magnet. If a bar magnet is cut in half, it is not the case that one half has the north pole and the other half has the south pole. Instead, each piece has its own north and south poles. A magnetic monopole cannot be created from normal matter such as atoms and electrons, but would instead be a new elementary particle.*

be created in particle accelerators (see below), and also too rare in the Universe to enter a particle detector with much probability.[15]

Some condensed matter systems propose a structure superficially similar to a magnetic monopole, known as a flux tube. The ends of a flux tube form a magnetic dipole, but since they move independently, they can be treated for many purposes as independent magnetic monopole quasiparticles. Since 2009, numerous news reports from the popular media[16][17] have incorrectly described these systems as the long-awaited discovery of the magnetic monopoles, but the two phenomena are only superficially related to one another.[18][19] These condensed-matter systems continue to be an area of active research. (See "Monopoles" in condensed-matter systems below.)

25.2   Poles and magnetism in ordinary matter

Main article: Magnetism

All matter ever isolated to date—including every atom on the periodic table and every particle in the standard model—has zero magnetic monopole charge. Therefore, the ordinary phenomena of magnetism and magnets have nothing to do with magnetic monopoles.

Instead, magnetism in ordinary matter comes from two sources. First, electric currents create magnetic fields according to Ampère's law. Second, many elementary particles have an "intrinsic" magnetic moment, the most important of which is the electron magnetic dipole moment. (This magnetism is related to quantum-mechanical "spin".)

Mathematically, the magnetic field of an object is often described in terms of a multipole expansion. This is an expression of the field as the sum of component fields with specific mathematical forms. The first term in the expansion is called the "monopole" term, the second is called "dipole", then "quadrupole", then "octupole", and so on. Any of these terms can be present in the multipole expansion of an electric field, for example. However, in the multipole expansion of a *magnetic* field, the "monopole" term is always exactly zero (for ordinary matter). A magnetic monopole, if it exists, would have the defining property of producing a magnetic field whose "monopole" term is nonzero.

A magnetic dipole is something whose magnetic field is predominantly or exactly described by the magnetic dipole term of the multipole expansion. The term "dipole" means "two poles", corresponding to the fact that a dipole magnet typically contains a "north pole" on one side and a "south pole" on the other side. This is analogous to an electric dipole, which has positive charge on one side and negative charge on the other. However, an electric dipole and magnetic dipole are fundamentally quite different. In an electric dipole made of ordinary matter, the positive charge is made of protons and the negative charge is made of electrons, but a magnetic dipole does *not* have different types of matter creating the north pole and south pole. Instead, the two magnetic poles arise simultaneously from the aggregate effect of all the currents and intrinsic moments throughout the magnet. Because of this, the two poles of a magnetic dipole must always have equal and opposite strength, and the two poles cannot be separated from each other.

25.3   Maxwell's equations

Maxwell's equations of electromagnetism relate the electric and magnetic fields to each other and to the motions of electric charges. The standard equations provide for electric charges, but they posit no magnetic charges. Except for this difference, the equations are symmetric under the interchange of the electric and magnetic fields.[20] In fact, symmetric Maxwell's equations can be written when all charges (and hence electric currents) are zero, and this is how the electromagnetic wave equation is derived.

Fully symmetric Maxwell's equations can also be written if one allows for the possibility of "magnetic charges" analogous to electric charges.[21] With the inclusion of a variable for the density of these magnetic charges, say ϱ_m, there will also be a "magnetic current density" variable in the equations, $\mathbf{j}_m$.

If magnetic charges do not exist – or if they do exist but are not present in a region of space – then the new terms in Maxwell's equations are all zero, and the extended equations reduce to the conventional equations of electromagnetism such as $\nabla \cdot \mathbf{B} = 0$ (where $\nabla \cdot$ is divergence and $\mathbf{B}$ is the magnetic $\mathbf{B}$ field).

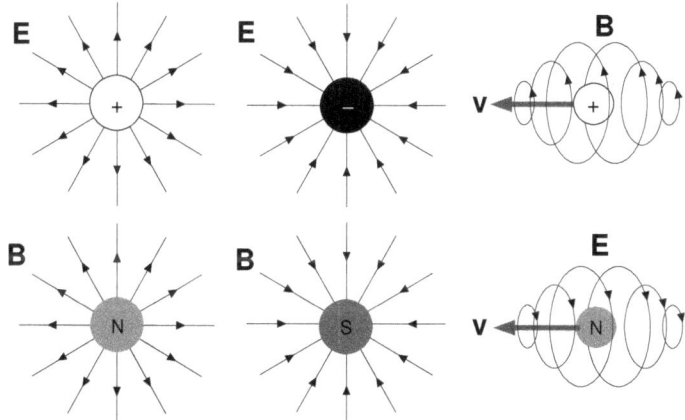

Left: Fields due to stationary electric and magnetic monopoles. **Right:** In motion (velocity **v**), an *electric* charge induces a **B** field while a *magnetic* charge induces an **E** field. Conventional current is used.

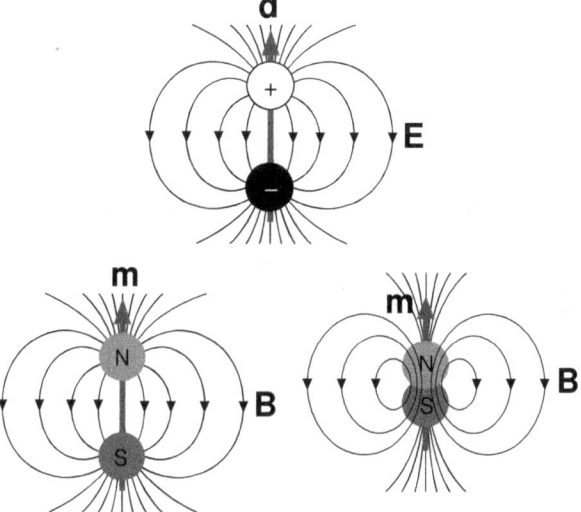

Top: E field due to an electric dipole moment **d**. **Bottom left: B** field due to a *mathematical* magnetic dipole **m** formed by two magnetic monopoles. **Bottom right: B** field due to a natural magnetic dipole moment **m** found in ordinary matter (*not* from monopoles).

The **E** fields and **B** fields due to electric charges (black/white) and magnetic poles (red/blue).[22][23]

25.3.1   In Gaussian cgs units

The extended Maxwell's equations are as follows, in Gaussian cgs units:[24]

In these equations ϱ_m is the *magnetic charge density*, $\mathbf{j}_m$ is the *magnetic current density*, and q_m is the *magnetic charge* of a test particle, all defined analogously to the related quantities of electric charge and current; **v** is the particle's velocity and c is the speed of light. For all other definitions and details, see Maxwell's equations. For the equations in nondimensionalized form, remove the factors of c.

25.3.2 In SI units

In SI units, there are two conflicting units in use for magnetic charge q_m: webers (Wb) and ampere·meters (A·m). The conversion between them is $q_m(\text{Wb}) = \mu_0 q_m(\text{A·m})$, since the units are 1 Wb = 1 H·A = (1 H·m^{-1})·(1 A·m) by dimensional analysis (H is the henry – the SI unit of inductance).

Maxwell's equations then take the following forms (using the same notation above):[26]

25.3.3 Tensor formulation

Maxwell's equations in the language of tensors makes Lorentz covariance clear. The generalized equations are:[27][28]

where

- $F^{\alpha\beta}$ is the electromagnetic tensor, $^{\alpha\beta} = 1/2\varepsilon^{\alpha\beta\gamma\delta}F\gamma\delta$ is the dual electromagnetic tensor,

- for a particle with electric charge q_e and magnetic charge q_m; v is the four-velocity and p the four-momentum,

- for an electric and magnetic charge distribution; $J_e = (\varrho_e, \mathbf{j}_e)$ is the electric four-current and $J_m = (\varrho_m, \mathbf{j}_m)$ the magnetic four-current.

For a particle having only electric charge, one can express its field using a four-potential, according to the standard covariant formulation of classical electromagnetism:

$$F_{\alpha\beta} = \partial_\alpha A_\beta - \partial_\beta A_\gamma$$

However, this formula is inadequate for a particle that has both electric and magnetic charge, and we must add a term involving another potential P.[29][30]

$$F_{\alpha\beta} = \partial_\alpha A_\beta - \partial_\beta A_\alpha + \partial^\mu(\varepsilon_{\alpha\beta\mu\nu}P^\nu),$$

This formula for the fields is often called the Cabibbo-Ferrari relation, though Shanmugadhasan proposed it earlier.[30] The quantity $\varepsilon^{\alpha\beta\gamma\delta}$ is the Levi-Civita symbol, and the indices (as usual) behave according to the Einstein summation convention.

25.3.4 Duality transformation

The generalized Maxwell's equations possess a certain symmetry, called a *duality transformation*. One can choose any real angle ξ, and simultaneously change the fields and charges everywhere in the universe as follows (in Gaussian units):[31]

where the primed quantities are the charges and fields before the transformation, and the unprimed quantities are after the transformation. The fields and charges after this transformation still obey the same Maxwell's equations. The matrix is a two-dimensional rotation matrix.

Because of the duality transformation, one cannot uniquely decide whether a particle has an electric charge, a magnetic charge, or both, just by observing its behavior and comparing that to Maxwell's equations. For example, it is merely a convention, not a requirement of Maxwell's equations, that electrons have electric charge but not magnetic charge; after a $\xi = \pi/2$ transformation, it would be the other way around. The key empirical fact is that all particles ever observed have the same ratio of magnetic charge to electric charge.[31] Duality transformations can change the ratio to any arbitrary numerical value, but cannot change the fact that all particles have the same ratio. Since this is the case, a duality transformation can be made that sets this ratio to be zero, so that all particles have no magnetic charge. This choice underlies the "conventional" definitions of electricity and magnetism.[31]

25.4 Dirac's quantization

One of the defining advances in quantum theory was Paul Dirac's work on developing a relativistic quantum electro-magnetism. Before his formulation, the presence of electric charge was simply "inserted" into the equations of quantum mechanics (QM), but in 1931 Dirac showed that a discrete charge naturally "falls out" of QM. That is to say, we can maintain the form of Maxwell's equations and still have magnetic charges.

Consider a system consisting of a single stationary electric monopole (an electron, say) and a single stationary magnetic monopole. Classically, the electromagnetic field surrounding them has a momentum density given by the Poynting vector, and it also has a total angular momentum, which is proportional to the product $q_e q_m$, and independent of the distance between them.

Quantum mechanics dictates, however, that angular momentum is quantized in units of $\hbar$, so therefore the product $q_e q_m$ must also be quantized. This means that if even a single magnetic monopole existed in the universe, and the form of Maxwell's equations is valid, all electric charges would then be quantized.

What are the units in which magnetic charge would be quantized? Although it would be possible simply to integrate over all space to find the total angular momentum in the above example, Dirac took a different approach. This led him to new ideas. He considered a point-like magnetic charge whose magnetic field behaves as q_m / r^2 and is directed in the radial direction, located at the origin. Because the divergence of **B** is equal to zero almost everywhere, except for the locus of the magnetic monopole at $r = 0$, one can locally define the vector potential such that the curl of the vector potential **A** equals the magnetic field **B**.

However, the vector potential cannot be defined globally precisely because the divergence of the magnetic field is proportional to the Dirac delta function at the origin. We must define one set of functions for the vector potential on the "northern hemisphere" (the half-space $z > 0$ above the particle), and another set of functions for the "southern hemisphere". These two vector potentials are matched at the "equator" (the plane $z = 0$ through the particle), and they differ by a gauge transformation. The wave function of an electrically-charged particle (a "probe charge") that orbits the "equator" generally changes by a phase, much like in the Aharonov–Bohm effect. This phase is proportional to the electric charge q_e of the probe, as well as to the magnetic charge q_m of the source. Dirac was originally considering an electron whose wave function is described by the Dirac equation.

Because the electron returns to the same point after the full trip around the equator, the phase φ of its wave function $e^{i\varphi}$ must be unchanged, which implies that the phase φ added to the wave function must be a multiple of 2π:

where ε_0 is the vacuum permittivity, $\hbar = h/2\pi$ is the reduced Planck's constant, c is the speed of light, and $\mathbb{Z}$ is the set of integers.

This is known as the **Dirac quantization condition**. The hypothetical existence of a magnetic monopole would imply that the electric charge must be quantized in certain units; also, the existence of the electric charges implies that the magnetic charges of the hypothetical magnetic monopoles, if they exist, must be quantized in units inversely proportional to the elementary electric charge.

At the time it was not clear if such a thing existed, or even had to. After all, another theory could come along that would explain charge quantization without need for the monopole. The concept remained something of a curiosity. However, in the time since the publication of this seminal work, no other widely accepted explanation of charge quantization has appeared. (The concept of local gauge invariance—see gauge theory below—provides a natural explanation of charge quantization, without invoking the need for magnetic monopoles; but only if the U(1) gauge group is compact, in which case we will have magnetic monopoles anyway.)

If we maximally extend the definition of the vector potential for the southern hemisphere, it will be defined everywhere except for a semi-infinite line stretched from the origin in the direction towards the northern pole. This semi-infinite line is called the Dirac string and its effect on the wave function is analogous to the effect of the solenoid in the Aharonov–Bohm effect. The quantization condition comes from the requirement that the phases around the Dirac string are trivial, which means that the Dirac string must be unphysical. The Dirac string is merely an artifact of the coordinate chart used and should not be taken seriously.

The Dirac monopole is a singular solution of Maxwell's equation (because it requires removing the worldline from spacetime); in more complicated theories, it is superseded by a smooth solution such as the 't Hooft–Polyakov monopole.

25.5 Topological interpretation

25.5.1 Dirac string

Main article: Dirac string

A gauge theory like electromagnetism is defined by a gauge field, which associates a group element to each path in space time. For infinitesimal paths, the group element is close to the identity, while for longer paths the group element is the successive product of the infinitesimal group elements along the way.

In electrodynamics, the group is U(1), unit complex numbers under multiplication. For infinitesimal paths, the group element is $1 + iA_\mu dx^\mu$ which implies that for finite paths parametrized by s, the group element is:

$$\prod_s \left(1 + ieA_\mu \frac{dx^\mu}{ds}ds\right) = \exp\left(ie\int A \cdot dx\right).$$

The map from paths to group elements is called the Wilson loop or the holonomy, and for a U(1) gauge group it is the phase factor which the wavefunction of a charged particle acquires as it traverses the path. For a loop:

$$e\oint_{\partial D} A \cdot dx = e\int_D (\nabla \times A)dS = e\int_D B\, dS.$$

So that the phase a charged particle gets when going in a loop is the magnetic flux through the loop. When a small solenoid has a magnetic flux, there are interference fringes for charged particles which go around the solenoid, or around different sides of the solenoid, which reveal its presence.

But if all particle charges are integer multiples of e, solenoids with a flux of $2\pi/e$ have no interference fringes, because the phase factor for any charged particle is $e^{2\pi i} = 1$. Such a solenoid, if thin enough, is quantum-mechanically invisible. If such a solenoid were to carry a flux of $2\pi/e$, when the flux leaked out from one of its ends it would be indistinguishable from a monopole.

Dirac's monopole solution in fact describes an infinitesimal line solenoid ending at a point, and the location of the solenoid is the singular part of the solution, the Dirac string. Dirac strings link monopoles and antimonopoles of opposite magnetic charge, although in Dirac's version, the string just goes off to infinity. The string is unobservable, so you can put it anywhere, and by using two coordinate patches, the field in each patch can be made nonsingular by sliding the string to where it cannot be seen.

25.5.2 Grand unified theories

Main article: 't Hooft–Polyakov monopole

In a U(1) gauge group with quantized charge, the group is a circle of radius $2\pi/e$. Such a U(1) gauge group is called compact. Any U(1) which comes from a Grand Unified Theory is compact – because only compact higher gauge groups make sense. The size of the gauge group is a measure of the inverse coupling constant, so that in the limit of a large-volume gauge group, the interaction of any fixed representation goes to zero.

The case of the U(1) gauge group is a special case because all its irreducible representations are of the same size – the charge is bigger by an integer amount, but the field is still just a complex number – so that in U(1) gauge field theory it is possible to take the decompactified limit with no contradiction. The quantum of charge becomes small, but each charged particle has a huge number of charge quanta so its charge stays finite. In a non-compact U(1) gauge group theory, the charges of particles are generically not integer multiples of a single unit. Since charge quantization is an experimental certainty, it is clear that the U(1) gauge group of electromagnetism is compact.

GUTs lead to compact U(1) gauge groups, so they explain charge quantization in a way that seems to be logically independent from magnetic monopoles. However, the explanation is essentially the same, because in any GUT which breaks down into a U(1) gauge group at long distances, there are magnetic monopoles.

The argument is topological:

1. The holonomy of a gauge field maps loops to elements of the gauge group. Infinitesimal loops are mapped to group elements infinitesimally close to the identity.

2. If you imagine a big sphere in space, you can deform an infinitesimal loop which starts and ends at the north pole as follows: stretch out the loop over the western hemisphere until it becomes a great circle (which still starts and ends at the north pole) then let it shrink back to a little loop while going over the eastern hemisphere. This is called *lassoing the sphere*.

3. Lassoing is a sequence of loops, so the holonomy maps it to a sequence of group elements, a continuous path in the gauge group. Since the loop at the beginning of the lassoing is the same as the loop at the end, the path in the group is closed.

4. If the group path associated to the lassoing procedure winds around the U(1), the sphere contains magnetic charge. During the lassoing, the holonomy changes by the amount of magnetic flux through the sphere.

5. Since the holonomy at the beginning and at the end is the identity, the total magnetic flux is quantized. The magnetic charge is proportional to the number of windings N, the magnetic flux through the sphere is equal to $2\pi N/e$. This is the Dirac quantization condition, and it is a topological condition which demands that the long distance U(1) gauge field configurations be consistent.

6. When the U(1) gauge group comes from breaking a compact Lie group, the path which winds around the U(1) group enough times is topologically trivial in the big group. In a non-U(1) compact Lie group, the covering space is a Lie group with the same Lie algebra, but where all closed loops are contractible. Lie groups are homogenous, so that any cycle in the group can be moved around so that it starts at the identity, then its lift to the covering group ends at P, which is a lift of the identity. Going around the loop twice gets you to P^2, three times to P^3, all lifts of the identity. But there are only finitely many lifts of the identity, because the lifts can't accumulate. This number of times one has to traverse the loop to make it contractible is small, for example if the GUT group is SO(3), the covering group is SU(2), and going around any loop twice is enough.

7. This means that there is a continuous gauge-field configuration in the GUT group allows the U(1) monopole configuration to unwind itself at short distances, at the cost of not staying in the U(1). In order to do this with as little energy as possible, you should leave only the U(1) gauge group in the neighborhood of one point, which is called the **core** of the monopole. Outside the core, the monopole has only magnetic field energy.

Hence, the Dirac monopole is a topological defect in a compact U(1) gauge theory. When there is no GUT, the defect is a singularity – the core shrinks to a point. But when there is some sort of short-distance regulator on space time, the monopoles have a finite mass. Monopoles occur in lattice U(1), and there the core size is the lattice size. In general, they are expected to occur whenever there is a short-distance regulator.

25.5.3 String theory

In the universe, quantum gravity provides the regulator. When gravity is included, the monopole singularity can be a black hole, and for large magnetic charge and mass, the black hole mass is equal to the black hole charge, so that the mass of the magnetic black hole is not infinite. If the black hole can decay completely by Hawking radiation, the lightest charged particles cannot be too heavy.[33] The lightest monopole should have a mass less than or comparable to its charge in natural units.

So in a consistent holographic theory, of which string theory is the only known example, there are always finite-mass monopoles. For ordinary electromagnetism, the mass bound is not very useful because it is about same size as the Planck mass.

25.5.4 Mathematical formulation

In mathematics, a (classical) gauge field is defined as a connection over a principal G-bundle over spacetime. G is the gauge group, and it acts on each fiber of the bundle separately.

A *connection* on a G bundle tells you how to glue fibers together at nearby points of M. It starts with a continuous symmetry group G which acts on the fiber F, and then it associates a group element with each infinitesimal path. Group multiplication along any path tells you how to move from one point on the bundle to another, by having the G element associated to a path act on the fiber F.

In mathematics, the definition of bundle is designed to emphasize topology, so the notion of connection is added on as an afterthought. In physics, the connection is the fundamental physical object. One of the fundamental observations in the theory of characteristic classes in algebraic topology is that many homotopical structures of nontrivial principal bundles may be expressed as an integral of some polynomial over **any** connection over it. Note that a connection over a trivial bundle can never give us a nontrivial principal bundle.

If space time is $\mathbf{R}^4$ the space of all possible connections of the G-bundle is connected. But consider what happens when we remove a timelike worldline from spacetime. The resulting spacetime is homotopically equivalent to the topological sphere S^2.

A principal G-bundle over S^2 is defined by covering S^2 by two charts, each homeomorphic to the open 2-ball such that their intersection is homeomorphic to the strip $S^1 \times I$. 2-balls are homotopically trivial and the strip is homotopically equivalent to the circle S^1. So a topological classification of the possible connections is reduced to classifying the transition functions. The transition function maps the strip to G, and the different ways of mapping a strip into G are given by the first homotopy group of G.

So in the G-bundle formulation, a gauge theory admits Dirac monopoles provided G is not simply connected, whenever there are paths that go around the group that cannot be deformed to a constant path (a path whose image consists of a single point). U(1), which has quantized charges, is not simply connected and can have Dirac monopoles while $\mathbf{R}$, its universal covering group, **is** simply connected, doesn't have quantized charges and does not admit Dirac monopoles. The mathematical definition is equivalent to the physics definition provided that, following Dirac, gauge fields are allowed which are defined only patch-wise and the gauge field on different patches are glued after a gauge transformation.

The total magnetic flux is none other than the first Chern number of the principal bundle, and depends only upon the choice of the principal bundle, and not the specific connection over it. In other words, it's a topological invariant.

This argument for monopoles is a restatement of the lasso argument for a pure U(1) theory. It generalizes to $d + 1$ dimensions with $d \geq 2$ in several ways. One way is to extend everything into the extra dimensions, so that U(1) monopoles become sheets of dimension $d - 3$. Another way is to examine the type of topological singularity at a point with the homotopy group $\pi d_{-2}(G)$.

25.6 Grand unified theories

In more recent years, a new class of theories has also suggested the existence of magnetic monopoles.

During the early 1970s, the successes of quantum field theory and gauge theory in the development of electroweak theory and the mathematics of the strong nuclear force led many theorists to move on to attempt to combine them in a single theory known as a Grand Unified Theory (GUT). Several GUTs were proposed, most of which implied the presence of a real magnetic monopole particle. More accurately, GUTs predicted a range of particles known as dyons, of which the most basic state was a monopole. The charge on magnetic monopoles predicted by GUTs is either 1 or 2 *gD*, depending on the theory.

The majority of particles appearing in any quantum field theory are unstable, and they decay into other particles in a variety of reactions that must satisfy various conservation laws. Stable particles are stable because there are no lighter particles into which they can decay and still satisfy the conservation laws. For instance, the electron has a lepton number of one and an electric charge of one, and there are no lighter particles that conserve these values. On the other hand, the muon, essentially a heavy electron, can decay into the electron plus two quanta of energy, and hence it is not stable.

The dyons in these GUTs are also stable, but for an entirely different reason. The dyons are expected to exist as a side effect of the "freezing out" of the conditions of the early universe, or a symmetry breaking. In this scenario, the dyons arise due to the configuration of the vacuum in a particular area of the universe, according to the original Dirac theory. They remain stable not because of a conservation condition, but because there is no simpler *topological* state into which they can decay.

The length scale over which this special vacuum configuration exists is called the *correlation length* of the system. A correlation length cannot be larger than causality would allow, therefore the correlation length for making magnetic monopoles must be at least as big as the horizon size determined by the metric of the expanding universe. According to that logic, there should be at least one magnetic monopole per horizon volume as it was when the symmetry breaking took place.

Cosmological models of the events following the big bang make predictions about what the horizon volume was, which lead to predictions about present-day monopole density. Early models predicted an enormous density of monopoles, in clear contradiction to the experimental evidence.[34][35] This was called the "monopole problem". Its widely accepted resolution was not a change in the particle-physics prediction of monopoles, but rather in the cosmological models used to infer their present-day density. Specifically, more recent theories of cosmic inflation drastically reduce the predicted number of magnetic monopoles, to a density small enough to make it unsurprising that humans have never seen one.[36] This resolution of the "monopole problem" was regarded as a success of cosmic inflation theory. (However, of course, it is only a noteworthy success if the particle-physics monopole prediction is correct.[37]) For these reasons, monopoles became a major interest in the 1970s and 80s, along with the other "approachable" predictions of GUTs such as proton decay.

Many of the other particles predicted by these GUTs were beyond the abilities of current experiments to detect. For instance, a wide class of particles known as the X and Y bosons are predicted to mediate the coupling of the electroweak and strong forces, but these particles are extremely heavy and well beyond the capabilities of any reasonable particle accelerator to create.

25.7   Searches for magnetic monopoles

A number of attempts have been made to detect magnetic monopoles. One of the simpler ones is to use a loop of superconducting wire to look for even tiny magnetic sources, a so-called "superconducting quantum interference device", or SQUID. Given the predicted density, loops small enough to fit on a lab bench would expect to see about one monopole event per year. Although there have been tantalizing events recorded, in particular the event recorded by Blas Cabrera on the night of February 14, 1982 (thus, sometimes referred to as the "Valentine's Day Monopole"[38]), there has never been reproducible evidence for the existence of magnetic monopoles.[13] The lack of such events places a limit on the number of monopoles of about one monopole per 10^{29} nucleons.

Another experiment in 1975 resulted in the announcement of the detection of a moving magnetic monopole in cosmic rays by the team led by P. Buford Price.[12] Price later retracted his claim, and a possible alternative explanation was offered by Alvarez.[39] In his paper it was demonstrated that the path of the cosmic ray event that was claimed to be due to a magnetic monopole could be reproduced by the path followed by a platinum nucleus decaying first to osmium, and then to tantalum.

Other experiments rely on the strong coupling of monopoles with photons, as is the case for any electrically-charged particle as well. In experiments involving photon exchange in particle accelerators, monopoles should be produced in reasonable numbers, and detected due to their effect on the scattering of the photons. The probability of a particle being created in such experiments is related to their mass – with heavier particles being less likely to be created – so by examining the results of such experiments, limits on the mass of a magnetic monopole can be calculated. The most recent such experiments suggest that monopoles with masses below 600 GeV/c^2 do not exist, while upper limits on their mass due to the very existence of the universe – which would have collapsed by now if they were too heavy – are about 10^{17} GeV/c^2.

The MoEDAL experiment, installed at the Large Hadron Collider, is currently searching for magnetic monopoles and large supersymmetric particles using layers of special plastic sheets attached to the walls around LHCb's VELO detector. The particles it is looking for will damage the sheets along their path, with various identifying features.

The Russian astrophysicist Igor Novikov claims the fields of macroscopic black holes to be potential magnetic monopoles, representing the entrance to an Einstein–Rosen bridge.[40]

25.8 "Monopoles" in condensed-matter systems

Since around 2003, various condensed-matter physics groups have used the term "magnetic monopole" to describe a different and largely unrelated phenomenon.[18][19]

A true magnetic monopole would be a new elementary particle, and would violate the law $\nabla \cdot \mathbf{B} = 0$. A monopole of this kind, which would help to explain the law of charge quantization as formulated by Paul Dirac in 1931,[41] has never been observed in experiments.

The monopoles studied by condensed-matter groups have none of these properties. They are not a new elementary particle, but rather are an emergent phenomenon in systems of everyday particles (protons, neutrons, electrons, photons); in other words, they are quasi-particles. They are not sources for the $\mathbf{B}$-field (i.e., they do not violate $\nabla \cdot \mathbf{B} = 0$); instead, they are sources for other fields, for example the $\mathbf{H}$-field,[5] or the "$\mathbf{B}$*-field" (related to superfluid vorticity).[6] They are not directly relevant to grand unified theories or other aspects of particle physics, and do not help explain charge quantization—except insofar as studies of analogous situations can help confirm that the mathematical analyses involved are sound.[42]

There are a number of examples in condensed-matter physics where collective behavior leads to emergent phenomena that resemble magnetic monopoles in certain respects,[17][43][44][45] including most prominently the spin ice materials.[5][46] While these should not be confused with hypothetical elementary monopoles existing in the vacuum, they nonetheless have similar properties and can be probed using similar techniques.

Some researchers use the term **magnetricity** to describe the manipulation of magnetic monopole quasiparticles in spin ice,[46][47] in analogy to the word "electricity".

One example of the work on magnetic monopole quasiparticles is a paper published in the journal *Science* in September 2009, in which researchers Jonathan Morris and Alan Tennant from the Helmholtz-Zentrum Berlin für Materialien und Energie (HZB) along with Santiago Grigera from Instituto de Física de Líquidos y Sistemas Biológicos (IFLYSIB, CONICET) and other colleagues from Dresden University of Technology, University of St. Andrews and Oxford University described the observation of quasiparticles resembling magnetic monopoles. A single crystal of the spin ice material dysprosium titanate was cooled to a temperature between 0.6 kelvin and 2.0 kelvin. Using observations of neutron scattering, the magnetic moments were shown to align into interwoven tubelike bundles resembling Dirac strings. At the defect formed by the end of each tube, the magnetic field looks like that of a monopole. Using an applied magnetic field to break the symmetry of the system, the researchers were able to control the density and orientation of these strings. A contribution to the heat capacity of the system from an effective gas of these quasiparticles was also described.[16][48]

This research went on to win the 2012 Europhysics Prize for condensed matter physics.

Another example is a paper in the February 11, 2011 issue of *Nature Physics* which describes creation and measurement of long-lived magnetic monopole quasiparticle currents in spin ice. By applying a magnetic-field pulse to crystal of dysprosium titanate at 0.36 K, the authors created a relaxing magnetic current that lasted for several minutes. They measured the current by means of the electromotive force it induced in a solenoid coupled to a sensitive amplifier, and quantitatively described it using a chemical kinetic model of point-like charges obeying the Onsager–Wien mechanism of carrier dissociation and recombination. They thus derived the microscopic parameters of monopole motion in spin ice and identified the distinct roles of free and bound magnetic charges.[47]

In superfluids, there is a field **B***, related to superfluid vorticity, which is mathematically analogous to the magnetic **B**-field. Because of the similarity, the field **B*** is called a "synthetic magnetic field". In January 2014, it was reported that monopole quasiparticles[49] for the **B*** field were created and studied in a spinor Bose–Einstein condensate.[6] This constitutes the first example of a magnetic monopole observed within a system governed by quantum field theory.[42]

25.9 Further descriptions in particle physics

In physics the phrase "magnetic monopole" usually denoted a Yang–Mills potential A and Higgs field ϕ whose equations of motion are determined by the Yang–Mills action

$$\int (F_A, F_A) + (D_A\phi, D_A\phi) - \lambda(1 - \|\phi\|^2)^2.$$

In mathematics, the phrase customarily refers to a static solution to these equations in the Bogomolny–Parasad–Sommerfeld limit $\lambda \to \phi$ which realizes, within topological class, the absolutes minimum of the functional

$$\int_{R^3} (F_A, F_A) + (D_A\phi, D_A\phi).$$

This means that it in a connection A on a principal G-bundle over $\mathbf{R}^3$ (c.f. also Connections on a manifold; principal G-object) and a section ϕ of the associated adjoint bundle of Lie algebras such that the curvature FA and covariant derivative $DA\,\phi$ satisfy the Bogomolny equations

$$F_A = *D_A\phi$$

and the boundary conditions.

$$\|\phi\| = 1 - \frac{m}{r} + \theta(r^2), \quad \|D_A\phi\| = \mathcal{O}(r^2)$$

Pure mathematical advances in the theory of monopoles from the 1980s onwards have often proceeded on the basis of physically motived questions.

The equations themselves are invariant under gauge transformation and orientation-preserving symmetries. When γ is large, $\phi/\|\phi\|$ defines a mapping from a 2-sphere of radius γ in $\mathbf{R}^3$ to an adjoint orbit G/k and the homotopy class of this mapping is called the magnetic charge. Most work has been done in the case G = SU(2), where the charge is a positive integer k. The absolute minimum value of the functional is then $8\pi k$ and the coefficient m in the asymptotic expansion of $\phi/\|\phi\|$ is $k/2$.

The first SU(2) solution was found by E. B. Bogomolny, J. K. Parasad and C. M. Sommerfeld in 1975. It is spherically symmetric of charge 1 and has the form

$$A = \left(\frac{1}{\sinh\gamma} - \frac{1}{\gamma}\right) \epsilon_{ijk}\frac{x_j}{\gamma}\sigma_k\,dx_i,$$
$$\phi = \left(\frac{1}{\tanh\gamma} - \frac{1}{\gamma}\right) \frac{x_j}{\gamma}\sigma_i$$

In 1980, C.H.Taubes[50] showed by a gluing construction that there exist solutions for all large k and soon after explicit axially-symmetric solutions were found. The first exact solution in the general case was given in 1981 by R.S.Ward for k = 2 in terms of elliptic functions.

There are two ways of solving the Bogomolny equations. The first is by twistor methods. In the formulation of N.J. Hitchin,[51] an arbitrary solution corresponds to a holomorphic vector bundle over the complex surface TP^1, the tangent bundle of the projective line. This is naturally isomorphic to the space of oriented straight lines in $\mathbf{R}^3$.

The boundary condition show that the holomorphic bundle is an extension of line bundles determined by a compact algebraic curve of genus $(k-1)^2$ (the spectral curve) in TP^1, satisfying certain constraints.

The second method, due to W.Nahm,[52] involves solving an eigen value problem for the coupled Dirac operator and transforming the equations with their boundary conditions into a system of ordinary differential equations, the Nahm equations.

$$\frac{dT_1}{ds} = [T_2, T_3], \quad \frac{dT_2}{ds} = [T_3, T_1], \quad \frac{dT_3}{ds} = [T_1, T_2]$$

where $Ti(s)$ is a $k \times k$ -matrix valued function on (0,2).

Both constructions are based on analogous procedures for instantons, the key observation due to N.S.Manton being of the self-dual Yang–Mills equations (c.f. also Yang–Mills field) in $\mathbf{R}^4$.

The equivalence of the two methods for SU(2) and their general applicability was established in[53] (see also[54]). Explicit formulas for A and ϕ are difficult to obtain by either method, despite some exact solutions of Nahm's equations in symmetric situations.[55]

Maximally imbedded spherically symmetric magnetic monopole solutions in the Bogolomony-Parasad-Sommerfield limit for the gauge group SU(n) were exhibited by Bais.[56][57] Gannoulis, Goddard and Olive,[58] and Farwell and Minami [59] showed that maximally imbedded spherically symmetric magnetic monopole solutions in the Bogolomony-Parasad-Sommerfield limit for an an arbitrary simple gauge group G corresponding to a Lie Algebra with Cartan matrix K and level vector[60] R, are solutions to the Toda molecule[61][62] equation:

$$\frac{d^2\theta_i}{dr^2} = \exp K_{ij}\theta_j, \quad where: B_r(r, \hat{k}) \quad = \frac{1}{2}\{\frac{d^2\theta_i}{dr^2} - \frac{\bar{R}_i}{r^2}\}H_i, \quad \phi(r, \hat{k}) = \frac{1}{2}\{\frac{d\theta_i}{dr} + \frac{\bar{R}_i}{r}\}H_i$$

Non-singular solutions have a magnetic field vanishes at the origin. Explicit finite energy solutions for the Lie Algebras A_n, B_n and C_n have been obtained using this method.

The case of a more generalLie group$G$, where the stabilizer ofϕat infinity is a maximal torus, was treated by M.K.Murray from the twistor point of view, where the single spectral curve of an SU(2)-monopole is replaced by a collection of curve sindexed by the vertices of the Dynkin diagram of G. The corresponding Nahm construction was designed by J. Hustubise and Murray.[64]

The moduli space (c.f. also Moduli theory) of all SU(2) monopoles of charge k up to gauge equivalence was shown by Taubes[65] to be a smooth non-compact manifold of dimension $4k - 1$. Restricting to gauge transformations that preserve the connection at infinity gives a $4k$-dimensional manifold Mk, which is a circle bundle over the true moduli space and carries a natural complete hyper-Kähler metric[66] (c.f. also Kähler–Einstein manifold). With suspected to any of the complex structures of the hyper-Kähler family, this manifold is holomorphically equivalent to the space of based rational mapping of degree k from P_1 to itself.[67]

The metric is known in twistor terms,[66] and its Kähler potential can be written using the Riemann theta functions of the spectral curve,[54] but only the case $k = 2$ is known in a more conventional and usable form[66] (as of 2000). This Atiyah–Hitchin manifold, the Einstein Taub-NUT metric and $\mathbf{R}^4$ are the only 4-dimensional complete hyper-Kähler manifolds with a non-triholomorphic SU(2) action. Its geodesics have been studied and a programme of Manton concerning monopole dynamics put into effect. Further dynamical features have been elucidated by numerical and analytical techniques.

A cyclic k-fold conering of Mk splits isometrically is a product $Mk \times S^1 \times \mathbf{R}^3$, where Mk is the space of strongly centred monopoles. This space features in an application of S-duality in theoretical physics, and in[68] G.B.Segal and A.Selby studied its topology and the L^2 harmonic forms defined on it, partially confirming the physical prediction.

Magnetic monopole on hyperbolic three-space were investigated from the twistor point of view by M. F. Atiyah[69] (replacing the complex surface TP^1 by the complement of the anti-diagonal in $P^1 \times P^1$) and in terms of discrete Nahm equations by Murray and M. A. Singer.[70]

25.10   See also

- Bogomolny equations

- Dirac string

- Dyon

- Felix Ehrenhaft

- Gauss's law for magnetism

- Halbach array

- Instanton

- Meron

- Soliton

- 't Hooft–Polyakov monopole

- Wu–Yang monopole

25.11   Notes

[1] Dark Cosmos: In Search of Our Universe's Missing Mass and Energy, by Dan Hooper, p192

[2] Particle Data Group summary of magnetic monopole search

[3] Wen, Xiao-Gang; Witten, Edward, *Electric and magnetic charges in superstring models*, Nuclear Physics B, Volume 261, p. 651–677

[4] S. Coleman, *The Magnetic Monopole 50 years Later*, reprinted in *Aspects of Symmetry*

[5] C. Castelnovo, R. Moessner and S. L. Sondhi (January 3, 2008). "Magnetic monopoles in spin ice". *Nature* **451**: 42–45. arXiv:0710.5515. Bibcode:2008Natur.451...42C. doi:10.1038/nature06433.

[6] Ray, M.W.; Ruokokoski, E.; Kandel, S.; Möttönen, M.; Hall, D. S. (2014). "Observation of Dirac monopoles in a synthetic magnetic field". *Nature* **505** (7485): 657–660. arXiv:1408.3133. Bibcode:2014Natur.505..657R. doi:10.1038/nature12954. ISSN 0028-0836.

[7] The encyclopædia britannica, Volume 17, p352

[8] Principles of Physics by William Francis Magie, p424

[9] Pierre Curie, *Sur la possibilité d'existence de la conductibilité magnétique et du magnétisme libre* (*On the possible existence of magnetic conductivity and free magnetism*), Séances de la Société Française de Physique (Paris), p76 (1894). (French)Free access online copy.

[10] Paul Dirac, "Quantised Singularities in the Electromagnetic Field". Proc. Roy. Soc. (London) **A 133**, 60 (1931). Journal Site, Free Access .

[11] Lecture notes by Robert Littlejohn, University of California, Berkeley, 2007–8

[12] P. B. Price; E. K. Shirk; W. Z. Osborne; L. S. Pinsky (August 25, 1975). "Evidence for Detection of a Moving Magnetic Monopole". *Physical Review Letters* (American Physical Society) **35** (8): 487–490. Bibcode:1975PhRvL..35..487P. doi:10.1103/PhysRevLett.35.487.

[13] Blas Cabrera (May 17, 1982). "First Results from a Superconductive Detector for Moving Magnetic Monopoles". *Physical Review Letters* (American Physical Society) **48** (20): 1378–1381. Bibcode:1982PhRvL..48.1378C. doi:10.1103/PhysRevLett.48.1378.

[14] Milton p.60

[15] Polchinski, arXiv 2003

[16] "Magnetic Monopoles Detected in a Real Magnet for the First Time". Science Daily. September 4, 2009. Retrieved September 4, 2009.

[17] Making magnetic monopoles, and other exotica, in the lab, Symmetry Breaking, January 29, 2009. Retrieved January 31, 2009.

[18] Magnetic monopoles spotted in spin ices, September 3, 2009. "Oleg Tchernyshyov at Johns Hopkins University [a researcher in this field] cautions that the theory and experiments are specific to spin ices, and are not likely to shed light on magnetic monopoles as predicted by Dirac."

[19] Elizabeth Gibney (29 January 2014). "Quantum cloud simulates magnetic monopole". *Nature (news section)*. doi:10.1038/na612. "This is not the first time that physicists have created monopole analogues. In 2009, physicists observed magnetic monopoles in a crystalline material called spin ice, which, when cooled to near-absolute zero, seems to fill with atom-sized, classical monopoles. These are magnetic in a true sense, but cannot be studied individually. Similar analogues have also been seen in other materials, such as in superfluid helium.... Steven Bramwell, a physicist at University College London who pioneered work on monopoles in spin ices, says that the [2014 experiment led by David Hall] is impressive, but that what it observed is not a Dirac monopole in the way many people might understand it. "There's a mathematical analogy here, a neat and beautiful one. But they're not magnetic monopoles."

[20] The fact that the electric and magnetic fields can be written in a symmetric way is specific to the fact that space is three-dimensional. When the equations of electromagnetism are extrapolated to other dimensions, the magnetic field is described as being a rank-two antisymmetric tensor, whereas the electric field remains a true vector. In dimensions other than three, these two mathematical objects do not have the same number of components.

[21] http://www.ieeeghn.org/wiki/index.php/STARS:Maxwell%27s_Equations

[22] Parker, C.B. (1994). *McGraw-Hill Encyclopaedia of Physics* (2nd ed.). McGraw-Hill. ISBN 0-07-051400-3.

[23] M. Mansfield, C. O'Sullivan (2011). *Understanding Physics* (4th ed.). John Wiley & Sons. ISBN 978-0-47-0746370.

[24] F. Moulin (2001). "Magnetic monopoles and Lorentz force". *Nuovo Cimento B* **116** (8): 869–877. arXiv:math-ph/0203043. Bibcode:2001NCimB.116..869M.

[25] Wolfgang Rindler (November 1989). "Relativity and electromagnetism: The force on a magnetic monopole". *American Journal of Physics* (American Journal of Physics) **57** (11): 993–994. Bibcode:1989AmJPh..57..993R. doi:10.1119/1.15782.

[26] For the convention where magnetic charge has units of webers, see Jackson 1999. In particular, for Maxwell's equations, see section 6.11, equation (6.150), page 273, and for the Lorentz force law, see page 290, exercise 6.17(a). For the convention where magnetic charge has units of ampere-meters, see (for example) arXiv:physics/0508099v1, eqn (4).

[27] J.A. Heras, G. Baez (2009). "The covariant formulation of Maxwell's equations expressed in a form independent of specific units". arXiv:0901.0194.

[28] F. Moulin (2002). "Magnetic monopoles and Lorentz force". arXiv:math-ph/0203043.

[29] Shanmugadhasan, S (1952). "The Dynamical Theory of Magnetic Monopoles". *Canadian Journal of Physics* **30**: 218. Bibcod18S. doi:10.1139/p52-021.

[30] Fryberger, D (1989). "On Generalized Electromagnetism and Dirac Algebra" (PDF). *Foundations of Physics* **19**: 125. Bib9..125F. doi:10.1007/bf00734522.

[31] Jackson 1999, section 6.11.

[32] Jackson 1999, section 6.11, equation (6.153), page 275

[33] Nima Arkani-Hamed, Lubos Motl, Alberto Nicolis, Cumrun Vafa: The String Landscape, Black Holes and Gravity as the Weakest Force (arXiv:hep-th/0601001, JHEP 0706:060,2007)

[34] Zel'dovich, Ya. B.; Khlopov, M. Yu. (1978). "On the concentration of relic monopoles in the universe". *Phys. Lett.* **B79** (3): 239–41. Bibcode:1978PhLB...79..239Z. doi:10.1016/0370-2693(78)90232-0.

[35] Preskill, John (1979). "Cosmological production of superheavy magnetic monopoles". *Phys. Rev. Lett.* **43** (19): 1365. Bibcode:1979PhRvL..43.1365P. doi:10.1103/PhysRevLett.43.1365.

[36] Preskill, John (1984). "Magnetic Monopoles". *Ann. Rev. Nucl. Part. Sci.* **34**: 461. Bibcode:1984ARNPS..34..461P. doi:10.1146/annurev.ns.34.120184.002333.

[37] Rees, Martin. (1998). *Before the Beginning* (New York: Basic Books) p. 185 ISBN 0-201-15142-1

[38] http://www.nature.com/nature/journal/v429/n6987/full/429010a.html

[39] Alvarez, Luis W. "Analysis of a Reported Magnetic Monopole". In ed. Kirk, W. T. *Proceedings of the 1975 international symposium on lepton and photon interactions at high energies.* International symposium on lepton and photon interactions at high energies, Aug 21, 1975. p. 967.

[40] „If the structures of the magnetic fields appear to be magnetic monopoles, that are macroscopic in size, then this is a wormhole." Taken from All About Space, issue No. 24, April 2014, item „Could wormholes really exist?"

[41] "Quantised Singularities in the Electromagnetic Field" Paul Dirac, *Proceedings of the Royal Society*, May 29, 1931. Retrieved February 1, 2014.

[42] Elizabeth Gibney (29 January 2014). "Quantum cloud simulates magnetic monopole".*Nature (news section)*.doi:10.1038/nature2.

[43] Zhong, Fang; Nagosa, Naoto; Takahashi, Mei S.; Asamitsu, Atsushi; Mathieu, Roland; Ogasawara, Takeshi; Yamada, Hiroyuki; Kawasaki, Masashi; Tokura, Yoshinori; Terakura, Kiyoyuki (2003). "The Anomalous Hall Effect and Magnetic Monopoles in Momentum Space". *Science* **302** (5642): 92–95. arXiv:cond-mat/0310232. Bibcode:2003Sci...302...92F. doi:10.1126/science.1089408.

[44] Inducing a Magnetic Monopole with Topological Surface States, American Association for the Advancement of Science (AAAS) *Science Express* magazine, Xiao-Liang Qi, Rundong Li, Jiadong Zang, Shou-Cheng Zhang, January 29, 2009. Retrieved January 31, 2009.

[45]   *Artificial Magnetic Monopoles Discovered*

[46] S. T. Bramwell, S. R. Giblin, S. Calder, R. Aldus, D. Prabhakaran, T. Fennell (15 October 2009). "Measurement of the charge and current of magnetic monopoles in spin ice". *Nature* **461** (7266): 956–959. arXiv:0907.0956. Bibcode:2009Natur.461..956B. doi:10.1038/nature08500. PMID 19829376.

[47] S. R. Giblin, S. T. Bramwell, P. C. W. Holdsworth, D. Prabhakaran & I. Terry (February 13, 2011). "Creation and measurement of long-lived magnetic monopole currents in spin ice" **7** (3). Nature Physics. Bibcode:2011NatPh...7..252G. doi:18/nphys1896. Retrieved February 28, 2011.

[48] D.J.P. Morris, D.A. Tennant, S.A. Grigera, B. Klemke, C. Castelnovo, R. Moessner, C. Czter-nasty, M. Meissner, K.C. Rule, J.-U. Hoffmann, K. Kiefer, S. Gerischer, D. Slobinsky, and R.S. Perry (September 3, 2009) [2009-07-09]. "Dirac Strings and Magnetic Monopoles in Spin Ice $Dy_2Ti_2O_7$". *Science* **326** (5951): 411–4. arXiv:1011.1174. Bibcode:2009Sci...326..411M. doi:10.1126/science.1178868. PMID 19729617.

[49] Pietilä, Ville; Möttönen, Mikko (2009). "Creation of Dirac Monopoles in Spinor Bose–Einstein Condensates". *Phys. Rev. Lett.* **103**: 030401. arXiv:0903.4732. Bibcode:2009PhRvL.103c0401P. doi:10.1103/physrevlett.103.030401.

[50] A.Jaffe, C.H.Taubes (1980). *Vortices and monopoles.*

[51] N.J. Hitchin (1982). *Monopoles and geodesics.*

[52] W.Nahm (1982). *The construction of all self-dual monopoles by the ADHM method.*

[53] N.J. Hitchin (1983). *On the construction of monopoles.*

[54] N.J. Hitchin (1999). *Integrable sustems in Riemannian geometry* (K.Uhlenbeck ed.). C-L.Terng (ed.).

[55] N.J. Hitchin, N.S. Manton, M.K. Murray (1995). *Symmetric Monopoles.*

[56] F.A. Bais and H. Weldon, (1978). *Exact Monopole Solutions in SU(N) Gauge Theory*, Phys. Rev. Let. 41, 601.

[57] D. Wilkinson and F.A. Bais, (1979). *Exact SU(N) monopole solutions with spherical symmetry*, Phys. Rev D. 19, 2410

[58] N. Ganoulis, P. Goddard, D. Olive, (1982).*Self dual Monopoles and Toda Molecules*, Nucl. Phys. B205, 601

[59] Farwell, Ruth and Minami, Masatsugu, (1983). *One-dimensional Toda Molecule. 2. The Solutions Applied To Bogomolny Monopoles With Spherical Symmetry*, Prog. Theor. Phys. 70 710.

[60] R. Slansky,(1981). *Group theory for unified model building", Physics Reports, 79, 1. (See table 10 pg. 84 of http://citeseerx.ist. psu.edu/viewdoc/download?doi=10.1.1.126.1581&rep=rep1&type=pdf)*

[61] M. Toda, (1975). *Studies of a non-linear lattice*, Phys. Rep., 8, 1.

[62] B. Kostant, (1979).*The solution to a generalized Toda lattice and representation theory.*, Adv. in Math. 34, 195.

[63] M.K.Murray (1983). *Monopoles and spectral curves for arbitrary Lie groups.*

[64] Hurtubise, Jacques; Murray, Michael K. (1989). "On the construction of monopoles for the classical groups". *Communications in Mathematical Physics* **122** (1): 35–89. Bibcode:1989CMaPh.122...35H. doi:10.1007/bf01221407. MR 994495.

[65] C.H.Taubes (1983). *Stability in Yang–Mills theories.*

[66] M.F. Atiyah; N.J. Hitchin (1988). *The geometry and dynamics of magnetic monopoles.* Princeton Univ.Press.

[67] S.K.Donaldson (1984). *Nahm's equations and the classification of monopoles.*

[68] G.B.Segal, A.Selby (1996). *The cohomology of the space of magnetic monopoles.*

[69] M.F.Atiyah (1987). *Magnetic monopoles in hyperbolic space, Vector bundles on algebraic varieties.* Oxford University Press.

[70] M.K.Murray (2000). *On the complete integrability of the discrete Nahm equations.*

25.12   References

- Brau, Charles A. (2004). *Modern Problems in Classical Electrodynamics.* Oxford University Press. ISBN 0-19-514665-4.

- Hitchin, N.J.; Murray, M.K. (1988). *Spectral curves and the ADHM method.*

- Jackson, John David (1999). *Classical Electrodynamics* (3rd ed.). New York: Wiley. ISBN 0-471-30932-X.

- Milton, Kimball A. (June 2006). "Theoretical and experimental status of magnetic monopoles". *Reports on Progress in Physics* **69** (6): 1637–1711. arXiv:hep-ex/0602040.Bibcode:2006RPPh...69.1637M.doi:10.1088/0/69/6/R02.

- Shnir, Yakov M. (2005). *Magnetic Monopoles.* Springer-Verlag. ISBN 3-540-25277-0.

- Sutcliffe, P.M. (1997). *BPS monopoles.*

- Vonsovsky, Sergey V. (1975). *Magnetism of Elemetary Particles.* Mir Publishers.

25.13   External links

- Magnetic Monopole Searches (lecture notes)

- Particle Data Group summary of magnetic monopole search

- 'Race for the Pole' Dr David Milstead Freeview 'Snapshot' video by the Vega Science Trust and the BBC/OU.

- Interview with Jonathan Morris about magnetic monopoles and magnetic monopole quasiparticles. Drillingsraum, April 16, 2010

- *Nature,* 2009

- *Sciencedaily,* 2009

- H. Kadowaki, N. Doi, Y. Aoki, Y.Tabata, T.J. Sato, J.W. Lynn, K. Matsuhira, Z. Hiroi (2009). "Observation of Magnetic Monopoles in Spin Ice". arXiv:0908.3568.

- *Video of lecture by Paul Dirac on magnetic monopoles,* 1975 on YouTube

Chapter 26

Atiyah–Singer index theorem

In differential geometry, the **Atiyah–Singer index theorem**, proved by Michael Atiyah and Isadore Singer (1963), states that for an elliptic differential operator on a compact manifold, the **analytical index** (related to the dimension of the space of solutions) is equal to the **topological index** (defined in terms of some topological data). It includes many other theorems, such as the Riemann–Roch theorem, as special cases, and has applications in theoretical physics.

26.1   History

The index problem for elliptic differential operators was posed by Israel Gel'fand (1960). He noticed the homotopy invariance of the index, and asked for a formula for it by means of topological invariants. Some of the motivating examples included the Riemann–Roch theorem and its generalization the Hirzebruch–Riemann–Roch theorem, and the Hirzebruch signature theorem. Hirzebruch and Borel had proved the integrality of the Â genus of a spin manifold, and Atiyah suggested that this integrality could be explained if it were the index of the Dirac operator (which was rediscovered by Atiyah and Singer in 1961).

The Atiyah–Singer theorem was announced by Atiyah & Singer (1963). The proof sketched in this announcement was never published by them, though it appears in the book (Palais 1965). Their first published proof (Atiyah & Singer 1968a) replaced the cobordism theory of the first proof with K-theory, and they used this to give proofs of various generalizations in the papers Atiyah and Singer (1968a, 1968b, 1971a, 1971b).

- **1965:** S.P. Novikov (Novikov 1965) published his results on the topological invariance of the rational Pontrjagin classes on smooth manifolds.

- Kirby and Siebenmann's results (Kirby & Siebenmann 1969), combined with René Thom's paper (Thom 1956) proved the existence of rational Pontryagin classes on topological manifolds. The rational Pontrjagin classes are essential ingredients of the index theorem on smooth and topological manifolds.

- **1969:** M.F. Atiyah (Atiyah 1970) defines abstract elliptic operators on arbitrary metric spaces. Abstract elliptic operators became protagonists in Kasparov's theory and Connes's noncommutative differential geometry.

- **1971:** I.M. Singer (Singer 1971) proposes a comprehensive program for future extensions of index theory.

- **1972:** G.G. Kasparov (Kasparov 1972) publishes his work on the realization of the K-homology by abstract elliptic operators.

- Atiyah, Bott, and Patodi (1973) gave a new proof of the index theorem using the heat equation, described in (Melrose 1993).

- **1977:** D. Sullivan (Sullivan 1979) establishes his theorem on the existence and uniqueness of Lipschitz and quasi-conformal structures on topological manifolds of dimension different from 4.

- Getzler (1983) motivated by ideas of Witten (1982) and Alvarez-Gaume, gave a short proof of the local index theorem for operators that are locally Dirac operators; this covers many of the useful cases.

- **1983:** N. Teleman (Teleman 1983) proves that the analytical indices of signature operators with values in vector bundles are topological invariants.

- **1984:** N. Teleman (Teleman 1984) establishes the index theorem on topological manifolds.

- **1986:** A. Connes (Connes 1986) publishes his fundamental paper on non-commutative geometry.

- **1989:** S.K. Donaldson and D. Sullivan (Donaldson & Sullivan 1989) study Yang–Mills theory on quasiconformal manifolds of dimension 4. They introduce the signature operator S defined on differential forms of degree two.

- **1990:** A. Connes and H. Moscovici (Connes & Moscovici 1990) prove the local index formula in the context of non-commutative geometry.

- **1994:** A. Connes, D. Sullivan and N. Teleman (Connes, Sullivan & Teleman 1994) prove the index theorem for signature operators on quasiconformal manifolds.

26.2   Notation

- X is a compact smooth manifold (without boundary).

- E and F are smooth vector bundles over X.

- D is an elliptic differential operator from E to F. So in local coordinates it acts as a differential operator, taking smooth sections of E to smooth sections of F.

26.3   Symbol of a differential operator

If D is a differential operator on a euclidean space of order n in k variables

$$x_1, ..., xk,$$

then its symbol is the function of $2k$ variables

$$x_1, ... , xk, y_1, ..., yk,$$

given by dropping all terms of order less than n and replacing $\partial/\partial xi$ by yi. So the symbol is homogeneous in the variables y, of degree n. The symbol is well defined even though $\partial/\partial xi$ does not commute with xi because we only keep the highest order terms and differential operators commute "up to lower-order terms". The operator is called **elliptic** if the symbol is nonzero whenever at least one y is nonzero.

Example: The Laplace operator in k variables has symbol $y_1{}^2 + ... + yk^2$, and so is elliptic as this is nonzero whenever any the yi's are nonzero. The wave operator has symbol $-y_1{}^2 + ... + yk^2$, which is not elliptic if $k \geq 2$, as the symbol vanishes for some non-zero values of the ys.

The symbol of a differential operator of order n on a smooth manifold X is defined in much the same way using local coordinate charts, and is a function on the cotangent bundle of X, homogeneous of degree n on each cotangent space. (In general, differential operators transform in a rather complicated way under coordinate transforms (see jet bundle); however, the highest order terms transform like tensors so we get well defined homogeneous functions on the cotangent spaces that are independent of the choice of local charts.) More generally, the symbol of a differential operator between two vector bundles E and F is a section of the pullback of the bundle $Hom(E, F)$ to the cotangent space of X. The differential operator is called *elliptic* if the element of $Hom(Ex, Fx)$ is invertible for all non-zero cotangent vectors at any point x of X.

A key property of elliptic operators is that they are almost invertible; this is closely related to the fact that their symbols are almost invertible. More precisely, an elliptic operator D on a compact manifold has a (non-unique) **parametrix** (or **pseudoinverse**) D' such that $DD'-1$ and $D'D-1$ are both compact operators. An important consequence is that the kernel of D is finite-dimensional, because all eigenspaces of compact operators, other than the kernel, are finite-dimensional. (The pseudoinverse of an elliptic differential operator is almost never a differential operator. However, it is an elliptic pseudodifferential operator.)

26.4   Analytical index

As the elliptic differential operator D has a pseudoinverse, it is a Fredholm operator. Any Fredholm operator has an *index*, defined as the difference between the (finite) dimension of the kernel of D (solutions of $Df = 0$), and the (finite) dimension of the cokernel of D (the constraints on the right-hand-side of an inhomogeneous equation like $Df = g$, or equivalently the kernel of the adjoint operator). In other words,

$$\text{Index}(D) = \dim \text{Ker(D)} - \dim \text{Coker}(D) = \dim \text{Ker(D)} - \dim \text{Ker}(D^*).$$

This is sometimes called the **analytical index** of D.

Example: Suppose that the manifold is the circle (thought of as **R/Z**), and D is the operator $d/dx - \lambda$ for some complex constant λ. (This is the simplest example of an elliptic operator.) Then the kernel is the space of multiples of $\exp(\lambda x)$ if λ is an integral multiple of $2\pi i$ and is 0 otherwise, and the kernel of the adjoint is a similar space with λ replaced by its complex conjugate. So D has index 0. This example shows that the kernel and cokernel of elliptic operators can jump discontinuously as the elliptic operator varies, so there is no nice formula for their dimensions in terms of continuous topological data. However the jumps in the dimensions of the kernel and cokernel are the same, so the index, given by the difference of their dimensions, does vary continuously, and can be given in terms of topological data by the index theorem.

26.5   Topological index

The **topological index** of an elliptic differential operator D between smooth vector bundles E and F on an n-dimensional compact manifold X is given by

$$\text{ch}(D)\text{Td}(X)[X],$$

in other words the value of the top dimensional component of the mixed cohomology class $\text{ch}(D)\text{Td}(X)$ on the fundamental homology class of the manifold X. Here,

- $\text{Td}(X)$ is the Todd class of the complexified tangent bundle of X.

- $\text{ch}(D)$ is equal to $\varphi^{-1}(\text{ch}(\text{d}(p^*E, p^*F, \sigma(D))))$, where

  - φ is the Thom isomorphism from $H^k(X, \mathbf{Q})$ to $H^{n+k}(B(X)/S(X), \mathbf{Q})$

- $B(X)$ is the unit ball bundle of the cotangent bundle of X, and $S(X)$ is its boundary, and p is the projection to X.

- ch is the Chern character from K-theory $K(X)$ to the rational cohomology ring $H(X, \mathbf{Q})$.

- $d(p^*E, p^*F, \sigma(D))$ is the "difference element" of $K(B(X)/S(X))$ associated to two vector bundles p^*E and p^*F on $B(X)$ and an isomorphism $\sigma(D)$ between them on the subspace $S(X)$.

- $\sigma(D)$ is the symbol of D

One can also define the topological index using only K theory (and this alternative definition is compatible in a certain sense with the Chern-character construction above). If X is a compact submanifold of a manifold Y then there is a pushforward (or "shriek") map from $K(TX)$ to $K(TY)$. The topological index of an element of $K(TX)$ is defined to be the image of this operation with Y some Euclidean space, for which $K(TY)$ can be naturally identified with the integers $\mathbf{Z}$ (as a consequence of Bott-periodicity). This map is independent of the embedding of X in Euclidean space. Now a differential operator as above naturally defines an element of $K(TX)$, and the image in $\mathbf{Z}$ under this map "is" the topological index.

As usual, D is an elliptic differential operator between vector bundles E and F over a compact manifold X.

The *index problem* is the following: compute the (analytical) index of D using only the symbol s and *topological* data derived from the manifold and the vector bundle. The Atiyah–Singer index theorem solves this problem, and states:

> **The analytical index of D is equal to its topological index.**

In spite of its formidable definition, the topological index is usually straightforward to evaluate explicitly. So this makes it possible to evaluate the analytical index. (The cokernel and kernel of an elliptic operator are in general extremely hard to evaluate individually; the index theorem shows that we can usually at least evaluate their **difference**.) Many important invariants of a manifold (such as the signature) can be given as the index of suitable differential operators, so the index theorem allows us to evaluate these invariants in terms of topological data.

Although the analytical index is usually hard to evaluate directly, it is at least obviously an integer. The topological index is by definition a rational number, but it is usually not at all obvious from the definition that it is also integral. So the Atiyah–Singer index theorem implies some deep integrality properties, as it implies that the topological index is integral.

The index of an elliptic differential operator obviously vanishes if the operator is self adjoint. It also vanishes if the manifold X has odd dimension, though there are **pseudodifferential** elliptic operators whose index does not vanish in odd dimensions.

26.6 Extensions of the Atiyah–Singer index theorem

26.6.1 Teleman index theorem (Teleman 1983), (Teleman 1984)

> **For any abstract elliptic operator (Atiyah 1970) on a closed, oriented, topological manifold, the analytical index equals the topological index.**

The proof of this result goes through specific considerations, including the extension of Hodge theory on combinatorial and Lipschitz manifolds (Teleman 1980), (Teleman 1983), the extension of Atiyah–Singer's signature operator to Lipschitz manifolds (Teleman 1983), Kasparov's K-homology (Kasparov 1972) and topological cobordism (Kirby & Siebenmann 1977).

This result shows that the index theorem is not merely a differentiable statement, but rather a topological statement.

26.6.2 Connes–Donaldson–Sullivan–Teleman index theorem (Donaldson & Sullivan 1989), (Connes, Sullivan & Teleman 1994)

> **For any quasiconformal manifold there exists a local construction of the Hirzebruch–Thom characteristic classes.**

This theory is based on a signature operator S, defined on middle degree differential forms on even-dimensional quasi-conformal manifolds (compare (Donaldson & Sullivan 1989)).

Using topological cobordism and K-homology one may provide a full statement of an index theorem on quasiconformal manifolds (see page 678 of (Connes, Sullivan & Teleman 1994)). The work (Connes, Sullivan & Teleman 1994) "provides local constructions for characteristic classes based on higher dimensional relatives of the measurable Riemann mapping in dimension two and the Yang–Mills theory in dimension four."

These results constitute significant advances along the lines of Singer's program *Prospects in Mathematics* (Singer 1971). At the same time, they provide, also, an effective construction of the rational Pontrjagin classes on topological manifolds. The paper (Teleman 1985) provides a link between Thom's original construction of the rational Pontrjagin classes (Thom 1956) and index theory.

It is important to mention that the index formula is a topological statement. The obstruction theories due to Milnor, Kervaire, Kirby, Siebenmann, Sullivan, Donaldson show that only a minority of topological manifolds possess differentiable structures and these are not necessarily unique. Sullivan's result on Lipschitz and quasiconformal structures (Sullivan 1979) shows that any topological manifold in dimension different from 4 possesses such a structure which is unique (up to isotopy close to identity).

The quasiconformal structures (Connes, Sullivan & Teleman 1994) and more generally the L^p-structures, $p > n(n+1)/2$, introduced by M. Hilsum (Hilsum 1999), are the weakest analytical structures on topological manifolds of dimension n for which the index theorem is known to hold.

26.7 Examples

26.7.1 Euler characteristic

Suppose that M is a compact oriented manifold. If we take E to be the sum of the even exterior powers of the cotangent bundle, and F to be the sum of the odd powers, define $D = d + d^*$, considered as a map from E to F. Then the topological index of D is the Euler characteristic of the Hodge cohomology of M, and the analytical index is the Euler class of the manifold. The index formula for this operator yields the Chern-Gauss-Bonnet theorem.

26.7.2 Hirzebruch–Riemann–Roch theorem

Take X to be a complex manifold with a complex vector bundle V. We let the vector bundles E and F be the sums of the bundles of differential forms with coefficients in V of type $(0,i)$ with i even or odd, and we let the differential operator D be the sum

$$\bar{\partial} + \bar{\partial}^*$$

restricted to E. Then the analytical index of D is the holomorphic Euler characteristic of V:

$$\text{index}(D) = \Sigma(-1)^p \dim H^p(X,V).$$

The topological index of D is given by

$$\text{index}(D) = \text{ch}(V)\text{Td}(X)[X],$$

the product of the Chern character of V and the Todd class of X evaluated on the fundamental class of X. By equating the topological and analytical indices we get the Hirzebruch–Riemann–Roch theorem. In fact we get a generalization of it to all complex manifolds: Hirzebruch's proof only worked for **projective** complex manifolds X.

This derivation of the Hirzebruch–Riemann–Roch theorem is more natural if we use the index theorem for elliptic complexes rather than elliptic operators. We can take the complex to be

$$0 \to V \to V \otimes \Lambda^{0,1} T^*(X) \to V \otimes \Lambda^{0,2} T^*(X)...$$

with the differential given by $\overline{\partial}$. Then the i'th cohomology group is just the coherent cohomology group $H^i(X, V)$, so the analytical index of this complex is the holomorphic Euler characteristic $\Sigma (-1)^i \dim(H^i(X, V))$. As before, the topological index is $\mathrm{ch}(V)\mathrm{Td}(X)[X]$.

26.7.3   Hirzebruch signature theorem

The Hirzebruch signature theorem states that the signature of a compact smooth manifold X of dimension $4k$ is given by the L genus of the manifold. This follows from the Atiyah–Singer index theorem applied to the following **signature operator**.

The bundles E and F are given by the $+1$ and -1 eigenspaces of the operator on the bundle of differential forms of X, that acts on k-forms as

$$i^{k(k-1)}$$

times the Hodge $*$ operator. The operator D is the Hodge Laplacian

$$D \equiv \Delta := (\mathbf{d} + \mathbf{d}^*)^2$$

restricted to E, where $\mathbf{d}$ is the Cartan exterior derivative and $\mathbf{d}^*$ is its adjoint.

The analytic index of D is the signature of the manifold X, and its topological index is the L genus of X, so these are equal.

26.7.4   Â genus and Rochlin's theorem

The Â genus is a rational number defined for any manifold, but is in general not an integer. Borel and Hirzebruch showed that it is integral for spin manifolds, and an even integer if in addition the dimension is 4 mod 8. This can be deduced from the index theorem, which implies that the Â genus for spin manifolds is the index of a Dirac operator. The extra factor of 2 in dimensions 4 mod 8 comes from the fact that in this case the kernel and cokernel of the Dirac operator have a quaternionic structure, so as complex vector spaces they have even dimensions, so the index is even.

In dimension 4 this result implies Rochlin's theorem that the signature of a 4-dimensional spin manifold is divisible by 16: this follows because in dimension 4 the Â genus is minus one eighth of the signature.

26.8   Proof techniques

26.8.1   Pseudodifferential operators

Main article: pseudodifferential operator

Pseudodifferential operators can be explained easily in the case of constant coefficient operators on Euclidean space. In this case, constant coefficient differential operators are just the Fourier transforms of multiplication by polynomials, and constant coefficient pseudodifferential operators are just the Fourier transforms of multiplication by more general functions.

Many proofs of the index theorem use pseudodifferential operators rather than differential operators. The reason for this is that for many purposes there are not enough differential operators. For example, a pseudoinverse of an elliptic differential operator of positive order is not a differential operator, but is a pseudodifferential operator. Also, there is a direct correspondence between data representing elements of K(B(X), S(X)) (clutching functions) and symbols of elliptic pseudodifferential operators.

Pseudodifferential operators have an order, which can be any real number or even $-\infty$, and have symbols (which are no longer polynomials on the cotangent space), and elliptic differential operators are those whose symbols are invertible for sufficiently large cotangent vectors. Most version of the index theorem can be extended from elliptic differential operators to elliptic pseudodifferential operators.

26.8.2   Cobordism

The initial proof was based on that of the Hirzebruch–Riemann–Roch theorem (1954), and involved cobordism theory and pseudodifferential operators.

The idea of this first proof is roughly as follows. Consider the ring generated by pairs (X, V) where V is a smooth vector bundle on the compact smooth oriented manifold X, with relations that the sum and product of the ring on these generators are given by disjoint union and product of manifolds (with the obvious operations on the vector bundles), and any boundary of a manifold with vector bundle is 0. This is similar to the cobordism ring of oriented manifolds, except that the manifolds also have a vector bundle. The topological and analytical indices are both reinterpreted as functions from this ring to the integers. Then one checks that these two functions are in fact both ring homomorphisms. In order to prove they are the same, it is then only necessary to check they are the same on a set of generators of this ring. Thom's cobordism theory gives a set of generators; for example, complex vector spaces with the trivial bundle together with certain bundles over even dimensional spheres. So the index theorem can be proved by checking it on these particularly simple cases.

26.8.3   K theory

Atiyah and Singer's first published proof used K theory rather than cobordism. If i is any inclusion of compact manifolds from X to Y, they defined a 'pushforward' operation $i!$ on elliptic operators of X to elliptic operators of Y that preserves the index. By taking Y to be some sphere that X embeds in, this reduces the index theorem to the case of spheres. If Y is a sphere and X is some point embedded in Y, then any elliptic operator on Y is the image under $i!$ of some elliptic operator on the point. This reduces the index theorem to the case of a point, when it is trivial.

26.8.4   Heat equation

Atiyah, Bott, and Patodi (1973) gave a new proof of the index theorem using the heat equation, described in (Melrose 1993) and (Gilkey 1994). Berline, Getzler & Vergne (2004) describe a simpler heat equation proof exploiting supersymmetry.
If D is a differential operator with adjoint D^*, then D^*D and DD^* are self adjoint operators whose non-zero eigenvalues have the same multiplicities. However their zero eigenspaces may have different multiplicities, as these multiplicities are the dimensions of the kernels of D and D^*. Therefore the index of D is given by

$$\text{Index}(D) = \dim \text{Ker}(D) - \dim \text{Ker}(D^*) = \text{Tr}(e^{-tD^*D}) - \text{Tr}(e^{-tDD^*})$$

for any positive t. The right hand side is given by the trace of the difference of the kernels of two heat operators. These have an asymptotic expansion for small positive t, which can be used to evaluate the limit as t tends to 0, giving a proof of the Atiyah–Singer index theorem. The asymptotic expansions for small t appear very complicated, but invariant theory shows that there are huge cancellations between the terms, which makes it possible to find the leading terms explicitly. These cancellations were later explained using supersymmetry.

26.9   Generalizations

- The Atiyah–Singer theorem applies to elliptic pseudodifferential operators in much the same way as for elliptic differential operators. In fact, for technical reasons most of the early proofs worked with pseudodifferential rather than differential operators: their extra flexibility made some steps of the proofs easier.

- Instead of working with an elliptic operator between two vector bundles, it is sometimes more convenient to work with an *elliptic complex*

$$0 \to E_0 \to E_1 \to E_2 \to ... \to Em \to 0$$

of vector bundles. The difference is that the symbols now form an exact sequence (off the zero section). In the case when there are just two non-zero bundles in the complex this implies that the symbol is an isomorphism off the zero section, so an elliptic complex with 2 terms is essentially the same as an elliptic operator between two vector bundles. Conversely the index theorem for an elliptic complex can easily be reduced to the case of an elliptic operator: the two vector bundles are given by the sums of the even or odd terms of the complex, and the elliptic operator is the sum of the operators of the elliptic complex and their adjoints, restricted to the sum of the even bundles.

- If the manifold is allowed to have boundary, then some restrictions must be put on the domain of the elliptic operator in order to ensure a finite index. These conditions can be local (like demanding that the sections in the domain vanish at the boundary) or more complicated global conditions (like requiring that the sections in the domain solve some differential equation). The local case was worked out by Atiyah and Bott, but they showed that many interesting operators (e.g., the signature operator) do not admit local boundary conditions. To handle these operators, Atiyah, Patodi and Singer introduced global boundary conditions equivalent to attaching a cylinder to the manifold along the boundary and then restricting the domain to those sections that are square integrable along the cylinder. This point of view is adopted in the proof of Melrose (1993) of the Atiyah–Patodi–Singer index theorem.

- Instead of just one elliptic operator, one can consider a family of elliptic operators parameterized by some space Y. In this case the index is an element of the K-theory of Y, rather than an integer. If the operators in the family are real, then the index lies in the real K-theory of Y. This gives a little extra information, as the map from the real K theory of Y to the complex K theory is not always injective.

- If there is a group action of a group G on the compact manifold X, commuting with the elliptic operator, then one replaces ordinary K theory with equivariant K-theory. Moreover, one gets generalizations of the Lefschetz fixed point theorem, with terms coming from fixed point submanifolds of the group G. See also: equivariant index theorem.

- Atiyah (1976) showed how to extend the index theorem to some non-compact manifolds, acted on by a discrete group with compact quotient. The kernel of the elliptic operator is in general infinite dimensional in this case, but it is possible to get a finite index using the dimension of a module over a von Neumann algebra; this index is in general real rather than integer valued. This version is called the L^2 **index theorem**, and was used by Atiyah & Schmid (1977) to rederive properties of the discrete series representations of semisimple Lie groups.

- The Callias index theorem is an index theorem for a Dirac operator on a noncompact odd-dimensional space. The Atiyah–Singer index is only defined on compact spaces, and vanishes when their dimension is odd. In 1978 Constantine Callias, at the suggestion of his Ph.D. advisor Roman Jackiw, used the axial anomaly to derive this index theorem on spaces equipped with a Hermitian matrix called the Higgs field. As presented in his paper Index Theorems on Open Spaces the index of the Dirac operator is a topological invariant which measures the winding of the Higgs field on a sphere at infinity. If U is the unit matrix in the direction of the Higgs field, then the index is proportional to the integral of $U(dU)^{n-1}$ over the $(n-1)$-sphere at infinity. If n is even, it is always zero. The topological interpretation of this invariant and its relation to the Hörmander index proposed by Boris Fedosov, as generalized by Lars Hörmander, was published by Raoul Bott and Robert Thomas Seeley in the article Some Remarks on the Paper of Callias in the same issue of Communications in Mathematical Physics as Callias' article.

26.10 References

26.10.1 Theoretical references

The papers by Atiyah are reprinted in volumes 3 and 4 of his collected works, (Atiyah 1988a, 1988b)

- Atiyah, M. F. (1970), "Global Theory of Elliptic Operators", *Proc. Int. Conf. on Functional Analysis and Related Topics (Tokyo, 1969)*, University of Tokio, Zbl 0193.43601

- Atiyah, M. F. (1976), "Elliptic operators, discrete groups and von Neumann algebras", *Colloque "Analyse et Topologie" en l'Honneur de Henri Cartan (Orsay, 1974)*, Asterisque, 32–33, Soc. Math. France, Paris, pp. 43–72, MR 0420729

- Atiyah, M. F.; Segal, G. B. (1968), "The Index of Elliptic Operators: II", *Annals of Mathematics*, Second Series **87** (3): 531–545, doi:10.2307/1970716, JSTOR 1970716 This reformulates the result as a sort of Lefschetz fixed point theorem, using equivariant K theory.

- Atiyah, Michael F.; Singer, Isadore M. (1963), "The Index of Elliptic Operators on Compact Manifolds", *Bull. Amer. Math. Soc.* **69** (3): 422–433, doi:10.1090/S0002-9904-1963-10957-X An announcement of the index theorem.

- Atiyah, Michael F.; Singer, Isadore M. (1968a), "The Index of Elliptic Operators I", *Annals of Mathematics* **87** (3): 484–530, doi:10.2307/1970715, JSTOR 1970715 This gives a proof using K theory instead of cohomology.

- Atiyah, Michael F.; Singer, Isadore M. (1968b), "The Index of Elliptic Operators III", *Annals of Mathematics*, Second Series **87** (3): 546–604, doi:10.2307/1970717, JSTOR 1970717 This paper shows how to convert from the K-theory version to a version using cohomology.

- Atiyah, Michael F.; Singer, Isadore M. (1971), "The Index of Elliptic Operators IV", *Annals of Mathematics*, Second Series **93** (1): 119–138, doi:10.2307/1970756, JSTOR 1970756 This paper studies families of elliptic operators, where the index is now an element of the K-theory of the space parametrizing the family.

- Atiyah, Michael F.; Singer, Isadore M. (1971), "The Index of Elliptic Operators V", *Annals of Mathematics*, Second Series **93** (1): 139–149, doi:10.2307/1970757, JSTOR 1970757. This studies families of real (rather than complex) elliptic operators, when one can sometimes squeeze out a little extra information.

- Atiyah, M. F.; Bott, R. (1966), "A Lefschetz Fixed Point Formula for Elliptic Differential Operators", *Bull. Am. Math. Soc.* **72** (2): 245–50, doi:10.1090/S0002-9904-1966-11483-0. This states a theorem calculating the Lefschetz number of an endomorphism of an elliptic complex.

- Atiyah, M. F.; Bott, R. (1967), "A Lefschetz Fixed Point Formula for Elliptic Complexes: I", *Annals of Mathematics*, Second series **86** (2): 374–407, doi:10.2307/1970694, JSTOR 1970694 and Atiyah, M. F.; Bott, R. (1968), "A Lefschetz Fixed Point Formula for Elliptic Complexes: II. Applications", *Annals of Mathematics*, Second Series **88** (3): 451–491, doi:10.2307/1970721, JSTOR 1970721 These give the proofs and some applications of the results announced in the previous paper.

- Atiyah, M.; Bott, R.; Patodi, V. K. (1973), "On the heat equation and the index theorem", *Invent. Math.* **19** (4): 279–330, Bibcode:1973InMat..19..279A, doi:10.1007/BF01425417, MR 0650828. "Errata", *Invent. Math.* **28** (3), 1975: 277–280, Bibcode:1975InMat..28..277A, doi:10.1007/BF01425562, MR 0650829

- Atiyah, Michael; Schmid, Wilfried (1977), "A geometric construction of the discrete series for semisimple Lie groups", *Invent. Math.* **42**: 1–62, Bibcode:1977InMat..42....1A, doi:10.1007/BF01389783, MR 0463358, Atiyah, Michael; Schmid, Wilfried (1979), "Erratum:", *Invent. Math.* **54** (2): 189–192, Bibcode:1979InMat..54..189A, doi:10.1007/BF01408936, MR 0550183

- Atiyah, Michael (1988a), *Collected works. Vol. 3. Index theory: 1*, Oxford Science Publications, New York: The Clarendon Press, Oxford University Press, ISBN 0-19-853277-6, MR 0951894

- Atiyah, Michael (1988b), *Collected works. Vol. 4. Index theory: 2*, Oxford Science Publications, New York: The Clarendon Press, Oxford University Press, ISBN 0-19-853278-4, MR 0951895

- Baum, P.; Fulton, W.; Macpherson, R. (1979), "Riemann-Roch for singular varieties", *Acta Mathematica* **143**: 155–191, doi:10.1007/BF02684299, Zbl 0332.14003

- Berline, Nicole; Getzler, Ezra; Vergne, Michèle (2004), *Heat Kernels and Dirac Operators*, Berlin: Springer, ISBN 3-540-20062-2 This gives an elementary proof of the index theorem for the Dirac operator, using the heat equation and supersymmetry.

- Bismut, Jean-Michel (1984), "The Atiyah–Singer Theorems: A Probabilistic Approach. I. The index theorem" (PDF), *J. Funct. Analysis* **57**: 56–99, doi:10.1016/0022-1236(84)90101-0 Bismut proves the theorem for elliptic complexes using probabilistic methods, rather than heat equation methods.

- Connes, A. (1986), "Non-commutative differential geometry", *Publications Mathematiques* (Paris) **62**: 257–360, doi:10.1007/BF02698807, Zbl 0592.46056

- Connes, A.(1994),*Noncommutative Geometry*, San Diego: Academic Press,ISBN978-0-12-185860-5,Zbl0818.466

- Connes, A.; Moscovici, H. (1990), "Cyclic cohomology, the Novikov conjecture and hyperbolic groups" (PDF), *Topology* **29** (3): 345–388, doi:10.1016/0040-9383(90)90003-3, Zbl 0759.58047

- Connes, A.; Sullivan, D.; Teleman, N. (1994), "Quasiconformal mappings, operators on Hilbert space and local formulae for characteristic classes", *Topology* **33** (4): 663–681, doi:10.1016/0040-9383(94)90003-5, Zbl 0840.57013

- Donaldson, S.K.;Sullivan, D.(1989), "Quasiconformal 4-manifolds",*Acta Mathematica***163**: 181–252,doi:10.1007/6, Zbl 0704.57008

- Gel'fand, I. M. (1960), "On elliptic equations", *Russ. Math.Surv.* **15** (3): 113–123, Bibcode:1960RuMaS..15..113G, doi:10.1070/rm1960v015n03ABEH004094 reprinted in volume 1 of his collected works, p. 65–75, ISBN 0-387-13619-3. On page 120 Gel'fand suggests that the index of an elliptic operator should be expressible in terms of topological data.

- Getzler, E. (1983), "Pseudodifferential operators on supermanifolds and the Atiyah–Singer index theorem", *Commun. Math. Phys.* **92** (2): 163–178, Bibcode:1983CMaPh..92..163G, doi:10.1007/BF01210843

- Getzler, E. (1988), "A short proof of the local Atiyah–Singer index theorem",*Topology***25**: 111–117,doi:10.1016/40-9383(86)90008-X

- Gilkey, Peter B. (1994), *Invariance Theory, the Heat Equation, and the Atiyah–Singer Theorem*, ISBN 0-8493-7874-5 Free online textbook that proves the Atiyah–Singer theorem with a heat equation approach

- Higson, Nigel; Roe, John (2000), *Analytic K-homology*, Oxford University Press, ISBN 9780191589201

- Hilsum, M. (1999), "Structures riemaniennes$L_p$et$K$-homologie",*Annals of Mathematics***149**: 1007–1022,doi:21079

- Kasparov, G.G. (1972), "Topological invariance of elliptic operators, I: K-homology", *Math. USSR Izvestija (Engl. Transl.)* **9** (4): 751–792, Bibcode:1975IzMat...9..751K, doi:10.1070/IM1975v009n04ABEH001497

- Kirby, R.; Siebenmann, L.C. (1969), "On the triangulation of manifolds and the Hauptvermutung", *Bull. Amer. Math. Soc.* **75** (4): 742–749, doi:10.1090/S0002-9904-1969-12271-8

- Kirby, R.; Siebenmann, L.C. (1977), *Foundational Essays on Topological Manifolds, Smoothings and Triangulations*, Annals of Mathematics Studies in Mathematics **88**, Princeton: Princeton University Press and Tokio University Press

- Melrose, Richard B. (1993), *The Atiyah–Patodi–Singer Index Theorem*, Wellesley, Mass.: Peters, ISBN 1-56881-002-4 Free online textbook.

- Novikov, S.P. (1965), "Topological invariance of the rational Pontrjagin classes" (PDF), *Doklady Akademii Nauk SSSR* **163**: 298–300

- Palais, Richard S. (1965), *Seminar on the Atiyah–Singer Index Theorem*, Annals of Mathematics Studies **57**, S.l.: Princeton Univ Press, ISBN 0-691-08031-3 This describes the original proof of the theorem (Atiyah and Singer never published their original proof themselves, but only improved versions of it.)

- Shanahan, P. (1978), *The Atiyah–Singer index theorem: an introduction*, Lecture Notes in Mathematics **638**, Springer, doi:10.1007/BFb0068264, ISBN 0-387-08660-9

- Singer, I.M. (1971), "Future extensions of index theory and elliptic operators", *Prospects in Mathematics*, Annals of Mathematics Studies in Mathematics **70**, pp. 171–185

- Sullivan, D. (1979), "Hyperbolic geometry and homeomorphisms", *J.C. Candrell, "Geometric Topology", Proc. Georgia Topology Conf. Athens, Georgia, 1977*, New York: Academic Press, pp. 543–595, ISBN 0-12-158860-2, Zbl 0478.57007

- Sullivan, D.; Teleman, N. (1983), "An analytic proof of Novikov's theorem on rational Pontrjagin classes", *Publications Mathematiques* (Paris) **58**: 291–293, Zbl 0531.58045

- Teleman, N. (1980), "Combinatorial Hodge theory and signature operator", *Inventiones Mathematicae* **61** (3): 227–249, Bibcode:1980InMat..61..227T, doi:10.1007/BF01390066

- Teleman, N. (1983), "The index of signature operators on Lipschitz manifolds", *Publications Mathematiques* (Paris) **58**: 251–290, doi:10.1007/BF02953772, Zbl 0531.58044

- Teleman, N. (1984), "The index theorem on topological manifolds", *Acta Mathematica* **153**: 117–152, doi:10.12376, Zbl 0.547.58036

- Teleman, N. (1985), "Transversality and the index theorem", *Integral Equations and Operator Theory* **8** (5): 693–719, doi:10.1007/BF01201710

- Thom, R. (1956), "Les classes caractéristiques de Pontrjagin de variétés triangulées", *Symp. Int. Top. Alg. Mexico*, pp. 54–67

- Witten, Edward (1982), "Supersymmetry and Morse theory", *J. Diff. Geom.* **17**: 661–692, MR 0683171

26.10.2   References on history

- Shing-Tung Yau, ed. (2009) [First published in 2005], *The Founders of Index Theory* (2nd ed.), Somerville, Mass.: International Press of Boston, ISBN 978-1571461377 - Personal accounts on Atiyah, Bott, Hirzebruch and Singer.

26.11   External links

26.11.1   Links on the theory

- Rafe Mazzeo: *The Atiyah–Singer Index Theorem: What it is and why you should care*. Pdf presentation.

- Voitsekhovskii, M.I.; Shubin, M.A. (2001), "Index formulas", in Hazewinkel, Michiel, *Encyclopedia of Mathematics*, Springer, ISBN 978-1-55608-010-4

- A. J. Wassermann, Lecture notes on the Atiyah–Singer Index Theorem

26.11.2   Links of interviews

- Raussen, Martin; Skau, Christian (2005), "Interview with Michael Atiyah and Isadore Singer" (PDF), *Notices of AMS*: 223–231

- R. R. Seeley and other (1999) Recollections from the early days of index theory and pseudo-differential operators - A partial transcript of informal post–dinner conversation during a symposium held in Roskilde, Denmark, in September 1998.

Chapter 27

Noncommutative geometry

Not to be confused with Anabelian geometry.

Noncommutative geometry (**NCG**) is a branch of mathematics concerned with a geometric approach to noncommutative algebras, and with the construction of *spaces* that are locally presented by noncommutative algebras of functions (possibly in some generalized sense). A noncommutative algebra is an associative algebra in which the multiplication is not commutative, that is, for which xy does not always equal yx ; or more generally an algebraic structure in which one of the principal binary operations is not commutative; one also allows additional structures, e.g. topology or norm, to be possibly carried by the noncommutative algebra of functions.

27.1   Motivation

The main motivation is to extend the commutative duality between spaces and functions to the noncommutative setting. In mathematics, *spaces*, which are geometric in nature, can be related to numerical functions on them. In general, such functions will form a commutative ring. For instance, one may take the ring $C(X)$ of continuous complex-valued functions on a topological space X. In many cases (*e.g.*, if X is a compact Hausdorff space), we can recover X from $C(X)$, and therefore it makes some sense to say that X has *commutative topology*.

More specifically, in topology, compact Hausdorff topological spaces can be reconstructed from the Banach algebra of functions on the space (Gel'fand-Neimark). In commutative algebraic geometry, algebraic schemes are locally prime spectra of commutative unital rings (A. Grothendieck), and schemes can be reconstructed from the categories of quasicoherent sheaves of modules on them (P. Gabriel-A. Rosenberg). For Grothendieck topologies, the cohomological properties of a site are invariant of the corresponding category of sheaves of sets viewed abstractly as a topos (A. Grothendieck). In all these cases, a space is reconstructed from the algebra of functions or its categorified version—some category of sheaves on that space.

Functions on a topological space can be multiplied and added pointwise hence they form a commutative algebra; in fact these operations are local in the topology of the base space, hence the functions form a sheaf of commutative rings over the base space.

The dream of noncommutative geometry is to generalize this duality to the duality between

- noncommutative algebras, or sheaves of noncommutative algebras, or sheaf-like noncommutative algebraic or operator-algebraic structures

- and geometric entities of certain kind,

and interact between the algebraic and geometric description of those via this duality.

155

Regarding that the commutative rings correspond to usual affine schemes, and commutative C*-algebras to usual topological spaces, the extension to noncommutative rings and algebras requires non-trivial generalization of topological spaces, as "non-commutative spaces". For this reason, some talk about non-commutative topology, though the term also has other meanings.

27.1.1 Applications in mathematical physics

Some applications in particle physics are described on the entries Noncommutative standard model and Noncommutative quantum field theory. Sudden rise in interest in noncommutative geometry in physics, follows after the speculations of its role in M-theory made in 1997.[1]

27.1.2 Motivation from ergodic theory

Some of the theory developed by Alain Connes to handle noncommutative geometry at a technical level has roots in older attempts, in particular in ergodic theory. The proposal of George Mackey to create a *virtual subgroup* theory, with respect to which ergodic group actions would become homogeneous spaces of an extended kind, has by now been subsumed.

27.2 Noncommutative C*-algebras, von Neumann algebras

(The formal duals of) non-commutative C*-algebras are often now called non-commutative spaces. This is by analogy with the Gelfand representation, which shows that commutative C*-algebras are dual to locally compact Hausdorff spaces. In general, one can associate to any C*-algebra S a topological space $\hat{S}$; see spectrum of a C*-algebra.

For the duality between σ-finite measure spaces and commutative von Neumann algebras, noncommutative von Neumann algebras are called *non-commutative measure spaces*.

27.3 Noncommutative differentiable manifolds

A smooth Riemannian manifold M is a topological space with a lot of extra structure. From its algebra of continuous functions $C(M)$ we only recover M topologically. The algebraic invariant that recovers the Riemannian structure is a spectral triple. It is constructed from a smooth vector bundle E over M, e.g. the exterior algebra bundle. The Hilbert space $L^2(M,E)$ of square integrable sections of E carries a representation of $C(M)$ by multiplication operators, and we consider an unbounded operator D in $L^2(M,E)$ with compact resolvent (e.g. the signature operator), such that the commutators $[D,f]$ are bounded whenever f is smooth. A recent deep theorem[2] states that M as a Riemannian manifold can be recovered from this data.

This suggests that one might define a noncommutative Riemannian manifold as a spectral triple (A,H,D), consisting of a representation of a C*-algebra A on a Hilbert space H, together with an unbounded operator D on H, with compact resolvent, such that $[D,a]$ is bounded for all a in some dense subalgebra of A. Research in spectral triples is very active, and many examples of noncommutative manifolds have been constructed.

27.4 Noncommutative affine and projective schemes

In analogy to the duality between affine schemes and commutative rings, we define a category of **noncommutative affine schemes** as the dual of the category of associative unital rings. There are certain analogues of Zariski topology in that context so that one can glue such affine schemes to more general objects.

There are also generalizations of the Cone and of the Proj of a commutative graded ring, mimicking a Serre's theorem on Proj. Namely the category of quasicoherent sheaves of O-modules on a Proj of a commutative graded algebra is equivalent to the category of graded modules over the ring localized on Serre's subcategory of graded modules of finite length; there is also analogous theorem for coherent sheaves when the algebra is Noetherian. This theorem is extended as a definition of **noncommutative projective geometry** by Michael Artin and J. J. Zhang,[3] who add also some general ring-theoretic conditions (e.g. Artin-Schelter regularity).

Many properties of projective schemes extend to this context. For example, there exist an analog of the celebrated Serre duality for noncommutative projective schemes of Artin and Zhang.[4]

A. L. Rosenberg has created a rather general relative concept of **noncommutative quasicompact scheme** (over a base category), abstracting the Grothendieck's study of morphisms of schemes and covers in terms of categories of quasicoherent sheaves and flat localization functors.[5] There is also another interesting approach via localization theory, due to Fred Van Oystaeyen, Luc Willaert and Alain Verschoren, where the main concept is that of a **schematic algebra**.[6]

27.5 Invariants for noncommutative spaces

Some of the motivating questions of the theory are concerned with extending known topological invariants to formal duals of noncommutative (operator) algebras and other replacements and candidates for noncommutative spaces. One of the main starting points of the Alain Connes' direction in noncommutative geometry is his discovery of a new homology theory associated to noncommutative associative algebras and noncommutative operator algebras, namely the cyclic homology and its relations to the algebraic K-theory (primarily via Connes-Chern character map).

The theory of characteristic classes of smooth manifolds has been extended to spectral triples, employing the tools of operator K-theory and cyclic cohomology. Several generalizations of now classical index theorems allow for effective extraction of numerical invariants from spectral triples. The fundamental characteristic class in cyclic cohomology, the JLO cocycle, generalizes the classical Chern character.

27.6 Examples of noncommutative spaces

- In the phase space formulation of quantum mechanics, the symplectic phase space of classical mechanics is deformed into a non-commutative phase space generated by the position and momentum operators.

- The standard model of particle physics is another example of a noncommutative geometry, cf noncommutative standard model.

- The noncommutative torus, deformation of the function algebra of the ordinary torus, can be given the structure of a spectral triple. This class of examples has been studied intensively and still functions as a test case for more complicated situations.

- **Snyder space**[7]

- Noncommutative algebras arising from foliations.

- Examples related to dynamical systems arising from number theory, such as the Gauss shift on continued fractions, give rise to noncommutative algebras that appear to have interesting noncommutative geometries.

27.7 See also

- Commutativity

- Phase space formulation

- Moyal product

- Fuzzy sphere

- Noncommutative algebraic geometry

27.8   Notes

[1] Alain Connes, Michael R. Douglas, Albert Schwarz, Noncommutative geometry and matrix theory: compactification on tori. J. High Energy Phys. 1998, no. 2, Paper 3, 35 pp. doi, hep-th/9711162

[2] Connes, Alain, On the spectral characterization of manifolds, arXiv:0810.2088v1

[3] M. Artin, J. J. Zhang, Noncommutative projective schemes, Adv. Math. 109 (1994), no. 2, 228--287, doi

[4] Amnon Yekutieli, James J. Zhang, Serre duality for noncommutative projective schemes, Proc. Amer. Math. Soc. 125, n. 3, 1997, 697-707, pdf

[5] A. L. Rosenberg, Noncommutative schemes, Compositio Mathematica 112 (1998) 93--125, doi; Underlying spaces of noncommutative schemes, preprint MPIM2003-111, dvi, ps; MSRI lecture *Noncommutative schemes and spaces* (Feb 2000): video

[6] Freddy van Oystaeyen, Algebraic geometry for associative algebras, ISBN 0-8247-0424-X - New York: Dekker, 2000.- 287 p. - (Monographs and textbooks in pure and applied mathematics , 232); F. van Oystaeyen, L. Willaert, Grothendieck topology, coherent sheaves and Serre's theorem for schematic algebras, J. Pure Appl. Alg. 104 (1995), p. 109--122

[7] H. S. Snyder, Quantized Space-Time, Phys. Rev. 71 (1947) 38

27.9   References

- Connes, Alain (1994), *Non-commutative geometry* (PDF), Boston, MA: Academic Press, ISBN 978-0-12-185860-5

- Connes, Alain; Marcolli, Matilde (2008), "A walk in the noncommutative garden", *An invitation to noncommutative geometry*, World Sci. Publ., Hackensack, NJ, pp. 1–128, arXiv:math/0601054, Bibcode:2006math......1054C, MR 2408150

- Connes, Alain; Marcolli, Matilde (2008), *Noncommutative geometry, quantum fields and motives* (PDF), American Mathematical Society Colloquium Publications **55**, Providence, R.I.: American Mathematical Society, ISBN 978-0-8218-4210-2, MR 2371808

- Gracia-Bondia, Jose M; Figueroa, Hector; Varilly, Joseph C (2000), *Elements of Non-commutative geometry*, Birkhauser, ISBN 978-0-8176-4124-5

- Landi, Giovanni (1997), *An introduction to noncommutative spaces and their geometries*, Lecture Notes in Physics. New Series m: Monographs **51**, Berlin, New York: Springer-Verlag, arXiv:hep-th/9701078, Bibcode:19971078L, ISBN 978-3-540-63509-3, MR 1482228

- Van Oystaeyen, Fred; Verschoren, Alain (1981), *Non-commutative algebraic geometry*, Lecture Notes in Mathematics **887**, Springer-Verlag, ISBN 978-3-540-11153-5

27.10   Further reading

- Consani, Caterina; Connes, Alain, eds. (2011), *Noncommutative geometry, arithmetic, and related topics. Proceedings of the 21st meeting of the Japan-U.S. Mathematics Institute (JAMI) held at Johns Hopkins University, Baltimore, MD, USA, March 23–26, 2009*, Baltimore, MD: Johns Hopkins University Press, ISBN 1-4214-0352-8, Zbl 1245.00040

- Grensing, Gerhard (2013). *Structural aspects of quantum field theory and noncommutative geometry*. Hackensack New Jersey: World Scientific. ISBN 978-981-4472-69-2.

27.11   External links

- Introduction to Quantum Geometry by Micho Đurđevich

- Lectures on Noncommutative Geometry by Victor Ginzburg

- Very Basic Noncommutative Geometry by Masoud Khalkhali

- Lectures on Arithmetic Noncommutative Geometry by Matilde Marcolli

- Noncommutative Geometry for Pedestrians by J. Madore

- An informal introduction to the ideas and concepts of noncommutative geometry by Thierry Masson (an easier introduction that is still rather technical)

- Noncommutative geometry on arxiv.org

- MathOverflow, Theories of Noncommutative Geometry

- S. Mahanta, On some approaches towards non-commutative algebraic geometry, math.QA/0501166

- G. Sardanashvily, *Lectures on Differential Geometry of Modules and Rings* (Lambert Academic Publishing, Saarbrücken, 2012); arXiv: 0910.1515

- Noncommutative geometry and particle physics

Chapter 28

Quantum group

In mathematics and theoretical physics, the term **quantum group** denotes various kinds of noncommutative algebra with additional structure. In general, a quantum group is some kind of Hopf algebra. There is no single, all-encompassing definition, but instead a family of broadly similar objects.

The term "quantum group" first appeared in the theory of quantum integrable systems, which was then formalized by Vladimir Drinfeld and Michio Jimbo as a particular class of Hopf algebra. The same term is also used for other Hopf algebras that deform or are close to classical Lie groups or Lie algebras, such as a 'bicrossproduct' class of quantum groups introduced by Shahn Majid a little after the work of Drinfeld and Jimbo.

In Drinfeld's approach, quantum groups arise as Hopf algebras depending on an auxiliary parameter q or h, which become universal enveloping algebras of a certain Lie algebra, frequently semisimple or affine, when $q = 1$ or $h = 0$. Closely related are certain dual objects, also Hopf algebras and also called quantum groups, deforming the algebra of functions on the corresponding semisimple algebraic group or a compact Lie group.

Just as groups often appear as symmetries, quantum groups act on many other mathematical objects and it has become fashionable to introduce the adjective *quantum* in such cases; for example there are quantum planes and quantum Grassmannians.

28.1 Intuitive meaning

The discovery of quantum groups was quite unexpected, since it was known for a long time that compact groups and semisimple Lie algebras are "rigid" objects, in other words, they cannot be "deformed". One of the ideas behind quantum groups is that if we consider a structure that is in a sense equivalent but larger, namely a group algebra or a universal enveloping algebra, then a group or enveloping algebra can be "deformed", although the deformation will no longer remain a group or enveloping algebra. More precisely, deformation can be accomplished within the category of Hopf algebras that are not required to be either commutative or cocommutative. One can think of the deformed object as an algebra of functions on a "noncommutative space", in the spirit of the noncommutative geometry of Alain Connes. This intuition, however, came after particular classes of quantum groups had already proved their usefulness in the study of the quantum Yang-Baxter equation and quantum inverse scattering method developed by the Leningrad School (Ludwig Faddeev, Leon Takhtajan, Evgenii Sklyanin, Nicolai Reshetikhin and Vladimir Korepin) and related work by the Japanese School.[1] The intuition behind the second, bicrossproduct, class of quantum groups was different and came from the search for self-dual objects as an approach to quantum gravity.[2]

28.2 Drinfeld-Jimbo type quantum groups

One type of objects commonly called a "quantum group" appeared in the work of Vladimir Drinfeld and Michio Jimbo as a deformation of the universal enveloping algebra of a semisimple Lie algebra or, more generally, a Kac–Moody algebra, in the category of Hopf algebras. The resulting algebra has additional structure, making it into a quasitriangular Hopf algebra.

Let $A = (a_{ij})$ be the Cartan matrix of the Kac–Moody algebra, and let q be a nonzero complex number distinct from 1, then the quantum group, $U_q(G)$, where G is the Lie algebra whose Cartan matrix is A, is defined as the unital associative algebra with generators k_λ (where λ is an element of the weight lattice, i.e. $2(\lambda, \alpha_i)/(\alpha_i, \alpha_i)$ is an integer for all i), and e_i and f_i (for simple roots, α_i), subject to the following relations:

- $k_0 = 1$,

- $k_\lambda k_\mu = k_{\lambda+\mu}$,

- $k_\lambda e_i k_\lambda^{-1} = q^{(\lambda, \alpha_i)} e_i$,

- $k_\lambda f_i k_\lambda^{-1} = q^{-(\lambda, \alpha_i)} f_i$,

- $[e_i, f_j] = \delta_{ij} \frac{k_i - k_i^{-1}}{q_i - q_i^{-1}}$,

- If $i \neq j$ then:

$$\sum_{n=0}^{1-a_{ij}} (-1)^n \frac{[1 - a_{ij}]_{q_i}!}{[1 - a_{ij} - n]_{q_i}![n]_{q_i}!} e_i^n e_j e_i^{1-a_{ij}-n} = 0,$$

$$\sum_{n=0}^{1-a_{ij}} (-1)^n \frac{[1 - a_{ij}]_{q_i}!}{[1 - a_{ij} - n]_{q_i}![n]_{q_i}!} f_i^n f_j f_i^{1-a_{ij}-n} = 0.$$

where $k_i = k_{\alpha_i}, q_i = q^{\frac{1}{2}(\alpha_i, \alpha_i)}, [0]_{q_i}! = 1, [n]_{q_i}! = \prod_{m=1}^{n} [m]_{q_i}$ for all positive integers n, and $[m]_{q_i} = \frac{q_i^m - q_i^{-m}}{q_i - q_i^{-1}}$. These are the q-factorial and q-number, respectively, the q-analogs of the ordinary factorial. The last two relations above are the q-Serre relations, the deformations of the Serre relations.

In the limit as $q \to 1$, these relations approach the relations for the universal enveloping algebra $U(G)$, where $k_\lambda \to 1$ and $\frac{k_\lambda - k_{-\lambda}}{q - q^{-1}} \to t_\lambda$ as $q \to 1$, where the element, t_λ, of the Cartan subalgebra satisfies $(t_\lambda, h) = \lambda(h)$ for all h in the Cartan subalgebra.

There are various coassociative coproducts under which these algebras are Hopf algebras, for example,

- $\Delta_1(k_\lambda) = k_\lambda \otimes k_\lambda$,
- $\Delta_1(e_i) = 1 \otimes e_i + e_i \otimes k_i$,
- $\Delta_1(f_i) = k_i^{-1} \otimes f_i + f_i \otimes 1$,
- $\Delta_2(k_\lambda) = k_\lambda \otimes k_\lambda$,
- $\Delta_2(e_i) = k_i^{-1} \otimes e_i + e_i \otimes 1$,
- $\Delta_2(f_i) = 1 \otimes f_i + f_i \otimes k_i$,
- $\Delta_3(k_\lambda) = k_\lambda \otimes k_\lambda$,
- $\Delta_3(e_i) = k_i^{-\frac{1}{2}} \otimes e_i + e_i \otimes k_i^{\frac{1}{2}}$,
- $\Delta_3(f_i) = k_i^{-\frac{1}{2}} \otimes f_i + f_i \otimes k_i^{\frac{1}{2}}$,

where the set of generators has been extended, if required, to include k_λ for λ which is expressible as the sum of an element of the weight lattice and half an element of the root lattice.

In addition, any Hopf algebra leads to another with reversed coproduct $T \circ \Delta$, where T is given by $T(x \otimes y) = y \otimes x$, giving three more possible versions.

The counit on $U_q(A)$ is the same for all these coproducts: $\varepsilon(k_\lambda) = 1$, $\varepsilon(e_i) = \varepsilon(f_i) = 0$, and the respective antipodes for the above coproducts are given by

- $S_1(k_\lambda) = k_{-\lambda},\ S_1(e_i) = -e_i k_i^{-1},\ S_1(f_i) = -k_i f_i$,
- $S_2(k_\lambda) = k_{-\lambda},\ S_2(e_i) = -k_i e_i,\ S_2(f_i) = -f_i k_i^{-1}$,
- $S_3(k_\lambda) = k_{-\lambda},\ S_3(e_i) = -q_i e_i,\ S_3(f_i) = -q_i^{-1} f_i.$

Alternatively, the quantum group $Uq(G)$ can be regarded as an algebra over the field $\mathbf{C}(q)$, the field of all rational functions of an indeterminate q over $\mathbf{C}$.

Similarly, the quantum group $Uq(G)$ can be regarded as an algebra over the field $\mathbf{Q}(q)$, the field of all rational functions of an indeterminate q over $\mathbf{Q}$ (see below in the section on quantum groups at $q = 0$). The center of quantum group can be described by quantum determinant.

28.2.1 Representation theory

Just as there are many different types of representations for Kac–Moody algebras and their universal enveloping algebras, so there are many different types of representation for quantum groups.

As is the case for all Hopf algebras, $Uq(G)$ has an adjoint representation on itself as a module, with the action being given by

$$\mathrm{Ad}_x \cdot y = \sum_{(x)} x_{(1)} y S(x_{(2)}),$$

where

$$\Delta(x) = \sum_{(x)} x_{(1)} \otimes x_{(2)}$$

Case 1: q is not a root of unity

One important type of representation is a weight representation, and the corresponding module is called a weight module. A weight module is a module with a basis of weight vectors. A weight vector is a nonzero vector v such that $k\lambda \cdot v = d\lambda v$ for all λ, where $d\lambda$ are complex numbers for all weights λ such that

- $d_0 = 1$,
- $d_\lambda d_\mu = d_{\lambda+\mu}$, for all weights λ and μ.

A weight module is called integrable if the actions of ei and fi are locally nilpotent (*i.e.* for any vector v in the module, there exists a positive integer k, possibly dependent on v, such that $e_i^k.v = f_i^k.v = 0$ for all i). In the case of integrable modules, the complex numbers $d\lambda$ associated with a weight vector satisfy $d_\lambda = c_\lambda q^{(\lambda,\nu)}$, where ν is an element of the weight lattice, and $c\lambda$ are complex numbers such that

- $c_0 = 1$,
- $c_\lambda c_\mu = c_{\lambda+\mu}$, for all weights λ and μ,
- $c_{2\alpha_i} = 1$ for all i.

Of special interest are highest weight representations, and the corresponding highest weight modules. A highest weight module is a module generated by a weight vector v, subject to $k\lambda \cdot v = d\lambda v$ for all weights μ, and $ei \cdot v = 0$ for all i. Similarly, a quantum group can have a lowest weight representation and lowest weight module, *i.e.* a module generated by a weight vector v, subject to $k\lambda \cdot v = d\lambda v$ for all weights λ, and $fi \cdot v = 0$ for all i.

Define a vector v to have weight ν if $k_\lambda.v = q^{(\lambda,\nu)}v$ for all λ in the weight lattice.

If G is a Kac–Moody algebra, then in any irreducible highest weight representation of $Uq(G)$, with highest weight ν, the multiplicities of the weights are equal to their multiplicities in an irreducible representation of $U(G)$ with equal highest weight. If the highest weight is dominant and integral (a weight μ is dominant and integral if μ satisfies the condition that $2(\mu,\alpha_i)/(\alpha_i,\alpha_i)$ is a non-negative integer for all i), then the weight spectrum of the irreducible representation is invariant under the Weyl group for G, and the representation is integrable.

Conversely, if a highest weight module is integrable, then its highest weight vector v satisfies $k_\lambda.v = c_\lambda q^{(\lambda,\nu)}v$, where $c\lambda \cdot v = d\lambda v$ are complex numbers such that

- $c_0 = 1$,
- $c_\lambda c_\mu = c_{\lambda+\mu}$, for all weights λ and μ,
- $c_{2\alpha_i} = 1$ for all i,

and ν is dominant and integral.

As is the case for all Hopf algebras, the tensor product of two modules is another module. For an element x of $Uq(G)$, and for vectors v and w in the respective modules, $x \cdot (v \boxtimes w) = \Delta(x) \boxtimes (v \boxtimes w)$, so that $k_\lambda.(v \otimes w) = k_\lambda.v \otimes k_\lambda.w$, and in the case of coproduct Δ_1, $e_i.(v \otimes w) = k_i.v \otimes e_i.w + e_i.v \otimes w$ and $f_i.(v \otimes w) = v \otimes f_i.w + f_i.v \otimes k_i^{-1}.w$.

The integrable highest weight module described above is a tensor product of a one-dimensional module (on which $k\lambda = c\lambda$ for all λ, and $ei = fi = 0$ for all i) and a highest weight module generated by a nonzero vector v_0, subject to $k_\lambda.v_0 = q^{(\lambda,\nu)}v_0$ for all weights λ, and $e_i.v_0 = 0$ for all i.

In the specific case where G is a finite-dimensional Lie algebra (as a special case of a Kac–Moody algebra), then the irreducible representations with dominant integral highest weights are also finite-dimensional.

In the case of a tensor product of highest weight modules, its decomposition into submodules is the same as for the tensor product of the corresponding modules of the Kac–Moody algebra (the highest weights are the same, as are their multiplicities).

Case 2: q is a root of unity

28.2.2 Quasitriangularity

Case 1: q is not a root of unity

Strictly, the quantum group $Uq(G)$ is not quasitriangular, but it can be thought of as being "nearly quasitriangular" in that there exists an infinite formal sum which plays the role of an R-matrix. This infinite formal sum is expressible in terms of generators ei and fi, and Cartan generators $t\lambda$, where $k\lambda$ is formally identified with $q^{i\lambda}$. The infinite formal sum is the product of two factors,

$$q^{\eta \sum_j t_{\lambda_j} \otimes t_{\mu_j}}$$

and an infinite formal sum, where λj is a basis for the dual space to the Cartan subalgebra, and μj is the dual basis, and $\eta = \pm 1$.

The formal infinite sum which plays the part of the R-matrix has a well-defined action on the tensor product of two irreducible highest weight modules, and also on the tensor product of two lowest weight modules. Specifically, if v has weight α and w has weight β, then

$$q^{\eta \sum_j t_{\lambda_j} \otimes t_{\mu_j}}.(v \otimes w) = q^{\eta(\alpha,\beta)}v \otimes w$$

and the fact that the modules are both highest weight modules or both lowest weight modules reduces the action of the other factor on $v \otimes W$ to a finite sum.

Specifically, if V is a highest weight module, then the formal infinite sum, R, has a well-defined, and invertible, action on $V \otimes V$, and this value of R (as an element of $\text{End}(V \otimes V)$) satisfies the Yang-Baxter equation, and therefore allows us to determine a representation of the braid group, and to define quasi-invariants for knots, links and braids.

Case 2: q is a root of unity

28.2.3   Quantum groups at $q = 0$

Main article: Crystal base

Masaki Kashiwara has researched the limiting behaviour of quantum groups as $q \to 0$, and found a particularly well behaved base called a crystal base.

28.2.4   Description and classification by root-systems and Dynkin diagrams

There has been considerable progress in describing finite quotients of quantum groups such as the above $Uq(\mathbf{g})$ for $q^n = 1$; one usually considers the class of **pointed** Hopf algebras, meaning that all subcoideals are 1-dimensional and thus there sum form a group called **coradical**:

- In 2002 H.-J. Schneider and N. Andruskiewitsch [3] finished their long-term classification effort of pointed Hopf algebras with coradical an abelian group (excluding primes 2, 3, 5, 7), especially as the above finite quotients of $Uq(\mathbf{g})$ Just like ordinary Semisimple Lie algebra they decompose into E's (Borel part), dual F's and K's (Cartan algebra):

$$(\mathfrak{B}(V) \otimes k[\mathbf{Z}^n] \otimes \mathfrak{B}(V^*))^\sigma$$

  Here, as in the classical theory V is a braided vector space of dimension n spanned by the E's, and σ (a so-called cocylce twist) creates the nontrivial **linking** between E's and F's. Note that in contrast to classical theory, more than two linked components may appear. The role of the **quantum Borel algebra** is taken by a Nichols algebra $\mathfrak{B}(V)$ of the braided vectorspace.

- A crucial ingredient was hence the classification of finite Nichols algebras for **abelian groups** by I. Heckenberger [4] in terms of generalized Dynkin diagrams. When small primes are present, some exotic examples, such as a triangle, occur (see also the Figure of a rank 3 Dankin diagram).

- In the meanwhile, Schneider and Heckenberger[5] have generally proven the existence of an **arithmetic** root system also in then nonabelian case, generating a PBW basis as proven by Kharcheko in the abelian case (without the assumption on finite dimension).This could recently be used[6] on the specific cases $Uq(\mathbf{g})$ and explains e.g. the numerical coincidence between certain coideal subalgebras of these quantum groups to the order of the Weyl group of the Lie algebra $\mathbf{g}$.

28.3   Compact matrix quantum groups

See also compact quantum group.

S.L. Woronowicz introduced compact matrix quantum groups. Compact matrix quantum groups are abstract structures on which the "continuous functions" on the structure are given by elements of a C*-algebra. The geometry of a compact matrix quantum group is a special case of a noncommutative geometry.

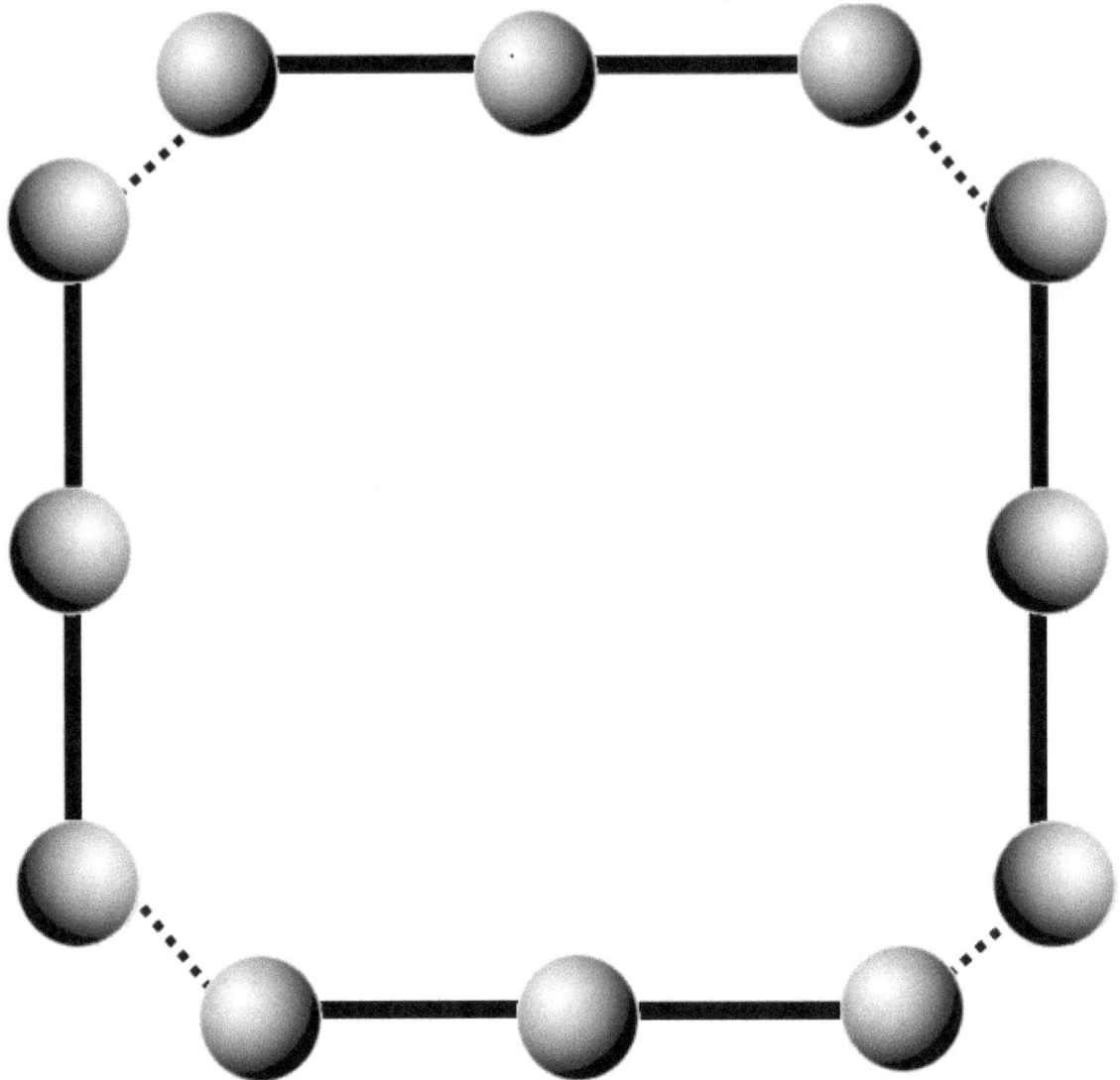

generalized Dynkin diagram for a pointed Hopf algebra linking four A3 copies

The continuous complex-valued functions on a compact Hausdorff topological space form a commutative C*-algebra. By the Gelfand theorem, a commutative C*-algebra is isomorphic to the C*-algebra of continuous complex-valued functions on a compact Hausdorff topological space, and the topological space is uniquely determined by the C*-algebra up to homeomorphism.

For a compact topological group, G, there exists a C*-algebra homomorphism Δ: $C(G) \to C(G) \otimes C(G)$ (where $C(G) \otimes C(G)$ is the C*-algebra tensor product - the completion of the algebraic tensor product of $C(G)$ and $C(G)$), such that $\Delta(f)(x, y) = f(xy)$ for all $f \in C(G)$, and for all $x, y \in G$ (where $(f \otimes g)(x, y) = f(x)g(y)$ for all $f, g \in C(G)$ and all $x, y \in G$). There also exists a linear multiplicative mapping κ: $C(G) \to C(G)$, such that $\kappa(f)(x) = f(x^{-1})$ for all $f \in C(G)$ and all $x \in G$. Strictly, this does not make $C(G)$ a Hopf algebra, unless G is finite. On the other hand, a finite-dimensional representation of G can be used to generate a *-subalgebra of $C(G)$ which is also a Hopf *-algebra. Specifically, if $g \mapsto (u_{ij}(g))_{i,j}$ is an n-dimensional representation of G, then for all i, j $uij \in C(G)$ and

$$\Delta(u_{ij}) = \sum_k u_{ik} \otimes u_{kj}.$$

It follows that the *-algebra generated by uij for all i, j and $\kappa(uij)$ for all i, j is a Hopf *-algebra: the counit is determined

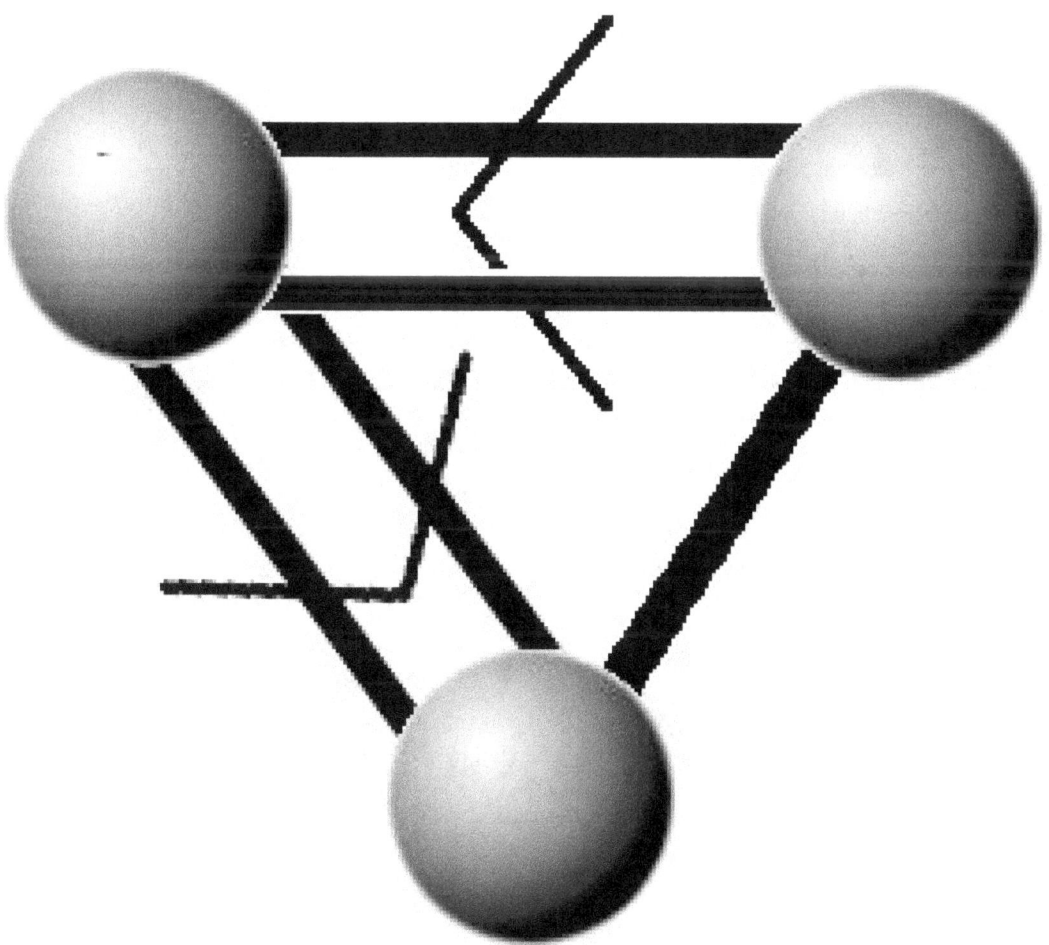

A rank 3 Dynkin diagram associated to a finite-dimensional Nichols algebra

by ε(*uij*) = δ*ij* for all *i*, *j* (where δ*ij* is the Kronecker delta), the antipode is κ, and the unit is given by

$$1 = \sum_k u_{1k}\kappa(u_{k1}) = \sum_k \kappa(u_{1k})u_{k1}.$$

As a generalization, a compact matrix quantum group is defined as a pair *(C, fu)*, where *C* is a C*-algebra and $u = (u_{ij})_{i,j=1,\dots,n}$ is a matrix with entries in *C* such that

- The *-subalgebra, C_0, of *C*, which is generated by the matrix elements of *u*, is dense in *C*;

- There exists a C*-algebra homomorphism called the comultiplication Δ: $C \to C \otimes C$ (where $C \text{ ⃝ } C$ is the C*-algebra tensor product - the completion of the algebraic tensor product of *C* and *C*) such that for all *i*, *j* we have:

$$\Delta(u_{ij}) = \sum_k u_{ik} \otimes u_{kj}$$

- There exists a linear antimultiplicative map κ: $C_0 \to C_0$ (the coinverse) such that κ(κ(*v**)*) = *v* for all $v \in C_0$ and

$$\sum_k \kappa(u_{ik})u_{kj} = \sum_k u_{ik}\kappa(u_{kj}) = \delta_{ij}I,$$

where I is the identity element of C. Since κ is antimultiplicative, then $\kappa(vw) = \kappa(w)\,\kappa(v)$ for all v, w in C_0.

As a consequence of continuity, the comultiplication on C is coassociative.

In general, C is not a bialgebra, and C_0 is a Hopf *-algebra.

Informally, C can be regarded as the *-algebra of continuous complex-valued functions over the compact matrix quantum group, and u can be regarded as a finite-dimensional representation of the compact matrix quantum group.

A representation of the compact matrix quantum group is given by a corepresentation of the Hopf *-algebra (a corepresentation of a counital coassociative coalgebra A is a square matrix $v = (v_{ij})_{i,j=1,\ldots,n}$ with entries in A (so v belongs to $M(n, A)$)) such that

$$\Delta(v_{ij}) = \sum_{k=1}^{n} v_{ik} \otimes v_{kj}$$

for all i, j and $\varepsilon(vij) = \delta ij$ for all i, j). Furthermore, a representation v, is called unitary if the matrix for v is unitary (or equivalently, if $\kappa(vij) = v*ij$ for all i, j).

An example of a compact matrix quantum group is SUμ(2), where the parameter μ is a positive real number. So SUμ(2) = (C(SUμ(2)), u), where C(SUμ(2)) is the C*-algebra generated by α and γ, subject to

$$\gamma\gamma^* = \gamma^*\gamma,$$

$$\alpha\gamma = \mu\gamma\alpha,$$

$$\alpha\gamma^* = \mu\gamma^*\alpha,$$

$$\alpha\alpha^* + \mu\gamma^*\gamma = \alpha^*\alpha + \mu^{-1}\gamma^*\gamma = I,$$

and

$$u = \begin{pmatrix} \alpha & \gamma \\ -\gamma^* & \alpha^* \end{pmatrix},$$

so that the comultiplication is determined by $\Delta(\alpha) = \alpha \otimes \alpha - \gamma \otimes \gamma^*$, $\Delta(\gamma) = \alpha \otimes \gamma + \gamma \otimes \alpha^*$, and the coinverse is determined by $\kappa(\alpha) = \alpha^*$, $\kappa(\gamma) = -\mu^{-1}\gamma$, $\kappa(\gamma^*) = -\mu\gamma^*$, $\kappa(\alpha^*) = \alpha$. Note that u is a representation, but not a unitary representation. u is equivalent to the unitary representation

$$v = \begin{pmatrix} \alpha & \sqrt{\mu}\gamma \\ -\frac{1}{\sqrt{\mu}}\gamma^* & \alpha^* \end{pmatrix}.$$

Equivalently, SUμ(2) = (C(SUμ(2)), w), where C(SUμ(2)) is the C*-algebra generated by α and β, subject to

$$\beta\beta^* = \beta^*\beta,$$

$$\alpha\beta = \mu\beta\alpha,$$

$$\alpha\beta^* = \mu\beta^*\alpha,$$

$$\alpha\alpha^* + \mu^2\beta^*\beta = \alpha^*\alpha + \beta^*\beta = I,$$

and

$$w = \begin{pmatrix} \alpha & \mu\beta \\ -\beta^* & \alpha^* \end{pmatrix},$$

so that the comultiplication is determined by $\Delta(\alpha) = \alpha \otimes \alpha - \mu\beta \otimes \beta^*$, $\Delta(\beta) = \alpha \otimes \beta + \beta \otimes \alpha^*$, and the coinverse is determined by $\kappa(\alpha) = \alpha^*$, $\kappa(\beta) = -\mu^{-1}\beta$, $\kappa(\beta^*) = -\mu\beta^*$, $\kappa(\alpha^*) = \alpha$. Note that w is a unitary representation. The realizations can be identified by equating $\gamma = \sqrt{\mu}\beta$.

When $\mu = 1$, then SUμ(2) is equal to the algebra $C(\mathrm{SU}(2))$ of functions on the concrete compact group SU(2).

28.4 Bicrossproduct quantum groups

Whereas compact matrix pseudogroups are typically versions of Drinfeld-Jimbo quantum groups in a dual function algebra formulation, with additional structure, the bicrossproduct ones are a distinct second family of quantum groups of increasing importance as deformations of solvable rather than semisimple Lie groups. They are associated to Lie splittings of Lie algebras or local factorisations of Lie groups and can be viewed as the cross product or Mackey quantisation of one of the factors acting on the other for the algebra and a similar story for the coproduct Δ with the second factor acting back on the first. The very simplest nontrivial example corresponds to two copies of **R** locally acting on each other and results in a quantum group (given here in an algebraic form) with generators p, K, K^{-1}, say, and coproduct

$$[p, K] = hK(K - 1)$$

$$\Delta p = p \otimes K + 1 \otimes p$$

$$\Delta K = K \otimes K$$

where h is the deformation parameter. This quantum group was linked to a toy model of Planck scale physics implementing Born reciprocity when viewed as a deformation of the Heisenberg algebra of quantum mechanics. Also, starting with any compact real form of a semisimple Lie algebra **g** its complexification as a real Lie algebra of twice the dimension splits into **g** and a certain solvable Lie algebra (the Iwasawa decomposition), and this provides a canonical bicrossproduct quantum group associated to **g**. For **su**(2) one obtains a quantum group deformation of the Euclidean group E(3) of motions in 3 dimensions.

28.5 See also

- Lie bialgebra
- Poisson–Lie group
- Affine quantum group
- Quantum affine algebras

28.6 Notes

[1] Schwiebert, Christian (1994), *Generalized quantum inverse scattering*, p. 12237, arXiv:hep-th/9412237v3, Bibcode:199437S

[2] Majid, Shahn (1988), "Hopf algebras for physics at the Planck scale", *Classical and Quantum Gravity* **5** (12): 1587–1607, Bibcode:1988CQGra...5.1587M, doi:10.1088/0264-9381/5/12/010

[3] Andruskiewitsch, Schneider: Pointed Hopf algebras, New directions in Hopf algebras, 1–68, Math. Sci. Res. Inst. Publ., 43, Cambridge Univ. Press, Cambridge, 2002.

[4] Heckenberger: Nichols algebras of diagonal type and arithmetic root systems, Habilitation thesis 2005.

[5] Heckenberger, Schneider: Root system and Weyl gruppoid for Nichols algebras, 2008.

[6] Heckenberger, Schneider: Right coideal subalgebras of Nichols algebras and the Duflo order of the Weyl grupoid, 2009.

28.7 References

- Jagannathan, R. (2001). "Some introductory notes on quantum groups, quantum algebras, and their applications". arXiv:math-ph/0105002v1.

- Kassel, Christian (1995), *Quantum groups*, Graduate Texts in Mathematics **155**, Berlin, New York: Springer-Verlag, ISBN 978-0-387-94370-1, MR 1321145

- Lusztig, George (2010) [1993]. *Introduction to Quantum Groups*. Cambridge, MA: Birkhäuser. ISBN 978-0-817-64716-2.

- Majid, Shahn (2002), *A quantum groups primer*, London Mathematical Society Lecture Note Series **292**, Cambridge University Press, ISBN 978-0-521-01041-2, MR 1904789

- Majid, Shahn (January 2006), "What Is...a Quantum Group?" (PDF), *Notices of the American Mathematical Society* **53** (1): 30–31, retrieved 2008-01-16

- Podles, P.; Muller, E. (1997), *Introduction to quantum groups*, p. 4002, arXiv:q-alg/9704002, Bibcode:14002P

- Shnider, Steven; Sternberg, Shlomo (1993). *Quantum groups: From coalgebras to Drinfeld algebras*. Graduate Texts in Mathematical Physics **2**. Cambridge, MA: International Press.

- Street, Ross (2007), *Quantum groups*, Australian Mathematical Society Lecture Series **19**, Cambridge University Press, ISBN 978-0-521-69524-4, MR 2294803

Chapter 29

Loop quantum gravity

Loop quantum gravity (**LQG**) is a theory that attempts to describe the quantum properties of the universe and gravity. It is also a theory of quantum space and quantum time because, according to general relativity, the geometry of spacetime is a manifestation of gravity. LQG is an attempt to merge and adapt standard quantum mechanics and standard general relativity. The main output of the theory is a physical picture of space where space is granular. The granularity is a direct consequence of the quantization. It has the same nature as the granularity of the photons in the quantum theory of electromagnetism or the discrete levels of the energy of the atoms. Here, it is space itself that is discrete. In other words, there is a minimum distance possible to travel through it.

More precisely, space can be viewed as an extremely fine fabric or network "woven" of finite loops. These networks of loops are called spin networks. The evolution of a spin network over time is called a spin foam. The predicted size of this structure is the Planck length, which is approximately 10^{-35} meters. According to the theory, there is no meaning to distance at scales smaller than the Planck scale. Therefore, LQG predicts that not just matter, but also space itself has an atomic structure.

Today LQG is a vast area of research, developing in several directions, which involves about 30 research groups worldwide. They all share the basic physical assumptions and the mathematical description of quantum space. The full development of the theory is being pursued in two directions: the more traditional canonical loop quantum gravity, and the newer covariant loop quantum gravity, more commonly called spin foam theory.

Research into the physical consequences of the theory is proceeding in several directions. Among these, the most well-developed is the application of LQG to cosmology, called loop quantum cosmology (LQC). LQC applies LQG ideas to the study of the early universe and the physics of the Big Bang. Its most spectacular consequence is that the evolution of the universe can be continued beyond the Big Bang. The Big Bang appears thus to be replaced by a sort of cosmic Big Bounce.

29.1   History

Main article: History of loop quantum gravity

In 1986, Abhay Ashtekar reformulated Einstein's general relativity in a language closer to that of the rest of fundamental physics. Shortly after, Ted Jacobson and Lee Smolin realized that the formal equation of quantum gravity, called the Wheeler–DeWitt equation, admitted solutions labelled by loops, when rewritten in the new Ashtekar variables, and Carlo Rovelli and Lee Smolin defined a nonperturbative and background-independent quantum theory of gravity in terms of these loop solutions. Jorge Pullin and Jerzy Lewandowski understood that the intersections of the loops are essential for the consistency of the theory, and the theory should be formulated in terms of intersecting loops, or graphs.

In 1994, Rovelli and Smolin showed that the quantum operators of the theory associated to area and volume have a discrete spectrum. That is, **geometry is quantized**. This result defines an explicit basis of states of quantum geometry, which turned out to be labelled by Roger Penrose's spin networks, which are graphs labelled by spins.

The canonical version of the dynamics was put on firm ground by Thomas Thiemann, who defined an anomaly-free Hamiltonian operator, showing the existence of a mathematically consistent background-independent theory. The covariant or spinfoam version of the dynamics developed during several decades, and crystallized in 2008, from the joint work of research groups in France, Canada, UK, Poland, and Germany, lead to the definition of a family of transition amplitudes, which in the classical limit can be shown to be related to a family of truncations of general relativity.[2] The finiteness of these amplitudes was proven in 2011.[3][4] It requires the existence of a positive cosmological constant, and this is consistent with observed acceleration in the expansion of the Universe.

29.2 General covariance and background independence

Main articles: General covariance, background-independent and diffeomorphism

In theoretical physics, general covariance is the invariance of the form of physical laws under arbitrary differentiable coordinate transformations. The essential idea is that coordinates are only artifices used in describing nature, and hence should play no role in the formulation of fundamental physical laws. A more significant requirement is the principle of general relativity that states that the laws of physics take the same form in all reference systems. This is a generalization of the principle of special relativity which states that the laws of physics take the same form in all inertial frames.

In mathematics, a diffeomorphism is an isomorphism in the category of smooth manifolds. It is an invertible function that maps one differentiable manifold to another, such that both the function and its inverse are smooth. These are the defining symmetry transformations of General Relativity since the theory is formulated only in terms of a differentiable manifold.

In general relativity, general covariance is intimately related to "diffeomorphism invariance". This symmetry is one of the defining features of the theory. However, it is a common misunderstanding that "diffeomorphism invariance" refers to the invariance of the physical predictions of a theory under arbitrary coordinate transformations; this is untrue and in fact every physical theory is invariant under coordinate transformations this way. Diffeomorphisms, as mathematicians define them, correspond to something much more radical; intuitively a way they can be envisaged is as simultaneously dragging all the physical fields (including the gravitational field) over the bare differentiable manifold while staying in the same coordinate system. Diffeomorphisms are the true symmetry transformations of general relativity, and come about from the assertion that the formulation of the theory is based on a bare differentiable manifold, but not on any prior geometry — the theory is background-independent (this is a profound shift, as all physical theories before general relativity had as part of their formulation a prior geometry). What is preserved under such transformations are the coincidences between the values the gravitational field take at such and such a "place" and the values the matter fields take there. From these relationships one can form a notion of matter being located with respect to the gravitational field, or vice versa. This is what Einstein discovered: that physical entities are located with respect to one another only and not with respect to the spacetime manifold. As Carlo Rovelli puts it: "No more fields on spacetime: just fields on fields.".[5] This is the true meaning of the saying "The stage disappears and becomes one of the actors"; space-time as a "container" over which physics takes place has no objective physical meaning and instead the gravitational interaction is represented as just one of the fields forming the world. This is known as the relationalist interpretation of space-time. The realization by Einstein that general relativity should be interpreted this way is the origin of his remark "Beyond my wildest expectations".

In LQG this aspect of general relativity is taken seriously and this symmetry is preserved by requiring that the physical states remain invariant under the generators of diffeomorphisms. The interpretation of this condition is well understood for purely spatial diffeomorphisms. However, the understanding of diffeomorphisms involving time (the Hamiltonian constraint) is more subtle because it is related to dynamics and the so-called "problem of time" in general relativity.[6] A generally accepted calculational framework to account for this constraint has yet to be found.[7][8] A plausible candidate for the quantum hamiltonian constraint is the operator introduced by Thiemann.[9]

LQG is formally background independent. The equations of LQG are not embedded in, or dependent on, space and time (except for its invariant topology). Instead, they are expected to give rise to space and time at distances which are large compared to the Planck length. The issue of background independence in LQG still has some unresolved subtleties. For example, some derivations require a fixed choice of the topology, while any consistent quantum theory of gravity should include topology change as a dynamical process.

29.3   Constraints and their Poisson bracket algebra

Main articles: Poisson bracket and Hamiltonian constraint

29.3.1   The constraints of classical canonical general relativity

Main article: Lie derivative

In the Hamiltonian formulation of ordinary classical mechanics the Poisson bracket is an important concept. A "canonical coordinate system" consists of canonical position and momentum variables that satisfy canonical Poisson-bracket relations,

$$\{q_i, p_j\} = \delta_{ij}$$

where the Poisson bracket is given by

$$\{f, g\} = \sum_{i=1}^{N} \left(\frac{\partial f}{\partial q_i} \frac{\partial g}{\partial p_i} - \frac{\partial f}{\partial p_i} \frac{\partial g}{\partial q_i} \right).$$

for arbitrary phase space functions $f(q_i, p_j)$ and $g(q_i, p_j)$. With the use of Poisson brackets, the Hamilton's equations can be rewritten as,

$$\dot{q}_i = \{q_i, H\},$$

$$\dot{p}_i = \{p_i, H\}.$$

These equations describe a "flow" or orbit in phase space generated by the Hamiltonian H. Given any phase space function $F(q, p)$, we have

$$\frac{d}{dt} F(q_i, p_i) = \{F, H\}.$$

Let us consider constrained systems, of which General relativity is an example. In a similar way the Poisson bracket between a constraint and the phase space variables generates a flow along an orbit in (the unconstrained) phase space generated by the constraint. There are three types of constraints in Ashtekar's reformulation of classical general relativity:

$SU(2)$ **Gauss gauge constraints**

The Gauss constraints

$$G_j(x) = 0.$$

This represents an infinite number of constraints one for each value of x. These come about from re-expressing General relativity as an SU(2) Yang–Mills type gauge theory (Yang–Mills is a generalization of Maxwell's theory where the gauge field transforms as a vector under Gauss transformations, that is, the Gauge field is of the form $A_a^i(x)$ where i is an internal index. See Ashtekar variables). These infinite number of Gauss gauge constraints can be smeared with test fields with internal indices, $\lambda^j(x)$,

$$G(\lambda) = \int d^3x G_j(x) \lambda^j(x).$$

which we demand vanish for any such function. These smeared constraints defined with respect to a suitable space of smearing functions give an equivalent description to the original constraints.

In fact Ashtekar's formulation may be thought of as ordinary SU(2) Yang–Mills theory together with the following special constraints, resulting from diffeomorphism invariance, and a Hamiltonian that vanishes. The dynamics of such a theory are thus very different from that of ordinary Yang–Mills theory.

Spatial diffeomorphisms constraints

The spatial diffeomorphism constraints

$$C_a(x) = 0$$

can be smeared by the so-called shift functions $\vec{N}(x)$ to give an equivalent set of smeared spatial diffeomorphism constraints,

$$C(\vec{N}) = \int d^3x C_a(x) N^a(x) .$$

These generate spatial diffeomorphisms along orbits defined by the shift function $N^a(x)$.

Hamiltonian constraints

The Hamiltonian

$$H(x) = 0$$

can be smeared by the so-called lapse functions $N(x)$ to give an equivalent set of smeared Hamiltonian constraints,

$$H(N) = \int d^3x H(x) N(x) .$$

These generate time diffeomorphisms along orbits defined by the lapse function $N(x)$.

In Ashtekar formulation the gauge field $A_a^i(x)$ is the configuration variable (the configuration variable being analogous to q in ordinary mechanics) and its conjugate momentum is the (densitized) triad (electrical field) $\tilde{E}_i^a(x)$. The constraints are certain functions of these phase space variables.

We consider the action of the constraints on arbitrary phase space functions. An important notion here is the Lie derivative, $\mathcal{L}_V$, which is basically a derivative operation that infinitesimally "shifts" functions along some orbit with tangent vector V .

29.3.2 The Poisson bracket algebra

Of particular importance is the Poisson bracket algebra formed between the (smeared) constraints themselves as it completely determines the theory. In terms of the above smeared constraints the constraint algebra amongst the Gauss' law reads,

$$\{G(\lambda), G(\mu)\} = G([\lambda, \mu])$$

where $[\lambda, \mu]^k = \lambda_i \mu_j \epsilon^{ijk}$. And so we see that the Poisson bracket of two Gauss' law is equivalent to a single Gauss' law evaluated on the commutator of the smearings. The Poisson bracket amongst spatial diffeomorphisms constraints reads

$$\{C(\vec{N}), C(\vec{M})\} = C(\mathcal{L}_{\vec{N}} \vec{M})$$

and we see that its effect is to "shift the smearing". The reason for this is that the smearing functions are not functions of the canonical variables and so the spatial diffeomorphism does not generate diffeomorphims on them. They do however generate diffeomorphims on everything else. This is equivalent to leaving everything else fixed while shifting the smearing .The action of the spatial diffeomorphism on the Gauss law is

$$\{C(\vec{N}), G(\lambda)\} = G(\mathcal{L}_{\vec{N}} \lambda) ,$$

again, it shifts the test field λ . The Gauss law has vanishing Poisson bracket with the Hamiltonian constraint. The spatial diffeomorphism constraint with a Hamiltonian gives a Hamiltonian with its smearing shifted,

$$\{C(\vec{N}), H(M)\} = H(\mathcal{L}_{\vec{N}} M) .$$

Finally, the poisson bracket of two Hamiltonians is a spatial diffeomorphism,

$$\{H(N), H(M)\} = C(K)$$

where K is some phase space function. That is, it is a sum over infinitesimal spatial diffeomorphisms constraints where the coefficients of proportionality are not constants but have non-trivial phase space dependence.

A (Poisson bracket) Lie algebra, with constraints C_I , is of the form

$$\{C_I, C_J\} = f_{IJ}^K C_K$$

where f_{IJ}^K are constants (the so-called structure constants). The above Poisson bracket algebra for General relativity does not form a true Lie algebra as we have structure functions rather than structure constants for the Poisson bracket between two Hamiltonians. This leads to difficulties.

29.3.3   Dirac observables

The constraints define a constraint surface in the original phase space. The gauge motions of the constraints apply to all phase space but have the feature that they leave the constraint surface where it is, and thus the orbit of a point in the hypersurface under gauge transformations will be an orbit entirely within it. Dirac observables are defined as phase space functions, O , that Poisson commute with all the constraints when the constraint equations are imposed,

$$\{G_j, O\}_{G_j = C_a = H = 0} = \{C_a, O\}_{G_j = C_a = H = 0} = \{H, O\}_{G_j = C_a = H = 0} = 0 \; ,$$

that is, they are quantities defined on the constraint surface that are invariant under the gauge transformations of the theory.

Then, solving only the constraint $G_j = 0$ and determining the Dirac observables with respect to it leads us back to the ADM phase space with constraints H, C_a . The dynamics of general relativity is generated by the constraints, it can be shown that six Einstein equations describing time evolution (really a gauge transformation) can be obtained by calculating the Poisson brackets of the three-metric and its conjugate momentum with a linear combination of the spatial diffeomorphism and Hamiltonian constraint. The vanishing of the constraints, giving the physical phase space, are the four other Einstein equations.[10]

29.4   Quantization of the constraints – the equations of quantum general relativity

29.4.1   Pre-history and Ashtekar new variables

Main articles: Frame fields in general relativity, Ashtekar variables and Self-dual Palatini action

Many of the technical problems in canonical quantum gravity revolve around the constraints. Canonical general relativity was originally formulated in terms of metric variables, but there seemed to be insurmountable mathematical difficulties in promoting the constraints to quantum operators because of their highly non-linear dependence on the canonical variables. The equations were much simplified with the introduction of Ashtekars new variables. Ashtekar variables describe canonical general relativity in terms of a new pair canonical variables closer to that of gauge theories. The first step consists of using densitized triads $\tilde{E}_i^a$ (a triad E_i^a is simply three orthogonal vector fields labeled by $i = 1, 2, 3$ and the densitized triad is defined by $\tilde{E}_i^a = \sqrt{\det(q)} E_i^a$) to encode information about the spatial metric,

$$\det(q) q^{ab} = \tilde{E}_i^a \tilde{E}_j^b \delta^{ij} \; .$$

(where δ^{ij} is the flat space metric, and the above equation expresses that q^{ab} , when written in terms of the basis E_i^a , is locally flat). (Formulating general relativity with triads instead of metrics was not new.) The densitized triads are not unique, and in fact one can perform a local in space rotation with respect to the internal indices i . The canonically conjugate variable is related to the extrinsic curvature by $K_a^i - K_{ab} \tilde{E}^{ai} / \sqrt{\det(q)}$. But problems similar to using the metric formulation arise when one tries to quantize the theory. Ashtekar's new insight was to introduce a new configuration variable,

$$A_a^i = \Gamma_a^i - i K_a^i$$

that behaves as a complex SU(2) connection where Γ_a^i is related to the so-called spin connection via $\Gamma_a^i = \Gamma_{ajk} \epsilon^{jki}$. Here A_a^i is called the chiral spin connection. It defines a covariant derivative $\mathcal{D}_a$. It turns out that $\tilde{E}_i^a$ is the conjugate momentum of A_a^i , and together these form Ashtekar's new variables.

The expressions for the constraints in Ashtekar variables; the Gauss's law, the spatial diffeomorphism constraint and the (densitized) Hamiltonian constraint then read:

$$G^i = \mathcal{D}_a \tilde{E}^a_i = 0$$

$$C_a = \tilde{E}^b_i F^i_{ab} - A^i_a (\mathcal{D}_b \tilde{E}^b_i) = V_a - A^i_a G^i = 0 \,,$$

$$\tilde{H} = \epsilon_{ijk} \tilde{E}^a_i \tilde{E}^b_j F^i_{ab} = 0$$

respectively, where F^i_{ab} is the field strength tensor of the connection A^i_a and where V_a is referred to as the vector constraint. The above-mentioned local in space rotational invariance is the original of the SU(2) gauge invariance here expressed by the Gauss law. Note that these constraints are polynomial in the fundamental variables, unlike as with the constraints in the metric formulation. This dramatic simplification seemed to open up the way to quantizing the constraints. (See the article Self-dual Palatini action for a derivation of Ashtekar's formulism).

With Ashtekar's new variables, given the configuration variable A^i_a , it is natural to consider wavefunctions $\Psi(A^i_a)$. This is the connection representation. It is analogous to ordinary quantum mechanics with configuration variable q and wavefunctions $\psi(q)$. The configuration variable gets promoted to a quantum operator via:

$$\hat{A}^i_a \Psi(A) = A^i_a \Psi(A) \,,$$

(analogous to $\hat{q}\psi(q) = q\psi(q)$) and the triads are (functional) derivatives,

$$\hat{\tilde{E}}^a_i \Psi(A) = -i \frac{\delta \Psi(A)}{\delta A^i_a} \,.$$

(analogous to $\hat{p}\psi(q) = -i\hbar d\psi(q)/dq$). In passing over to the quantum theory the constraints become operators on a kinematic Hilbert space (the unconstrained SU(2) Yang–Mills Hilbert space). Note that different ordering of the A 's and $\tilde{E}$'s when replacing the $\tilde{E}$'s with derivatives give rise to different operators - the choice made is called the factor ordering and should be chosen via physical reasoning. Formally they read

$$\hat{G}_j |\psi\rangle = 0$$

$$\hat{C}_a |\psi\rangle = 0$$

$$\hat{\tilde{H}} |\psi\rangle = 0 \,.$$

There are still problems in properly defining all these equations and solving them. For example the Hamiltonian constraint Ashtekar worked with was the densitized version instead of the original Hamiltonian, that is, he worked with $\tilde{H} = \sqrt{\det(q)} H$. There were serious difficulties in promoting this quantity to a quantum operator. Moreover, although Ashtekar variables had the virtue of simplifying the Hamiltonian, they are complex. When one quantizes the theory, it is difficult to ensure that one recovers real general relativity as opposed to complex general relativity.

29.4.2 Quantum constraints as the equations of quantum general relativity

We now move on to demonstrate an important aspect of the quantum constraints. We consider Gauss' law only. First we state the classical result that the Poisson bracket of the smeared Gauss' law $G(\lambda) = \int d^3x \lambda^j (D_a E^a)^j$ with the connections is

$$\{G(\lambda), A^i_a\} = \partial_a \lambda^i + g \epsilon^{ijk} A^j_a \lambda^k = (D_a \lambda)^i.$$

The quantum Gauss' law reads

$$\hat{G}_j \Psi(A) = -i D_a \frac{\delta \lambda \Psi[A]}{\delta A^j_a} = 0.$$

If one smears the quantum Gauss' law and study its action on the quantum state one finds that the action of the constraint on the quantum state is equivalent to shifting the argument of Ψ by an infinitesimal (in the sense of the parameter λ small) gauge transformation,

$$\left[1 + \int d^3x \lambda^j(x) \hat{G}_j\right] \Psi(A) = \Psi[A + D\lambda] = \Psi[A],$$

and the last identity comes from the fact that the constraint annihilates the state. So the constraint, as a quantum operator, is imposing the same symmetry that its vanishing imposed classically: it is telling us that the functions $\Psi[A]$ have to be gauge invariant functions of the connection. The same idea is true for the other constraints.

Therefore the two step process in the classical theory of solving the constraints $C_I = 0$ (equivalent to solving the admissibility conditions for the initial data) and looking for the gauge orbits (solving the `evolution' equations) is replaced by a one step process in the quantum theory, namely looking for solutions Ψ of the quantum equations $\hat{C}_I \Psi = 0$. This is because it obviously solves the constraint at the quantum level and it simultaneously looks for states that are gauge invariant because $\hat{C}_I$ is the quantum generator of gauge transformations (gauge invariant functions are constant along the gauge orbits and thus characterize them).[11] Recall that, at the classical level, solving the admissibility conditions and evolution equations was equivalent to solving all of Einstein's field equations, this underlines the central role of the quantum constraint equations in canonical quantum gravity.

29.4.3 Introduction of the loop representation

Main articles: Holonomy, Wilson loop and Knot invariant

It was in particular the inability to have good control over the space of solutions to the Gauss' law and spacial diffeomorphism constraints that led Rovelli and Smolin to consider a new representation - the loop representation in gauge theories and quantum gravity.[12]

We need the notion of a holonomy. A holonomy is a measure of how much the initial and final values of a spinor or vector differ after parallel transport around a closed loop; it is denoted

$h_\gamma[A]\lambda^p hi$.

Knowledge of the holonomies is equivalent to knowledge of the connection, up to gauge equivalence. Holonomies can also be associated with an edge; under a Gauss Law these transform as

$(h'_e)_{\alpha\beta} = U^{-1}_{\alpha\gamma}(x)(h_e)_{\gamma\sigma}U_{\sigma\beta}(y)$.

For a closed loop $x = y$ if we take the trace of this, that is, putting $\alpha = \beta$ and summing we obtain

$(h'_e)_{\alpha\alpha} = U^{-1}_{\alpha\gamma}(x)(h_e)_{\gamma\sigma}U_{\sigma\alpha}(x) = [U_{\sigma\alpha}(x)U^{-1}_{\alpha\gamma}(x)](h_e)_{\gamma\sigma} = \delta_{\sigma\gamma}(h_e)_{\gamma\sigma} = (h_e)_{\gamma\gamma}$

or

$\text{Tr}\, h'_\gamma = \text{Tr}\, h_\gamma..$

The trace of an holonomy around a closed loop is written

$W_\gamma[A]$

and is called a Wilson loop. Thus Wilson loops are gauge invariant. The explicit form of the Holonomy is

$h_\gamma[A] = \mathcal{P} \exp \left\{ - \int_{\gamma_0}^{\gamma_1} ds \dot{\gamma}^a A^i_a(\gamma(s)) T_i \right\}$

where γ is the curve along which the holonomy is evaluated, and s is a parameter along the curve, $\mathcal{P}$ denotes path ordering meaning factors for smaller values of s appear to the left, and T_i are matrices that satisfy the SU(2) algebra

$[T^i, T^j] = 2i\epsilon^{ijk}T^k$.

The Pauli matrices satisfy the above relation. It turns out that there are infinitely many more examples of sets of matrices that satisfy these relations, where each set comprises $(N + 1) \times (N + 1)$ matrices with $N = 1, 2, 3, \dots$, and where none of these can be thought to `decompose' into two or more examples of lower dimension. They are called different irreducible representations of the SU(2) algebra. The most fundamental representation being the Pauli matrices. The holonomy is labelled by a half integer $N/2$ according to the irreducible representation used.

The use of Wilson loops explicitly solves the Gauss gauge constraint. To handle the spatial diffeomorphism constraint we need to go over to the loop representation. As Wilson loops form a basis we can formally expand any Gauss gauge invariant function as,

$\Psi[A] = \sum_\gamma \Psi[\gamma]W_\gamma[A]$.

This is called the loop transform. We can see the analogy with going to the momentum representation in quantum mechanics(see Position and momentum space). There one has a basis of states $\exp(ikx)$ labelled by a number k and one expands

$\psi[x] = \int dk \psi(k) \exp(ikx)$.

and works with the coefficients of the expansion $\psi(k)$.

The inverse loop transform is defined by

$\Psi[\gamma] = \int [dA] \Psi[A] W_\gamma[A]$.

This defines the loop representation. Given an operator $\hat{O}$ in the connection representation,

$\Phi[A] = \hat{O} \Psi[A]$ $Eq\ 1$,

one should define the corresponding operator $\hat{O}'$ on $\Psi[\gamma]$ in the loop representation via,

$\Phi[\gamma] = \hat{O}' \Psi[\gamma]$ $Eq\ 2$,

where $\Phi[\gamma]$ is defined by the usual inverse loop transform,

$\Phi[\gamma] = \int [dA] \Phi[A] W_\gamma[A]$ $Eq\ 3.$.

A transformation formula giving the action of the operator $\hat{O}'$ on $\Psi[\gamma]$ in terms of the action of the operator $\hat{O}$ on $\Psi[A]$ is then obtained by equating the R.H.S. of $Eq\ 2$ with the R.H.S. of $Eq\ 3$ with $Eq\ 1$ substituted into $Eq\ 3$, namely

$\hat{O}' \Psi[\gamma] = \int [dA] W_\gamma[A] \hat{O} \Psi[A]$,

or

$\hat{O}' \Psi[\gamma] = \int [dA] (\hat{O}^\dagger W_\gamma[A]) \Psi[A]$,

where by $\hat{O}^\dagger$ we mean the operator $\hat{O}$ but with the reverse factor ordering (remember from simple quantum mechanics where the product of operators is reversed under conjugation). We evaluate the action of this operator on the Wilson loop as a calculation in the connection representation and rearranging the result as a manipulation purely in terms of loops (one should remember that when considering the action on the Wilson loop one should choose the operator one wishes to transform with the opposite factor ordering to the one chosen for its action on wavefunctions $\Psi[A]$). This gives the physical meaning of the operator $\hat{O}'$. For example if $\hat{O}^\dagger$ corresponded to a spatial diffeomorphism, then this can be thought of as keeping the connection field A of $W_\gamma[A]$ where it is while performing a spatial diffeomorphism on γ instead. Therefore the meaning of $\hat{O}'$ is a spatial diffeomorphism on γ , the argument of $\Psi[\gamma]$.

In the loop representation we can then solve the spatial diffeomorphism constraint by considering functions of loops $\Psi[\gamma]$ that are invariant under spatial diffeomorphisms of the loop γ . That is, we construct what mathematicians call knot invariants. This opened up an unexpected connection between knot theory and quantum gravity.

What about the Hamiltonian constraint? Let us go back to the connection representation. Any collection of non-intersecting Wilson loops satisfy Ashtekar's quantum Hamiltonian constraint. This can be seen from the following. With a particular ordering of terms and replacing $\tilde{E}_i^a$ by a derivative, the action of the quantum Hamiltonian constraint on a Wilson loop is

$\hat{\tilde{H}}^\dagger W_\gamma[A] = -\epsilon_{ijk} \hat{F}_{ab}^k \frac{\delta}{\delta A_a^i} \frac{\delta}{\delta A_b^j} W_\gamma[A]$.

When a derivative is taken it brings down the tangent vector, $\dot{\gamma}^a$, of the loop, γ . So we have something like

$\hat{F}_{ab}^i \dot{\gamma}^a \dot{\gamma}^b$.

However, as F_{ab}^i is anti-symmetric in the indices a and b this vanishes (this assumes that γ is not discontinuous anywhere and so the tangent vector is unique). Now let us go back to the loop representation.

We consider wavefunctions $\Psi[\gamma]$ that vanish if the loop has discontinuities and that are knot invariants. Such functions solve the Gauss law, the spatial diffeomorphism constraint and (formally) the Hamiltonian constraint. Thus we have identified an infinite set of exact (if only formal) solutions to all the equations of quantum general relativity![12] This generated a lot of interest in the approach and eventually led to LQG.

29.4.4   Geometric operators, the need for intersecting Wilson loops and spin network states

The easiest geometric quantity is the area. Let us choose coordinates so that the surface Σ is characterized by $x^3 = 0$. The area of small parallelogram of the surface Σ is the product of length of each side times $\sin\theta$ where θ is the angle between the sides. Say one edge is given by the vector $\vec{u}$ and the other by $\vec{v}$ then,

$$A = \|\vec{u}\|\|\vec{v}\|\sin\theta = \sqrt{\|\vec{u}\|^2\|\vec{v}\|^2(1-\cos^2\theta)} \quad = \sqrt{\|\vec{u}\|^2\|\vec{v}\|^2 - (\vec{u}\cdot\vec{v})^2}$$

From this we get the area of the surface Σ to be given by

$$A_\Sigma = \int_\Sigma dx^1 dx^2 \sqrt{\det(q^{(2)})}$$

where $\det(q^{(2)}) = q_{11}q_{22} - q_{12}^2$ and is the determinant of the metric induced on Σ. This can be rewritten as

$$\det(q^{(2)}) = \frac{\epsilon^{3ab}\epsilon^{3cd}q_{ac}q_{bc}}{2} .$$

The standard formula for an inverse matrix is

$$q^{ab} = \frac{\epsilon^{acd}\epsilon^{bef}q_{ce}q_{df}}{3!\det(q)}$$

Note the similarity between this and the expression for $\det(q^{(2)})$. But in Ashtekar variables we have $\tilde{E}_i^a \tilde{E}^{bi} = \det(q)q^{ab}$. Therefore

$$A_\Sigma = \int_\Sigma dx^1 dx^2 \sqrt{\tilde{E}_i^3 \tilde{E}^{3i}} .$$

According to the rules of canonical quantization we should promote the triads $\tilde{E}_i^3$ to quantum operators,

$$\hat{\tilde{E}}_i^3 \sim \frac{\delta}{\delta A_3^i} .$$

It turns out that the area A_Σ can be promoted to a well defined quantum operator despite the fact that we are dealing with product of two functional derivatives and worse we have a square-root to contend with as well.[13] Putting $N = 2J$, we talk of being in the J-th representation. We note that $\sum_i T^i T^i = J(J+1)1$. This quantity is important in the final formula for the area spectrum. We simply state the result below,

$$\hat{A}_\Sigma W_\gamma[A] = 8\pi\ell_{\text{Planck}}^2 \beta \sum_I \sqrt{j_I(j_I+1)} W_\gamma[A]$$

where the sum is over all edges I of the Wilson loop that pierce the surface Σ.

The formula for the volume of a region R is given by

$$V = \int_R d^3x \sqrt{\det(q)} = \frac{1}{6}\int_R dx^3 \sqrt{\epsilon_{abc}\epsilon^{ijk}\tilde{E}_i^a \tilde{E}_j^b \tilde{E}_k^c} .$$

The quantization of the volume proceeds the same way as with the area. As we take the derivative, and each time we do so we bring down the tangent vector $\dot{\gamma}^a$, when the volume operator acts on non-intersecting Wilson loops the result vanishes. Quantum states with non-zero volume must therefore involve intersections. Given that the anti-symmetric summation is taken over in the formula for the volume we would need at least intersections with three non-coplanar lines. Actually it turns out that one needs at least four-valent vertices for the volume operator to be non-vanishing.

We now consider Wilson loops with intersections. We assume the real representation where the gauge group is $SU(2)$. Wilson loops are an over complete basis as there are identities relating different Wilson loops. These come about from the fact that Wilson loops are based on matrices (the holonomy) and these matrices satisfy identities. Given any two $SU(2)$ matrices $\mathbb{A}$ and $\mathbb{B}$ it is easy to check that,

$$\text{Tr}(\mathbb{A})\text{Tr}(\mathbb{B}) = \text{Tr}(\mathbb{A}\mathbb{B}) + \text{Tr}(\mathbb{A}\mathbb{B}^{-1}) .$$

This implies that given two loops γ and η that intersect, we will have,

$$W_\gamma[A]W_\eta[A] = W_{\gamma\circ\eta}[A] + W_{\gamma\circ\eta^{-1}}[A]$$

where by η^{-1} we mean the loop η traversed in the opposite direction and $\gamma\circ\eta$ means the loop obtained by going around the loop γ and then along η. See figure below. Given that the matrices are unitary one has that $W_\gamma[A] = W_{\gamma^{-1}}[A]$. Also given the cyclic property of the matrix traces (i.e. $Tr(\mathbb{A}\mathbb{B}) = Tr(\mathbb{B}\mathbb{A})$) one has that $W_{\gamma\circ\eta}[A] = W_{\eta\circ\gamma}[A]$. These identities can be combined with each other into further identities of increasing complexity adding more loops. These identities are the so-called Mandelstam identities. Spin networks certain are linear combinations of intersecting Wilson loops designed to address the over completeness introduced by the Mandelstam identities (for trivalent intersections they eliminate the over-completeness entirely) and actually constitute a basis for all gauge invariant functions.

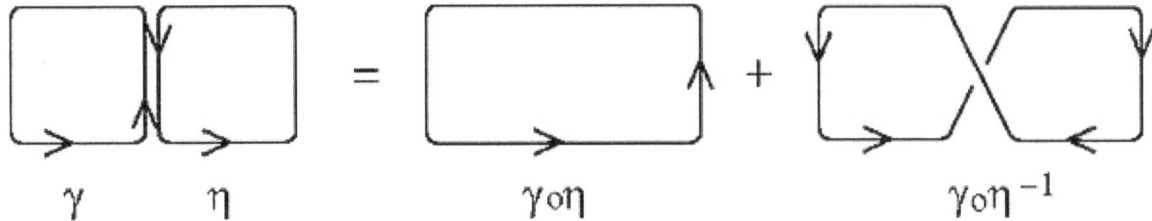

Graphical representation of the simplest non-trivial Mandestam identity relating different Wilson loops.

As mentioned above the holonomy tells you how to propagate test spin half particles. A spin network state assigns an amplitude to a set of spin half particles tracing out a path in space, merging and splitting. These are described by spin networks γ : the edges are labelled by spins together with `intertwiners' at the vertices which are prescription for how to sum over different ways the spins are rerouted. The sum over rerouting are chosen as such to make the form of the intertwiner invariant under Gauss gauge transformations.

29.4.5 Real variables, modern analysis and LQG

Main article: Hamiltonian constraint of LQG

Let us go into more detail about the technical difficulties associated with using Ashtekar's variables:

With Ashtekar's variables one uses a complex connection and so the relevant gauge group as actually $SL(2, \mathbb{C})$ and not $SU(2)$. As $SL(2, \mathbb{C})$ is non-compact it creates serious problems for the rigorous construction of the necessary mathematical machinery. The group $SU(2)$ is on the other hand is compact and the relevant constructions needed have been developed.

As mentioned above, because Ashtekar's variables are complex it results in complex general relativity. To recover the real theory one has to impose what are known as the reality conditions. These require that the densitized triad be real and that the real part of the Ashtekar connection equals the compatible spin connection (the compatibility condition being $\nabla_a e_b^I = 0$) determined by the desitized triad. The expression for compatible connection Γ_a^i is rather complicated and as such non-polynomial formula enters through the back door.

Before we state the next difficulty we should give a definition; a tensor density of weight W transforms like an ordinary tensor, except that in additional the W th power of the Jacobian,

$$J = \left| \frac{\partial x^a}{\partial x'^b} \right|$$

appears as a factor, i.e.

$$T'^{a\ldots}_{b\ldots} = J^W \frac{\partial x'^a}{\partial x^c} \ldots \frac{\partial x^d}{\partial x'^b} T^{c\ldots}_{d\ldots} .$$

It turns out that it is impossible, on general grounds, to construct a UV-finite, diffeomorphism non-violating operator corresponding to $\sqrt{\det(q)}H$. The reason is that the rescaled Hamiltonian constraint is a scalar density of weight two while it can be shown that only scalar densities of weight one have a chance to result in a well defined operator. Thus, one is forced to work with the original unrescaled, density one-valued, Hamiltonian constraint. However, this is non-polynomial and the whole virtue of the complex variables is questioned. In fact, all the solutions constructed for Ashtekar's Hamiltonian constraint only vanished for finite regularization (physics), however, this violates spatial diffeomorphism invariance.

Without the implementation and solution of the Hamiltonian constraint no progress can be made and no reliable predictions are possible!

To overcome the first problem one works with the configuration variable

$$A_a^i = \Gamma_a^i + \beta K_a^i$$

where β is real (as pointed out by Barbero, who introduced real variables some time after Ashtekar's variables[14][15]).

The Guass law and the spatial diffeomorphism constraints are the same. In real Ashtekar variables the Hamiltonian is

$$H = \frac{\epsilon_{ijk} F_{ab}^k \tilde{E}_i^a \tilde{E}_j^b}{\sqrt{\det(q)}} + 2\frac{\beta^2+1}{\beta^2} \frac{(\tilde{E}_i^a \tilde{E}_j^b - \tilde{E}_j^a \tilde{E}_i^b)}{\sqrt{\det(q)}} (A_a^i - \Gamma_a^i)(A_b^j - \Gamma_b^j) = H_E + H'.$$

The complicated relationship between Γ_a^i and the desitized triads causes serious problems upon quantization. It is with the choice $\beta = \pm i$ that the second more complicated term is made to vanish. However, as mentioned above Γ_a^i reappears in the reality conditions. Also we still have the problem of the $1/\sqrt{\det(q)}$ factor.

Thiemann was able to make it work for real β. First he could simplify the troublesome $1/\sqrt{\det(q)}$ by using the identity

$$\{A_c^k, V\} = \frac{\epsilon_{abc}\epsilon^{ijk} \tilde{E}_i^a \tilde{E}_j^b}{\sqrt{\det(q)}}$$

where V is the volume. The A_c^k and V can be promoted to well defined operators in the loop representation and the Poisson bracket is replaced by a commutator upon quantization; this takes care of the first term. It turns out that a similar trick can be used to treat the second term. One introduces the quantity

$$K = \int d^3x K_a^i \tilde{E}_i^a$$

and notes that

$$K_a^i = \{A_a^i, K\}.$$

We are then able to write

$$A_a^i - \Gamma_a^i = \beta K_a^i = \beta\{A_a^i, K\}.$$

The reason the quantity K is easier to work with at the time of quantization is that it can be written as

$$K = -\{V, \int d^3x H_E\}$$

where we have used that the integrated densitized trace of the extrinsic curvature, K, is the ``time derivative of the volume''.

In the long history of canonical quantum gravity formulating the Hamiltonian constraint as a quantum operator (Wheeler–DeWitt equation) in a mathematically rigorous manner has been a formidable problem. It was in the loop representation that a mathematically well defined Hamiltonian constraint was finally formulated in 1996.[9] We leave more details of its construction to the article Hamiltonian constraint of LQG. This together with the quantum versions of the Gauss law and spatial diffeomorphism constrains written in the loop representation are the central equations of LQG (modern canonical quantum General relativity).

Finding the states that are annihilated by these constraints (the physical states), and finding the corresponding physical inner product, and observables is the main goal of the technical side of LQG.

A very important aspect of the Hamiltonian operator is that it only acts at vertices (a consequence of this is that Thiemann's Hamiltonian operator, like Ashtekar's operator, annihilates non-intersecting loops except now it is not just formal and has rigorous mathematical meaning). More precisely, its action is non-zero on at least vertices of valence three and greater and results in a linear combination of new spin networks where the original graph has been modified by the addition of lines at each vertex together and a change in the labels of the adjacent links of the vertex.

29.4.6   Solving the quantum constraints

Main articles: spectrum, dual space and Rigged Hilbert space

We solve, at least approximately, all the quantum constraint equations and for the physical inner product to make physical predictions.

Before we move on to the constraints of LQG, lets us consider certain cases. We start with a kinematic Hilbert space $\mathcal{H}_{\text{Kin}}$ as so is equipped with an inner product—the kinematic inner product $\langle \phi, \psi \rangle_{\text{Kin}}$.

i) Say we have constraints $\hat{C}_I$ whose zero eigenvalues lie in their discrete spectrum. Solutions of the first constraint, $\hat{C}_1$, correspond to a subspace of the kinematic Hilbert space, $\mathcal{H}_1 \subset \mathcal{H}_{\text{Kin}}$. There will be a projection operator P_1 mapping

$\mathcal{H}_{\text{Kin}}$ onto $\mathcal{H}_1$. The kinematic inner product structure is easily employed to provide the inner product structure after solving this first constraint; the new inner product $\langle \phi, \psi \rangle_1$ is simply

$$\langle \phi, \psi \rangle_1 = \langle P\phi, P\psi \rangle_{\text{Kin}}$$

They are based on the same inner product and are states normalizable with respect to it.

ii) The zero point is not contained in the point spectrum of all the $\hat{C}_I$, there is then no non-trivial solution $\Psi \in \mathcal{H}_{\text{Kin}}$ to the system of quantum constraint equations $\hat{C}_I \Psi = 0$ for all I .

For example the zero eigenvalue of the operator

$$\hat{C} = \left(i \frac{d}{dx} - k \right)$$

on $L_2(\mathbb{R}, dx)$ lies in the continuous spectrum $\mathbb{R}$ but the formal ``eigenstate'' $\exp(-ikx)$ is not normalizable in the kinematic inner product,

$$\int_{-\infty}^{\infty} dx \psi^*(x)\psi(x) = \int_{-\infty}^{\infty} dx e^{ikx} e^{-ikx} = \int_{-\infty}^{\infty} dx = \infty$$

and so does not belong to the kinematic Hilbert space $\mathcal{H}_{\text{Kin}}$. In these cases we take a dense subset $\mathcal{S}$ of $\mathcal{H}_{\text{Kin}}$ (intuitively this means either any point in $\mathcal{S}$ is either in $\mathcal{H}_{\text{Kin}}$ or arbitrarily close to a point in $\mathcal{H}_{\text{Kin}}$) with very good convergence properties and consider its dual space $\mathcal{S}'$ (intuitively these map elements of $\mathcal{S}$ onto finite complex numbers in a linear manner), then $\mathcal{S} \subset \mathcal{H}_{\text{Kin}} \subset \mathcal{S}'$ (as $\mathcal{S}'$ contains distributional functions). The constraint operator is then implemented on this larger dual space, which contains distributional functions, under the adjoint action on the operator. One looks for solutions on this larger space. This comes at the price that the solutions must be given a new Hilbert space inner product with respect to which they are normalizable (see article on rigged Hilbert space). In this case we have a generalized projection operator on the new space of states. We cannot use the above formula for the new inner product as it diverges, instead the new inner product is given by the simply modification of the above,

$$\langle \phi, \psi \rangle_1 = \langle P\phi, \psi \rangle_{\text{Kin}}.$$

The generalized projector P is known as a rigging map.

Let us move to LQG, additional complications will arise from the fact the constraint algebra is not a Lie algebra due to the bracket between two Hamiltonian constraints.

The Gauss law is solved by the use of spin network states. They provide a basis for the Kinematic Hilbert space $\mathcal{H}_{\text{Kin}}$. The spatial diffeomorphism constraint has been solved. The induced inner product on $\mathcal{H}_{\text{Diff}}$ (we do not pursue the details) has a very simple description in terms of spin network states; given two spin networks s and s' , with associated spin network states ψ_s and $\psi_{s'}$, the inner product is 1 if s and s' are related to each other by a spatial diffeomorphism and zero otherwise.

The Hamiltonian constraint maps diffeomorphism invariant states onto non-diffeomorphism invaiant states as so does not preserve the diffeomorphism Hilbert space $\mathcal{H}_{\text{Diff}}$. This is an unavoidable consequence of the operator algebra, in particular the commutator:

$$[\hat{C}(\vec{N}), \hat{H}(M)] \propto \hat{H}(\mathcal{L}_{\vec{N}} M)$$

as can be seen by applying this to $\psi_s \in \mathcal{H}_{Diff}$,

$$(\vec{C}(\vec{N})\hat{H}(M) - \hat{H}(M)\vec{C}(\vec{N}))\psi_s \propto \hat{H}(\mathcal{L}_{\vec{N}} M)\psi_s$$

and using $\vec{C}(\vec{N})\psi_s = 0$ to obtain

$$\vec{C}(\vec{N})[\hat{H}(M)\psi_s] \propto \hat{H}(\mathcal{L}_{\vec{N}} M)\psi_s \neq 0$$

and so $\hat{H}(M)\psi_s$ is not in $\mathcal{H}_{Diff}$.

This means that you can't just solve the diffeomorphism constraint and then the Hamiltonian constraint. This problem can be circumvented by the introduction of the master constraint, with its trivial operator algebra, one is then able in principle to construct the physical inner product from $\mathcal{H}_{\text{Diff}}$.

29.5   Spin foams

Main articles: spin network, spin foam, BF model and Barrett–Crane model

In loop quantum gravity (LQG), a spin network represents a "quantum state" of the gravitational field on a 3-dimensional hypersurface. The set of all possible spin networks (or, more accurately, "s-knots" - that is, equivalence classes of spin networks under diffeomorphisms) is countable; it constitutes a basis of LQG Hilbert space.

In physics, a spin foam is a topological structure made out of two-dimensional faces that represents one of the configurations that must be summed to obtain a Feynman's path integral (functional integration) description of quantum gravity. It is closely related to loop quantum gravity.

29.5.1   Spin foam derived from the Hamiltonian constraint operator

The Hamiltonian constraint generates `time' evolution. Solving the Hamiltonian constraint should tell us how quantum states evolve in `time' from an initial spin network state to a final spin network state. One approach to solving the Hamiltonian constraint starts with what is called the Dirac delta function. This is a rather singular function of the real line, denoted $\delta(x)$, that is zero everywhere except at $x = 0$ but whose integral is finite and nonzero. It can be represented as a Fourier integral,

$\delta(x) = \int e^{ikx} dk$.

One can employ the idea of the delta function to impose the condition that the Hamiltonian constraint should vanish. It is obvious that

$\prod_{x \in \Sigma} \delta(\hat{H}(x))$

is non-zero only when $\hat{H}(x) = 0$ for all x in Σ. Using this we can `project' out solutions to the Hamiltonian constraint. With analogy to the Fourier integral given above, this (generalized) projector can formally be written as

$\int [dN] e^{i \int d^3 x N(x) \hat{H}(x)}$.

Interestingly, this is formally spatially diffeomorphism-invariant. As such it can be applied at the spatially diffeomorphism-invariant level. Using this the physical inner product is formally given by

$$\left\langle \int [dN] e^{i \int d^3 x N(x) \hat{H}(x)} s_{\text{int}} s_{\text{fin}} \right\rangle_{\text{Diff}}$$

where s_{int} are the initial spin network and s_{fin} is the final spin network.

The exponential can be expanded

$$\left\langle \int [dN] (1 + i \int d^3 x N(x) \hat{H}(x) + \tfrac{i^2}{2!} [\int d^3 x N(x) \hat{H}(x)][\int d^3 x' N(x') \hat{H}(x')] + \dots) s_{\text{int}}, s_{\text{fin}} \right\rangle_{\text{Diff}}$$

and each time a Hamiltonian operator acts it does so by adding a new edge at the vertex. The summation over different sequences of actions of $\hat{H}$ can be visualized as a summation over different histories of `interaction vertices' in the `time' evolution sending the initial spin network to the final spin network. This then naturally gives rise to the two-complex (a combinatorial set of faces that join along edges, which in turn join on vertices) underlying the spin foam description; we evolve forward an initial spin network sweeping out a surface, the action of the Hamiltonian constraint operator is to produce a new planar surface starting at the vertex. We are able to use the action of the Hamiltonian constraint on the vertex of a spin network state to associate an amplitude to each "interaction" (in analogy to Feynman diagrams). See figure below. This opens up a way of trying to directly link canonical LQG to a path integral description. Now just as a spin networks describe quantum space, each configuration contributing to these path integrals, or sums over history, describe `quantum space-time'. Because of their resemblance to soap foams and the way they are labeled John Baez gave these `quantum space-times' the name `spin foams'.

There are however severe difficulties with this particular approach, for example the Hamiltonian operator is not self-adjoint, in fact it is not even a normal operator (i.e. the operator does not commute with its adjoint) and so the spectral theorem cannot be used to define the exponential in general. The most serious problem is that the $\hat{H}(x)$'s are not mutually

The action of the Hamiltonian constraint translated to the path integral or so-called spin foam description. A single node splits into three nodes, creating a spin foam vertex. $N(x_n)$ is the value of N at the vertex and H_{nop} are the matrix elements of the Hamiltonian constraint $\hat{H}$.

commuting, it can then be shown the formal quantity $\int [dN] e^{i \int d^3 x N(x) \hat{H}(x)}$ cannot even define a (generalized) projector. The master constraint (see below) does not suffer from these problems and as such offers a way of connecting the canonical theory to the path integral formulation.

29.5.2 Spin foams from BF theory

It turns out there are alternative routes to formulating the path integral, however their connection to the Hamiltonian formalism is less clear. One way is to start with the BF theory. This is a simpler theory to general relativity. It has no local degrees of freedom and as such depends only on topological aspects of the fields. BF theory is what is known as a topological field theory. Surprisingly, it turns out that general relativity can be obtained from BF theory by imposing a constraint,[16] BF theory involves a field B_{ab}^{IJ} and if one chooses the field B to be the (anti-symmetric) product of two tetrads

$$B_{ab}^{IJ} = \tfrac{1}{2}(E_a^I E_b^J - E_b^I E_a^J)$$

(tetrads are like triads but in four spacetime dimensions), one recovers general relativity. The condition that the B field be given by the product of two tetrads is called the simplicity constraint. The spin foam dynamics of the topological field theory is well understood. Given the spin foam `interaction' amplitudes for this simple theory, one then tries to implement the simplicity conditions to obtain a path integral for general relativity. The non-trivial task of constructing a spin foam model is then reduced to the question of how this simplicity constraint should be imposed in the quantum theory. The first attempt at this was the famous Barrett–Crane model.[17] However this model was shown to be problematic, for example there did not seem to be enough degrees of freedom to ensure the correct classical limit.[18] It has been argued that the simplicity constraint was imposed too strongly at the quantum level and should only be imposed in the sense of expectation values just as with the Lorenz gauge condition $\partial_\mu \hat{A}^\mu$ in the Gupta–Bleuler formalism of quantum electrodynamics. New models have now been put forward, sometimes motivated by imposing the simplicity conditions in a weaker sense.

Another difficulty here is that spin foams are defined on a discretization of spacetime. While this presents no problems for a topological field theory as it has no local degrees of freedom, it presents problems for GR. This is known as the problem triangularization dependence.

29.5.3 Modern formulation of spin foams

Just as imposing the classical simplicity constraint recovers general relativity from BF theory, one expects an appropriate quantum simplicity constraint will recover quantum gravity from quantum BF theory.

Much progress has been made with regard to this issue by Engle, Pereira, and Rovelli[19] and Freidal and Krasnov[20] in defining spin foam interaction amplitudes with much better behaviour.

An attempt to make contact between EPRL-FK spin foam and the canonical formulation of LQG has been made.[21]

29.5.4   Spin foam derived from the master constraint operator

See below.

29.6   The semi-classical limit

29.6.1   What is the semiclassical limit?

Main articles: Correspondence principle and classical limit

The **classical limit** or **correspondence limit** is the ability of a physical theory to approximate or "recover" classical mechanics when considered over special values of its parameters.[22] The classical limit is used with physical theories that predict non-classical behavior.

In physics, the **correspondence principle** states that the behavior of systems described by the theory of quantum mechanics (or by the old quantum theory) reproduces classical physics in the limit of large quantum numbers. In other words, it says that for large orbits and for large energies, quantum calculations must agree with classical calculations.[23]

The principle was formulated by Niels Bohr in 1920,[24] though he had previously made use of it as early as 1913 in developing his model of the atom.[25]

There are two basic requirements in establishing the semi-classical limit of any quantum theory:

i) reproduction of the Poisson brackets (of the diffeomorphism constraints in the case of general relativity). This is extremely important because, as noted above, the Poisson bracket algebra formed between the (smeared) constraints themselves completely determines the classical theory. This is analogous to establishing Ehrenfest's theorem;

ii) the specification of a complete set of classical observables whose corresponding operators (see complete set of commuting observables for the quantum mechanical definition of a complete set of observables) when acted on by appropriate semi-classical states reproduce the same classical variables with small quantum corrections (a subtle point is that states that are semi-classical for one class of observables may not be semi-classical for a different class of observables[26]).

This may be easily done, for example, in ordinary quantum mechanics for a particle but in general relativity this becomes a highly non-trivial problem as we will see below.

29.6.2   Why might LQG not have general relativity as its semiclassical limit?

Any candidate theory of quantum gravity must be able to reproduce Einstein's theory of general relativity as a classical limit of a quantum theory. This is not guaranteed because of a feature of quantum field theories which is that they have different sectors, these are analogous to the different phases that come about in the thermodynamical limit of statistical systems. Just as different phases are physically different, so are different sectors of a quantum field theory. It may turn out that LQG belongs to an unphysical sector - one in which you do not recover general relativity in the semi classical limit (in fact there might not be any physical sector at all).

Theorems establishing the uniqueness of the loop representation as defined by Ashtekar et al. (i.e. a certain concrete realization of a Hilbert space and associated operators reproducing the correct loop algebra - the realization that everybody was using) have been given by two groups (Lewandowski, Okolow, Sahlmann and Thiemann)[27] and (Christian Fleischhack).[28] Before this result was established it was not known whether there could be other examples of Hilbert spaces with operators invoking the same loop algebra, other realizations, not equivalent to the one that had been used so far. These uniqueness theorems imply no others exist and so if LQG does not have the correct semiclassical limit then this would mean the end of the loop representation of quantum gravity altogether.

29.6.3 Difficulties checking the semiclassical limit of LQG

There are difficulties in trying to establish LQG gives Einstein's theory of general relativity in the semi classical limit. There are a number of particular difficulties in establishing the semi-classical limit

1. There is no operator corresponding to infinitesimal spacial diffeomorphisms (it is not surprising that the theory has no generator of infinitesimal spatial 'translations' as it predicts spatial geometry has a discrete nature, compare to the situation in condensed matter). Instead it must be approximated by finite spatial diffeomorphisms and so the Poisson bracket structure of the classical theory is not exactly reproduced. This problem can be circumvented with the introduction of the so-called master constraint (see below)[29]

2. There is the problem of reconciling the discrete combinatorial nature of the quantum states with the continuous nature of the fields of the classical theory.

3. There are serious difficulties arising from the structure of the Poisson brackets involving the spatial diffeomorphism and Hamiltonian constraints. In particular, the algebra of (smeared) Hamiltonian constraints does not close, it is proportional to a sum over infinitesimal spatial diffeomorphisms (which, as we have just noted, does not exist in the quantum theory) where the coefficients of proportionality are not constants but have non-trivial phase space dependence - as such it does not form a Lie algebra. However, the situation is much improved by the introduction of the master constraint.[29]

4. The semi-classical machinery developed so far is only appropriate to non-graph-changing operators, however, Thiemann's Hamiltonian constraint is a graph-changing operator - the new graph it generates has degrees of freedom upon which the coherent state does not depend and so their quantum fluctuations are not suppressed. There is also the restriction, so far, that these coherent states are only defined at the Kinematic level, and now one has to lift them to the level of $\mathcal{H}_{Diff}$ and $\mathcal{H}_{Phys}$. It can be shown that Thiemann's Hamiltonian constraint is required to be graph changing in order to resolve problem 3 in some sense. The master constraint algebra however is trivial and so the requirement that it be graph changing can be lifted and indeed non-graph changing master constraint operators have been defined.

5. Formulating observables for classical general relativity is a formidable problem by itself because of its non-linear nature and space-time diffeomorphism invariance. In fact a systematic approximation scheme to calculate observables has only been recently developed.[30][31]

Difficulties in trying to examine the semi classical limit of the theory should not be confused with it having the wrong semi classical limit.

29.6.4 Progress in demonstrating LQG has the correct semiclassical limit

Much details here to be written up...

Concerning issue number 2 above one can consider so-called weave states. Ordinary measurements of geometric quantities are macroscopic, and planckian discreteness is smoothed out. The fabric of a T-shirt is analogous. At a distance it is a smooth curved two-dimensional surface. But a closer inspection we see that it is actually composed of thousands of one-dimensional linked threads. The image of space given in LQG is similar, consider a very large spin network formed by a very large number of nodes and links, each of Planck scale. But probed at a macroscopic scale, it appears as a three-dimensional continuous metric geometry.

As far as the editor knows problem 4 of having semi-classical machinery for non-graph changing operators is as the moment still out of reach.

To make contact with familiar low energy physics it is mandatory to have to develop approximation schemes both for the physical inner product and for Dirac observables.

The spin foam models have been intensively studied can be viewed as avenues toward approximation schemes for the physical inner product.

Markopoulou et al. adopted the idea of noiseless subsystems in an attempt to solve the problem of the low energy limit in background independent quantum gravity theories[32][33][34] The idea has even led to the intriguing possibility of matter of the standard model being identified with emergent degrees of freedom from some versions of LQG (see section below: *LQG and related research programs*).

As Wightman emphasized in the 1950s, in Minkowski QFTs the $n-$ point functions

$$W(x_1, \ldots, x_n) = \langle 0 | \phi(x_n) \ldots \phi(x_1) | 0 \rangle \,,$$

completely determine the theory. In particular, one can calculate the scattering amplitudes from these quantities. As explained below in the section on the *Background independent scattering amplitudes*, in the background-independent context, the $n-$ point functions refer to a state and in gravity that state can naturally encode information about a specific geometry which can then appear in the expressions of these quantities. To leading order LQG calculations have been shown to agree in an appropriate sense with the $n-$ point functions calculated in the effective low energy quantum general relativity.

29.7   Improved dynamics and the master constraint

Main articles: Hamiltonian (quantum mechanics), Hamiltonian constraint of LQG and Friedrichs extension

29.7.1   The master constraint

Thiemann's master constraint should not be confused with the master equation which has to do with random processes. The Master Constraint Programme for Loop Quantum Gravity (LQG) was proposed as a classically equivalent way to impose the infinite number of Hamiltonian constraint equations

$$H(x) = 0$$

(x being a continuous index) in terms of a single master constraint,

$$M = \int d^3x \frac{[H(x)]^2}{\sqrt{\det(q(x))}} \,.$$

which involves the square of the constraints in question. Note that $H(x)$ were infinitely many whereas the master constraint is only one. It is clear that if M vanishes then so do the infinitely many $H(x)$'s. Conversely, if all the $H(x)$'s vanish then so does M , therefore they are equivalent. The master constraint M involves an appropriate averaging over all space and so is invariant under spatial diffeomorphisms (it is invariant under spatial "shifts" as it is a summation over all such spatial "shifts" of a quantity that transforms as a scalar). Hence its Poisson bracket with the (smeared) spacial diffeomorphism constraint, $C(\vec{N})$, is simple:

$$\{M, C(\vec{N})\} = 0 \,.$$

(it is $su(2)$ invariant as well). Also, obviously as any quantity Poisson commutes with itself, and the master constraint being a single constraint, it satisfies

$$\{M, M\} = 0 \,.$$

We also have the usual algebra between spatial diffeomorphisms. This represents a dramatic simplification of the Poisson bracket structure, and raises new hope in understanding the dynamics and establishing the semi-classical limit.[35]

An initial objection to the use of the master constraint was that on first sight it did not seem to encode information about the observables; because the Mater constraint is quadratic in the constraint, when you compute its Poisson bracket with any quantity, the result is proportional to the constraint, therefore it always vanishes when the constraints are imposed and as such does not select out particular phase space functions. However, it was realized that the condition

$$\{\{M, O\}, O\}_{M=0} = 0$$

is equivalent to O being a Dirac observable. So the master constraint does capture information about the observables. Because of its significance this is known as the Master equation.[35]

That the master constraint Poisson algebra is an honest Lie algebra opens up the possibility of using a certain method, known as group averaging, in order to construct solutions of the infinite number of Hamiltonian constraints, a physical inner product thereon and Dirac observables via what is known as refined algebraic quantization RAQ[36]

29.7.2 The quantum master constraint

Define the quantum master constraint (regularisation issues aside) as

$$\hat{M} := \int d^3x \left(\widehat{\frac{H}{\det(q(x))^{1/4}}} \right)^\dagger (x) \left(\widehat{\frac{H}{\det(q(x))^{1/4}}} \right)(x) \, .$$

Obviously,

$$\left(\widehat{\frac{H}{\det(q(x))^{1/4}}} \right)(x)\Psi = 0$$

for all x implies $\hat{M}\Psi = 0$. Conversely, if $\hat{M}\Psi = 0$ then

$$0 = < \Psi, \hat{M}\Psi > = \int d^3x \left\| \left(\widehat{\frac{H}{\det(q(x))^{1/4}}} \right)(x)\Psi \right\|^2 \quad Eq\ 4$$

implies

$$\left(\widehat{\frac{H}{\det(q(x))^{1/4}}} \right)(x)\Psi = 0 \, .$$

What is done first is, we are able to compute the matrix elements of the would-be operator $\hat{M}$, that is, we compute the quadratic form Q_M . It turns out that as Q_M is a graph changing, diffeomorphism invariant quadratic form it cannot exist on the kinematic Hilbert space H_{Kin} , and must be defined on H_{Diff} . The fact that the master constraint operator $\hat{M}$ is densely defined on H_{Diff} , it is obvious that $\hat{M}$ is a positive and symmetric operator in H_{Diff} . Therefore, the quadratic form Q_M associated with $\hat{M}$ is closable. The closure of Q_M is the quadratic form of a unique self-adjoint operator $\overline{\hat{M}}$, called the Friedrichs extension of $\hat{M}$. We relabel $\overline{\hat{M}}$ as $\hat{M}$ for simplicity. (Note that the presence of an inner product, viz Eq 4, means there are no superfluous solutions i.e. there are no Ψ such that $\left(\widehat{\frac{H}{\det(q(x))^{1/4}}} \right)(x)\Psi \neq 0$ but for which $\hat{M}\Psi = 0$).

It is also possible to construct a quadratic form Q_{M_E} for what is called the extended master constraint (discussed below) on H_{Kin} which also involves the weighted integral of the square of the spatial diffeomorphism constraint (this is possible because Q_{M_E} is not graph changing).

The spectrum of the master constraint may not contain zero due to normal or factor ordering effects which are finite but similar in nature to the infinite vacuum energies of background-dependent quantum field theories. In this case it turns out to be physically correct to replace $\hat{M}$ with $\hat{M}' := \hat{M} - min(spec(\hat{M}))\hat{1}$ provided that the "normal ordering constant" vanishes in the classical limit, that is, $\lim_{\hbar \to 0} min(spec(\hat{M})) = 0$, so that $\hat{M}'$ is a valid quantisation of M .

29.7.3 Testing the master constraint

The constraints in their primitive form are rather singular, this was the reason for integrating them over test functions to obtain smeared constraints. However, it would appear that the equation for the master constraint, given above, is even more singular involving the product of two primitive constraints (although integrated over space). Squaring the constraint is dangerous as it could lead to worsened ultraviolent behaviour of the corresponding operator and hence the master constraint programme must be approached with due care.

In doing so the master constraint programme has been satisfactorily tested in a number of model systems with non-trivial constraint algebras, free and interacting field theories.[37][38][39][40][41] The master constraint for LQG was established as a genuine positive self-adjoint operator and the physical Hilbert space of LQG was shown to be non-empty,[42] an obvious consistency test LQG must pass to be a viable theory of quantum General relativity.

29.7.4   Applications of the master constraint

The master constraint has been employed in attempts to approximate the physical inner product and define more rigorous path integrals.[43][44][45][46]

The Consistent Discretizations approach to LQG,[47][48] is an application of the master constraint program to construct the physical Hilbert space of the canonical theory.

29.7.5   Spin foam from the master constraint

It turns out that the master constraint is easily generalized to incorporate the other constraints. It is then referred to as the extended master constraint, denoted M_E. We can define the extended master constraint which imposes both the Hamiltonian constraint and spatial diffeomorphism constraint as a single operator,

$$M_E = \int_\Sigma d^3x \frac{H(x)^2 - q^{ab} V_a(x) V_b(x)}{\sqrt{det(q)}} \,.$$

Setting this single constraint to zero is equivalent to $H(x) = 0$ and $V_a(x) = 0$ for all x in Σ. This constraint implements the spatial diffeomorphism and Hamiltonian constraint at the same time on the Kinematic Hilbert space. The physical inner product is then defined as

$$\langle \phi, \psi \rangle_{\text{Phys}} = \lim_{T \to \infty} \left\langle \phi, \int_{-T}^{T} dt e^{it\hat{M}_E} \psi \right\rangle$$

(as $\delta(\hat{M}_E) = \lim_{T \to \infty} \int_{-T}^{T} dt e^{it\hat{M}_E}$). A spin foam representation of this expression is obtained by splitting the t -parameter in discrete steps and writing

$$e^{it\hat{M}_E} = \lim_{n \to \infty} [e^{it\hat{M}_E/n}]^n = \lim_{n \to \infty} [1 + it\hat{M}_E/n]^n.$$

The spin foam description then follows from the application of $[1 + it\hat{M}_E/n]$ on a spin network resulting in a linear combination of new spin networks whose graph and labels have been modified. Obviously an approximation is made by truncating the value of n to some finite integer. An advantage of the extended master constraint is that we are working at the kinematic level and so far it is only here we have access semi-classical coherent states. Moreover, one can find none graph changing versions of this master constraint operator, which are the only type of operators appropriate for these coherent states.

29.7.6   Algebraic quantum gravity

The master constraint programme has evolved into a fully combinatorial treatment of gravity known as Algebraic Quantum Gravity (AQG).[49] The non-graph changing master constraint operator is adapted in the framework of algebraic quantum gravity. While AQG is inspired by LQG, it differs drastically from it because in AQG there is fundamentally no topology or differential structure - it is background independent in a more generalized sense and could possibly have something to say about topology change. In this new formulation of quantum gravity AQG semiclassical states always control the fluctuations of all present degrees of freedom. This makes the AQG semiclassical analysis superior over that of LQG, and progress has been made in establishing it has the correct semiclassical limit and providing contact with familiar low energy physics.[50][51] See Thiemann's book for details.

29.8   Physical applications of LQG

29.8.1   Black hole entropy

Main articles: Black hole thermodynamics, Isolated horizon and Immirzi parameter

The Immirzi parameter (also known as the Barbero-Immirzi parameter) is a numerical coefficient appearing in loop quantum gravity. It may take real or imaginary values.

An artist depiction of two black holes merging, a process in which the laws of thermodynamics are upheld.

Black hole thermodynamics is the area of study that seeks to reconcile the laws of thermodynamics with the existence of black hole event horizons. The no hair conjecture of general relativity states that a black hole is characterized only by its mass, its charge, and its angular momentum; hence, it has no entropy. It appears, then, that one can violate the second law of thermodynamics by dropping an object with nonzero entropy into a black hole.[52] Work by Stephen Hawking and Jacob Bekenstein showed that one can preserve the second law of thermodynamics by assigning to each black hole a *black-hole entropy*

$$S_{\mathrm{BH}} = \frac{k_{\mathrm{B}} A}{4 \ell_{\mathrm{P}}^2},$$

where A is the area of the hole's event horizon, k_{B} is the Boltzmann constant, and $\ell_{\mathrm{P}} = \sqrt{G\hbar/c^3}$ is the Planck length.[53] The fact that the black hole entropy is also the maximal entropy that can be obtained by the Bekenstein bound (wherein the Bekenstein bound becomes an equality) was the main observation that led to the holographic principle.[52]

An oversight in the application of the no-hair theorem is the assumption that the relevant degrees of freedom accounting for the entropy of the black hole must be classical in nature; what if they were purely quantum mechanical instead and had non-zero entropy? Actually, this is what is realized in the LQG derivation of black hole entropy, and can be seen as a consequence of its background-independence – the classical black hole spacetime comes about from the semi-classical limit of the quantum state of the gravitational field, but there are many quantum states that have the same semiclassical limit. Specifically, in LQG[54] it is possible to associate a quantum geometrical interpretation to the microstates: These are the quantum geometries of the horizon which are consistent with the area, A, of the black hole and the topology of the horizon (i.e. spherical). LQG offers a geometric explanation of the finiteness of the entropy and of the proportionality of the area of the horizon.[55][56] These calculations have been generalized to rotating black holes.[57]

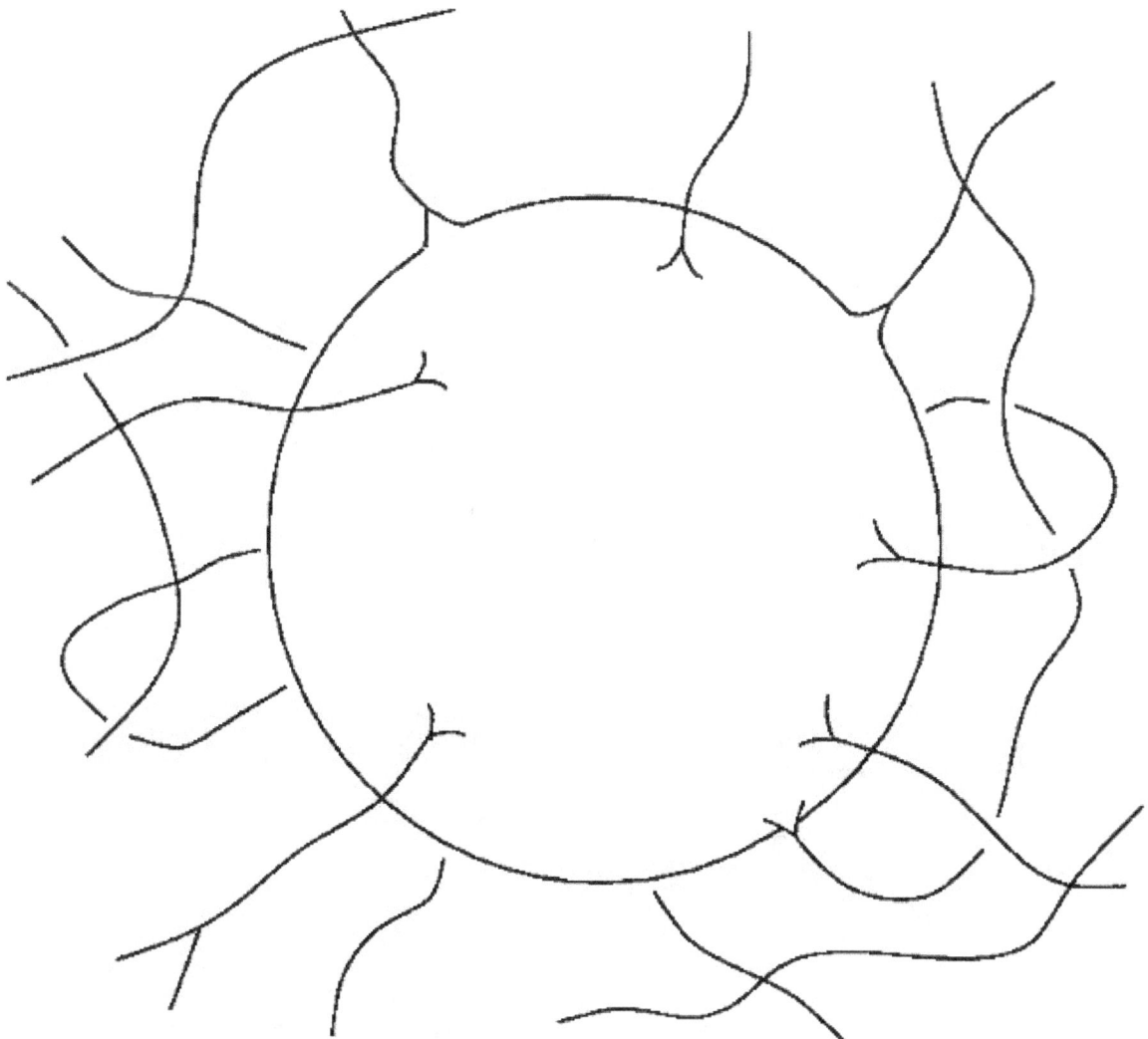

Representation of quantum geometries of the horizon. Polymer excitations in the bulk puncture the horizon, endowing it with quantized area. Intrinsically the horizon is flat except at punctures where it acquires a quantized deficit angle or quantized amount of curvature. These deficit angles add up to 4π .

It is possible to derive, from the covariant formulation of full quantum theory (Spinfoam) the correct relation between energy and area (1st law), the Unruh temperature and the distribution that yields Hawking entropy.[58] The calculation makes use of the notion of dynamical horizon and is done for non-extremal black holes.

A recent success of the theory in this direction is the computation of the entropy of all non singular black holes directly from theory and independent of Immirzi parameter.[59] The result is the expected formula $S = A/4$, where S is the entropy and A the area of the black hole, derived by Bekenstein and Hawking on heuristic grounds. This is the only known derivation of this formula from a fundamental theory, for the case of generic non singular black holes. Older attempts at this calculation had difficulties. The problem was that although Loop quantum gravity predicted that the entropy of a black hole is proportional to the area of the event horizon, the result depended on a crucial free parameter in the theory, the above-mentioned Immirzi parameter. However, there is no known computation of the Immirzi parameter, so it had to be fixed by demanding agreement with Bekenstein and Hawking's calculation of the black hole entropy.

29.8.2 Loop quantum cosmology

Main articles: loop quantum cosmology, Big bounce and inflation (cosmology)

The popular and technical literature makes extensive references to LQG-related topic of loop quantum cosmology. LQC was mainly developed by Martin Bojowald, it was popularized Loop quantum cosmology in *Scientific American* for predicting a Big Bounce prior to the Big Bang. Loop quantum cosmology (LQC) is a symmetry-reduced model of classical general relativity quantized using methods that mimic those of loop quantum gravity (LQG) that predicts a "quantum bridge" between contracting and expanding cosmological branches.

Achievements of LQC have been the resolution of the big bang singularity, the prediction of a Big Bounce, and a natural mechanism for inflation (cosmology).

LQC models share features of LQG and so is a useful toy model. However, the results obtained are subject to the usual restriction that a truncated classical theory, then quantized, might not display the true behaviour of the full theory due to artificial suppression of degrees of freedom that might have large quantum fluctuations in the full theory. It has been argued that singularity avoidance in LQC are by mechanisms only available in these restrictive models and that singularity avoidance in the full theory can still be obtained but by a more subtle feature of LQG.[60][61]

29.8.3 Loop quantum gravity phenomenology

Quantum gravity effects are notoriously difficult to measure because the Planck length is so incredibly small. However recently physicists have started to consider the possibility of measuring quantum gravity effects, mostly from astrophysical observations and gravitational wave detectors.

29.8.4 Background independent scattering amplitudes

Loop quantum gravity is formulated in a background-independent language. No spacetime is assumed a priori, but rather it is built up by the states of theory themselves - however scattering amplitudes are derived from n -point functions (Correlation function (quantum field theory)) and these, formulated in conventional quantum field theory, are functions of points of a background space-time. The relation between the background-independent formalism and the conventional formalism of quantum field theory on a given spacetime is far from obvious, and it is far from obvious how to recover low-energy quantities from the full background-independent theory. One would like to derive the n -point functions of the theory from the background-independent formalism, in order to compare them with the standard perturbative expansion of quantum general relativity and therefore check that loop quantum gravity yields the correct low-energy limit.

A strategy for addressing this problem has been suggested;[62] the idea is to study the boundary amplitude, namely a path integral over a finite space-time region, seen as a function of the boundary value of the field.[63] In conventional quantum field theory, this boundary amplitude is well–defined[64][65] and codes the physical information of the theory; it does so in quantum gravity as well, but in a fully background–independent manner.[66] A generally covariant definition of n -point functions can then be based on the idea that the distance between physical points –arguments of the n -point function is determined by the state of the gravitational field on the boundary of the spacetime region considered.

Progress has been made in calculating background independent scattering amplitudes this way with the use of spin foams. This is a way to extract physical information from the theory. Claims to have reproduced the correct behaviour for graviton scattering amplitudes and to have recovered classical gravity have been made. "We have calculated Newton's law starting from a world with no space and no time." - Carlo Rovelli.

29.9   Gravitons, string theory, supersymmetry, extra dimensions in LQG

Main articles: graviton, string theory, supersymmetry, Kaluza–Klein theory and supergravity

Some quantum theories of gravity posit a spin-2 quantum field that is quantized, giving rise to gravitons. In string theory one generally starts with quantized excitations on top of a classically fixed background. This theory is thus described as background dependent. Particles like photons as well as changes in the spacetime geometry (gravitons) are both described as excitations on the string worldsheet. While string theory is "background dependent", the choice of background, like a gauge fixing, does not affect the physical predictions. This is not the case, however, for quantum field theories, which give different predictions for different backgrounds. In contrast, loop quantum gravity, like general relativity, is manifestly background independent, eliminating the (in some sense) "redundant" background required in string theory. Loop quantum gravity, like string theory, also aims to overcome the nonrenormalizable divergences of quantum field theories.

LQG never introduces a background and excitations living on this background, so LQG does not use gravitons as building blocks. Instead one expects that one may recover a kind of semiclassical limit or weak field limit where something like "gravitons" will show up again. In contrast, gravitons play a key role in string theory where they are among the first (massless) level of excitations of a superstring.

LQG differs from string theory in that it is formulated in 3 and 4 dimensions and without supersymmetry or Kaluza-Klein extra dimensions, while the latter requires both to be true. There is no experimental evidence to date that confirms string theory's predictions of supersymmetry and Kaluza–Klein extra dimensions. In a 2003 paper A dialog on quantum gravity,[67] Carlo Rovelli regards the fact LQG is formulated in 4 dimensions and without supersymmetry as a strength of the theory as it represents the most parsimonious explanation, consistent with current experimental results, over its rival string/M-theory. Proponents of string theory will often point to the fact that, among other things, it demonstrably reproduces the established theories of general relativity and quantum field theory in the appropriate limits, which Loop Quantum Gravity has struggled to do. In that sense string theory's connection to established physics may be considered more reliable and less speculative, at the mathematical level. Peter Woit in Not Even Wrong and Lee Smolin in The Trouble with Physics regard string/M-theory to be in conflict with current known experimental results.

Since LQG has been formulated in 4 dimensions (with and without supersymmetry), and M-theory requires supersymmetry and 11 dimensions, a direct comparison between the two has not been possible. It is possible to extend mainstream LQG formalism to higher-dimensional supergravity, general relativity with supersymmetry and Kaluza–Klein extra dimensions should experimental evidence establish their existence. It would therefore be desirable to have higher-dimensional Supergravity loop quantizations at one's disposal in order to compare these approaches. In fact a series of recent papers have been published attempting just this.[68][69][70][71][72][73][74][75] Most recently, Thiemann (and alumni) have made progress toward calculating black hole entropy for supergravity in higher dimensions. It will be interesting to compare these results to the corresponding super string calculations.[76][77]

As of April 2013 LHC has failed to find evidence of supersymmetry or Kaluza–Klein extra dimensions, which has encouraged LQG researchers. Shaposhnikov in his paper "Is there a new physics between electroweak and Planck scales?" has proposed the neutrino minimal standard model,[78] which claims the most parsimonious theory is a standard model extended with neutrinos, plus gravity, and that extra dimensions, GUT physics, and supersymmetry, string/M-theory physics are unrealized in nature, and that any theory of quantum gravity must be four dimensional, like loop quantum gravity.

29.10   LQG and related research programs

Main articles: noncommutative geometry, twistor theory, entropic gravity, Sundance Bilson-Thompson, Asymptotic safety in quantum gravity, Causal dynamical triangulation, group field theory and consistent discretizations

Several research groups have attempted to combine LQG with other research programs: Johannes Aastrup, Jesper M. Grimstrup et al. research combines noncommutative geometry with loop quantum gravity,[79] Laurent Freidel, Simone Speziale, et al., spinors and twistor theory with loop quantum gravity,[80] and Lee Smolin et al. with Verlinde entropic gravity and loop gravity.[81] Stephon Alexander, Antonino Marciano and Lee Smolin have attempted to explain the origins of weak force chirality in terms of Ashketar's variables, which describe gravity as chiral,[82] and LQG with Yang–Mills theory fields[83] in four dimensions. Sundance Bilson-Thompson, Hackett et al.,[84][85] has attempted to introduce standard model via LQG"s degrees of freedom as an emergent property (by employing the idea noiseless subsystems a useful notion introduced in more general situation for constrained systems by Fotini Markopoulou-Kalamara et al.[86]) LQG has also drawn philosophical comparisons with causal dynamical triangulation[87] and asymptotically safe gravity,[88] and the spinfoam with group field theory and AdS/CFT correspondence.[89] Smolin and Wen have suggested combining LQG with String-net liquid, tensors, and Smolin and Fotini Markopoulou-Kalamara Quantum Graphity. There is the consistent discretizations approach. In addition to what has already mentioned above, Pullin and Gambini provide a framework to connect the path integral and canonical approaches to quantum gravity. They may help reconcile the spin foam and canonical loop representation approaches. Recent research by Chris Duston and Matilde Marcolli introduces topology change via topspin networks.[90]

29.11 Problems and comparisons with alternative approaches

Main article: List of unsolved problems in physics

Some of the major unsolved problems in physics are theoretical, meaning that existing theories seem incapable of explaining a certain observed phenomenon or experimental result. The others are experimental, meaning that there is a difficulty in creating an experiment to test a proposed theory or investigate a phenomenon in greater detail.

Can quantum mechanics and general relativity be realized as a fully consistent theory (perhaps as a quantum field theory)? Is spacetime fundamentally continuous or discrete? Would a consistent theory involve a force mediated by a hypothetical graviton, or be a product of a discrete structure of spacetime itself (as in loop quantum gravity)? Are there deviations from the predictions of general relativity at very small or very large scales or in other extreme circumstances that flow from a quantum gravity theory?

The theory of LQG is one possible solution to the problem of quantum gravity, as is string theory. There are substantial differences however. For example, string theory also addresses unification, the understanding of all known forces and particles as manifestations of a single entity, by postulating extra dimensions and so-far unobserved additional particles and symmetries. Contrary to this, LQG is based only on quantum theory and general relativity and its scope is limited to understanding the quantum aspects of the gravitational interaction. On the other hand, the consequences of LQG are radical, because they fundamentally change the nature of space and time and provide a tentative but detailed physical and mathematical picture of quantum spacetime.

Presently, no semiclassical limit recovering general relativity has been shown to exist. This means it remains unproven that LQG's description of spacetime at the Planck scale has the right continuum limit (described by general relativity with possible quantum corrections). Specifically, the dynamics of the theory is encoded in the Hamiltonian constraint, but there is no candidate Hamiltonian.[91] Other technical problems include finding off-shell closure of the constraint algebra and physical inner product vector space, coupling to matter fields of Quantum field theory, fate of the renormalization of the graviton in perturbation theory that lead to ultraviolet divergence beyond 2-loops (see One-loop Feynman diagram in Feynman diagram).[91]

While there has been a recent proposal relating to observation of naked singularities,[92] and doubly special relativity as a part of a program called loop quantum cosmology, there is no experimental observation for which loop quantum gravity makes a prediction not made by the Standard Model or general relativity (a problem that plagues all current theories of quantum gravity). Because of the above-mentioned lack of a semiclassical limit, LQG has not yet even reproduced the predictions made by general relativity.

An alternative criticism is that general relativity may be an effective field theory, and therefore quantization ignores the fundamental degrees of freedom.

29.12 See also

29.13 Notes

[1] Rovelli, Carlo (August 2008). "Loop Quantum Gravity" (PDF). *CERN*. Retrieved 14 September 2014.

[2] Rovelli, C. (2011). "Zakopane lectures on loop gravity". arXiv:1102.3660 [gr-qc].

[3] Muxin, H. (2011). "Cosmological constant in loop quantum gravity vertex amplitude". *Physical Review D* **84** (6): 064010. arXiv:1105.2212. Bibcode:2011PhRvD..84f4010H. doi:10.1103/PhysRevD.84.064010.

[4] Fairbairn, W. J.; Meusburger, C. (2011). "q-Deformation of Lorentzian spin foam models". arXiv:1112.2511 [gr-qc].

[5] Rovelli, C. (2004). *Quantum Gravity*. Cambridge Monographs on Mathematical Physics. p. 71. ISBN 978-0-521-83733-0.

[6] Kauffman, S.; Smolin, L. (7 April 1997). "A Possible Solution For The Problem Of Time In Quantum Cosmology". *Edge.org*. Retrieved 2014-08-20.

[7] Smolin, L. (2006). "The Case for Background Independence". In Rickles, D.; French, S.; Saatsi, J. T. *The Structural Foundations of Quantum Gravity*. Clarendon Press. pp. 196*ff*. ISBN 978-0-19-926969-3.

[8] Rovelli, C. (2004). *Quantum Gravity*. Cambridge Monographs on Mathematical Physics. p. 13ff. ISBN 978-0-521-83733-0.

[9] Thiemann, T. (1996). "Anomaly-free formulation of non-perturbative, four-dimensional Lorentzian quantum gravity". *Physics Letters B* **380**: 257–264. arXiv:gr-qc/9606088. Bibcode:1996PhLB..380..257T. doi:10.1016/0370-2693(96)00532-1.

[10] Baez, J.; de Muniain, J. P. (1994). *Gauge Fields, Knots and Quantum Gravity*. Series on Knots and Everything. Vol. 4. World Scientific. Part III, chapter 4. ISBN 978-981-02-1729-7.

[11] Thiemann, T. (2003). "Lectures on Loop Quantum Gravity". *Lecture Notes in Physics* **631**: 41–135. arXiv:gr-qc/0210094. Bibcode:2003LNP...631...41T. doi:10.1007/978-3-540-45230-0_3.

[12] Rovelli, C.; Smolin, L.(1988). "Knot Theory and Quantum Gravity". *Physical Review Letters* **61**(10): 1155–1958. Bibcode1.1155R. doi:10.1103/PhysRevLett.61.1155.

[13] Gambini, R.; Pullin, J. (2011). *A First Course in Loop Quantum Gravity*. Oxford University Press. Section 8.2. ISBN 978-0-19-959075-9.

[14] Fernando, J.; Barbero, G. (1995). "Reality Conditions and Ashtekar Variables: A Different Perspective". *Physical Review D* **51**: 5498–5506. arXiv:gr-qc/9410013. Bibcode:1995PhRvD..51.5498B. doi:10.1103/PhysRevD.51.5498.

[15] Fernando, J.; Barbero, G. (1995). "Real Ashtekar Variables for Lorentzian Signature Space-times". *Physical Review D* **51**: 5507–5520. arXiv:gr-qc/9410014. Bibcode:1995PhRvD..51.5507B. doi:10.1103/PhysRevD.51.5507.

[16] Bojowald, M.; Alejandro, P. "Spin Foam Quantization and Anomalies". arXiv:gr-qc/0303026 [gr-qc].

[17] Barrett, J.; Crane, L. (2000). "A Lorentzian signature model for quantum general relativity". *Classical and Quantum Gravity* **17**: 3101–3118. arXiv:gr-qc/9904025. Bibcode:2000CQGra..17.3101B. doi:10.1088/0264-9381/17/16/302..

[18] Rovelli, C.; Alesci, E. (2007). "The complete LQG propagator I. Difficulties with the Barrett–Crane vertex". *Physical Review D* **76**: 104012. arXiv:hep-th/0703074. Bibcode:2007PhRvD..76b4012B. doi:10.1103/PhysRevD.76.024012.

[19] Engle, J.; Pereira, R.; Rovelli, C. (2009). "Loop-Quantum-Gravity Vertex Amplitude". *Physical Review Letters* **99**: 161301. arXiv:0705.2388. Bibcode:2007PhRvL..99p1301E. doi:10.1103/physrevlett.99.161301.

[20] Freidal, L.; Krasnov, K. (2008). "A new spin foam model for 4D gravity". *Classical and Quantum Gravity* **25**: 125018. arXiv:0708.1595. Bibcode:2008CQGra..25l5018F. doi:10.1088/0264-9381/25/12/125018.

[21] Alesci, E.; Thiemann, T.; Zipfel, A. (2011). "Linking covariant and canonical LQG: new solutions to the Euclidean Scalar Constraint". arXiv:1109.1290.

[22] Bohm, D. (1989). *Quantum Theory*. Dover Publications. ISBN 978-0-486-65969-5.

[23] Tipler, P.; Llewellyn, R. (2008). *Modern Physics* (5th ed.). W. H. Freeman and Co. pp. 160–161. ISBN 978-0-7167-7550-8.

[24] Bohr, N. (1920). "Über die Serienspektra der Element". *Zeitschrift für Physik* **2** (5): 423–478. Bibcode:1920ZPhy....2..423B. doi:10.1007/BF01329978. (English translation in Bohr 1976, pp. 241–282)

[25] Jammer, M. (1989). *The Conceptual Development of Quantum Mechanics* (2nd ed.). Tomash Publishers. Section 3.2. ISBN 978-0-88318-617-6.

[26] Ashtekar, A.; Bombelli, L.; Corichi, A. (2005). "Semiclassical States for Constrained Systems". *Physical Review D* **72**: 025008. arXiv:hep-ph/0504114. Bibcode:2005PhRvD..72a5008C. doi:10.1103/PhysRevD.72.015008.

[27] Lewandowski, J.; Okołów, A.; Sahlmann, H.; Thiemann, T. (2005). "Uniqueness of Diffeomorphism Invariant States on Holonomy-Flux Algebras". *Communications in Mathematical Physics* **267**: 703–733. arXiv:gr-qc/0504147.Bibcode:207..703L. doi:10.1007/s00220-006-0100-7.

[28] Fleischhack, C. (2006). "Irreducibility of the Weyl algebra in loop quantum gravity". *Physical Review Letters* **97**: 061302. Bibcode:2006PhRvL..97f1302F. doi:10.1103/physrevlett.97.061302.

[29] Thiemann, T. (2008). *Modern Canonical General Relativity*. Cambridge Monographs on Mathematical Physics. Cambridge University Press. Section 10.6. ISBN 978-0-521-74187-3.

[30] "Partial and Complete Observables for Hamiltonian Constrained Systems". *General Relativity and Gravitation* **39**: 1891–1927. 2007. arXiv:gr-qc/0411013. Bibcode:2007GReGr..39.1891D. doi:10.1007/s10714-007-0495-2.

[31] "Partial and Complete Observables for Canonical General Relativity". *Classical and Quantum Gravity* **23**: 6155–6184. arXiv:gr-qc/0507106. Bibcode:2006CQGra..23.6155D. doi:10.1088/0264-9381/23/22/006.

[32] Dreyer, O.; Markopoulou, f.; Smolin, L. (2006). "Symmetry and entropy of black hole horizons". *Nuclear Physics B* **774**: 1–13. arXiv:hep-th/0409056. Bibcode:2006NuPhB.744....1D. doi:10.1016/j.nuclphysb.2006.02.045.

[33] Kribs, D. W.; Markopoulou, F. "Geometry from quantum particles". arXiv:gr-qc/0510052.

[34] Markopoulou, F.; Poulin, D. "Noiseless subsystems and the low energy limit of spin foam models" (unpublished).

[35] *The Phoenix Project: Master Constraint Programme for Loop Quantum Gravity*, Class.Quant.Grav.23:2211-2248,2006 or http://fr.arxiv.org/pdf/gr-qc/0305080

[36] *Modern Canonical Quantum General Relativity* by Thomas Thiemann

[37] *Testing the Master Constraint Programme for Loop Quantum Gravity I. General Framework*, Bianca Dittrich, Thomas Thiemann, Class.Quant.Grav. 23 (2006) 1025-1066.

[38] *Testing the Master Constraint Programme for Loop Quantum Gravity II. Finite Dimensional Systems*, Bianca Dittrich, Thomas Thiemann, Class.Quant.Grav. 23 (2006) 1067-1088.

[39] *Testing the Master Constraint Programme for Loop Quantum Gravity III. SL(2,R) Models*, Bianca Dittrich, Thomas Thiemann, Class.Quant.Grav. 23 (2006) 1089-1120.

[40] *Testing the Master Constraint Programme for Loop Quantum Gravity IV. Free Field Theories*, Bianca Dittrich, Thomas Thiemann, Class.Quant.Grav. 23 (2006) 1121-1142.

[41] *Testing the Master Constraint Programme for Loop Quantum Gravity V. Interacting Field Theories*, Bianca Dittrich, Thomas Thiemann, Class.Quant.Grav. 23 (2006) 1143-1162.

[42] *Quantum Spin Dynamics VIII. The Master Constraint*, Thomas Thiemann, Class.Quant.Grav. 23 (2006) 2249-2266.

[43] *Approximating the physical inner product of Loop Quantum Cosmology*, Benjamin Bahr, Thomas Thiemann, Class.Quant.Grav09-2138,2007.

[44] *On the Relation between Operator Constraint --, Master Constraint --, Reduced Phase Space --, and Path Integral Quantisation*, Muxin Han, Thomas Thiemann, Class.Quant.Grav.27:225019,2010.

[45] *On the Relation between Rigging Inner Product and Master Constraint Direct Integral Decomposition*, Muxin Han, Thomas Thiemann, J.Math.Phys.51:092501,2010.

[46] *A Path-integral for the Master Constraint of Loop Quantum Gravity*, Muxin Han, Class.Quant.Grav.27:215009,2010

[47] *Emergent diffeomorphism invariance in a discrete loop quantum gravity model*, Rodolfo Gambini, Jorge Pullin, Class.Quant.G2009

[48] Section 10.2.2 *A First Course in Loop quantum Gravity*, Rodolfo Gambinni, Jorge Pullin, Oxford University Press, first published 2011.

[49] *Algebraic Quantum Gravity (AQG) I. Conceptual Setup*, K. Giesel, T. Thiemann, Class.Quant.Grav.24:2465-2498,2007.

[50] *Algebraic Quantum Gravity (AQG) II. Semiclassical Analysis*, K. Giesel, T. Thiemann, Class.Quant.Grav.24:2499-2564,2007.

[51] *Algebraic Quantum Gravity (AQG) III. Semiclassical Perturbation Theory*, K. Giesel, T. Thiemann, Class.Quant.Grav.24:2565-2588,2007.

[52] Bousso, Raphael (2002). "The Holographic Principle". *Reviews of Modern Physics* **74** (3): 825–874. arXiv:hep-th/0203101. Bibcode:2002RvMP...74..825B. doi:10.1103/RevModPhys.74.825.

[53] Majumdar, Parthasarathi (1998). "Black Hole Entropy and Quantum Gravity". *ArXiv: General Relativity and Quantum Cosmology* **73**: 147. arXiv:gr-qc/9807045. Bibcode:1999InJPB..73..147M.

[54] See List of loop quantum gravity researchers

[55] Rovelli, Carlo (1996). "Black Hole Entropy from Loop Quantum Gravity". *Physical Review Letters* **77** (16): 3288–3291. arXiv:gr-qc/9603063. Bibcode:1996PhRvL..77.3288R. doi:10.1103/PhysRevLett.77.3288.

[56] Ashtekar, Abhay; Baez, John; Corichi, Alejandro; Krasnov, Kirill (1998). "Quantum Geometry and Black Hole Entropy". *Physical Review Letters* **80** (5): 904–907. arXiv:gr-qc/9710007. Bibcode:1998PhRvL..80..904A. doi:10.1103/PhysRevLett.80.904.

[57] *Quantum horizons and black hole entropy: Inclusion of distortion and rotation*, Abhay Ashtekar, Jonathan Engle, Chris Van Den Broeck, Class.Quant.Grav.22:L27-L34, 2005.

[58] Bianchi, Eugenio (2012). "Entropy of Non-Extremal Black Holes from Loop Gravity". arXiv:1204.5122.

[59] http://inspirehep.net/record/940357?ln=en. http://inspirehep.net/record/1111991.

[60] *On (Cosmological) Singularity Avoidance in Loop Quantum Gravity*, Johannes Brunnemann, Thomas Thiemann, Class.Quant.Grav. 23 (2006) 1395-1428.

[61] *Unboundedness of Triad -- Like Operators in Loop Quantum Gravity*, Johannes Brunnemann, Thomas Thiemann, Class.Quant.Grav. 23 (2006) 1429-1484.

[62] L. Modesto, C. Rovelli:*Particle scattering in loop quantum gravity*, Phys Rev Lett 95 (2005) 191301

[63] R Oeckl, *A 'general boundary' formulation for quantum mechanics and quantum gravity*, Phys Lett B575 (2003) 318-324 ; *Schrodinger's cat and the clock: lessons for quantum gravity*, Class Quant Grav 20 (2003) 5371-5380l

[64] F. Conrady, C. Rovelli *Generalized Schrodinger equation in Euclidean field theory*", Int J Mod Phys A 19, (2004) 1-32.

[65] L Doplicher, *Generalized Tomonaga-Schwinger equation from the Hadamard formula*, Phys Rev D70 (2004) 064037

[66] F. Conrady, L. Doplicher, R. Oeckl, C. Rovelli, M. Testa, *Minkowski vacuum in background independent quantum gravity*, Phys Rev D69 (2004) 064019.

[67] http://arxiv.org/abs/arXiv:hep-th/0310077

[68] *New Variables for Classical and Quantum Gravity in all Dimensions I. Hamiltonian Analysis*, Norbert Bodendorfer, Thomas Thiemann, Andreas Thurn, Class. Quantum Grav. 30 (2013) 045001

[69] *New Variables for Classical and Quantum Gravity in all Dimensions II. Lagrangian Analysis*, Norbert Bodendorfer, Thomas Thiemann, Andreas Thurn, Quantum Grav. 30 (2013) 045002

[70] *New Variables for Classical and Quantum Gravity in all Dimensions III. Quantum Theory*, Norbert Bodendorfer, Thomas Thiemann, Andreas Thurn, Class. Quantum Grav. 30 (2013) 045003

[71] *New Variables for Classical and Quantum Gravity in all Dimensions IV. Matter Coupling*, Norbert Bodendorfer, Thomas Thiemann, Andreas Thurn, Class. Quantum Grav. 30 (2013) 045004

[72] *On the Implementation of the Canonical Quantum Simplicity Constraint*, Norbert Bodendorfer, Thomas Thiemann, Andreas Thurn, Class. Quantum Grav. 30 (2013) 045005

[73] *Towards Loop Quantum Supergravity (LQSG) I. Rarita-Schwinger Sector*, Norbert Bodendorfer, Thomas Thiemann, Andreas Thurn, Class. Quantum Grav. 30 (2013) 045006

[74] *Towards Loop Quantum Supergravity (LQSG) II. p-Form Sector*, Norbert Bodendorfer, Thomas Thiemann, Andreas Thurn, Class. Quantum Grav. 30 (2013) 045007

[75] *Towards Loop Quantum Supergravity (LQSG)*, Norbert Bodendorfer, Thomas Thiemann, Andreas Thurn, Phys. Lett. B 711: 205-211 (2012)

[76] *New Variables for Classical and Quantum Gravity in all Dimensions V. Isolated Horizon Boundary Degrees of Freedom*, Norbert Bodendorfer, Thomas Thiemann, Andreas Thurn, http://uk.arxiv.org/pdf/1304.2679.

[77] *Black hole entropy from loop quantum gravity in higher dimensions*, Norbert Bodendorfer http://uk.arxiv.org/pdf/1307.5029

[78] http://arxiv.org/abs/0708.3550

[79] http://arxiv.org/abs/1203.6164

[80] http://arxiv.org/abs/1006.0199

[81] http://arxiv.org/abs/1001.3668

[82] http://arxiv.org/abs/1212.5246

[83] http://arxiv.org/abs/1105.3480

[84]*Quantum gravity and the standard model*, Sundance O. Bilson-Thompson, Fotini Markopoulou, Lee Smolin, Class.Quant.Grav.24:39-3994,2007.

[85] For a precise review and outlook of this research see: *Emergent Braided Matter of Quantum Geometry*, Sundance Bilson-Thompson, Jonathan Hackett, Louis Kauffman, Yidun Wan, SIGMA 8 (2012), 014, 43 pages.

[86] *Constrained Mechanics and Noiseless Subsystems*, Tomasz Konopka, Fotini Markopoulou, arXiv:gr-qc/0601028.

[87] http://www.perimeterinstitute.ca/people/renate-loll

[88] wwnpqft.inln.cnrs.fr/pdf/Bianchi.pdf

[89] http://arxiv.org/abs/0804.0632

[90] http://arxiv.org/abs/1308.2934

[91] Nicolai, Hermann; Peeters, Kasper; Zamaklar, Marija (2005). "Loop quantum gravity: an outside view". *Classical and Quantum Gravity* **22** (19): R193–R247. arXiv:hep-th/0501114. Bibcode:2005CQGra..22R.193N. doi:10.1088/0264-9381/22/19/R01.

[92] Goswami; Joshi, Pankaj S.; Singh, Parampreet et al. (2006). "Quantum evaporation of a naked singularity". *Physical Review Letters* **96** (3): 31302. arXiv:gr-qc/0506129. Bibcode:2006PhRvL..96c1302G. doi:10.1103/PhysRevLett.96.031302.

29.14   References

- Topical Reviews

  - Rovelli, Carlo (2011). "Zakopane lectures on loop gravity". arXiv:1102.3660.

  - Rovelli, Carlo (1998). "Loop Quantum Gravity". *Living Reviews in Relativity* **1**. Retrieved 2008-03-13.

  - Thiemann, Thomas (2003). "Lectures on Loop Quantum Gravity". *Lectures Notes in Physics*. Lecture Notes in Physics **631**: 41–135. arXiv:gr-qc/0210094. Bibcode:2003LNP...631...41T. doi:10.1007/978-3-540-45230-0_3. ISBN 978-3-540-40810-9.

  - Ashtekar, Abhay; Lewandowski, Jerzy (2004). "Background Independent Quantum Gravity: A Status Report". *Classical and Quantum Gravity* **21** (15): R53–R152. arXiv:gr-qc/0404018. Bibcode:2004CQGr53A. doi:10.1088/0264-9381/21/15/R01.

  - Carlo Rovelli and Marcus Gaul, *Loop Quantum Gravity and the Meaning of Diffeomorphism Invariance*, e-print available as gr-qc/9910079.

  - Lee Smolin, *The case for background independence*, e-print available as hep-th/0507235.

  - Alejandro Corichi, *Loop Quantum Geometry: A primer*, e-print available as .

  - Alejandro Perez, *Introduction to loop quantum gravity and spin foams*, e-print available as .

- Hermann Nicolai and Kasper Peeters *Loop and spin foam quantum gravity: A Brief guide for beginners.*, e-print available as .

- Popular books:

  - Lee Smolin, *Three Roads to Quantum Gravity*

  - Carlo Rovelli, *Che cos'è il tempo? Che cos'è lo spazio?*, Di Renzo Editore, Roma, 2004. French translation: *Qu'est ce que le temps? Qu'est ce que l'espace?*, Bernard Gilson ed, Brussel, 2006. English translation: *What is Time? What is space?*, Di Renzo Editore, Roma, 2006.

  - Julian Barbour, *The End of Time: The Next Revolution in Our Understanding of the Universe*

  - Musser, George (2008). "The Complete Idiot's Guide to String Theory". *The Physics Teacher* (Indianapolis: Alpha) **47** (2): 368. Bibcode:2009PhTea..47Q.128H. doi:10.1119/1.3072469. ISBN 978-1-59257-702-6. – Focuses on string theory but has an extended discussion of loop gravity as well.

- Magazine articles:

  - Lee Smolin, "Atoms of Space and Time", *Scientific American*, January 2004

  - Martin Bojowald, "Following the Bouncing Universe", *Scientific American*, October 2008

- Easier introductory, expository or critical works:

  - Abhay Ashtekar, *Gravity and the quantum*, e-print available as gr-qc/0410054 (2004)

  - John C. Baez and Javier Perez de Muniain, *Gauge Fields, Knots and Quantum Gravity*, World Scientific (1994)

  - Carlo Rovelli, *A Dialog on Quantum Gravity*, e-print available as hep-th/0310077 (2003)

  - Rodolfo Gambini and Jorge Pullin, *A First Course in Loop Quantum Gravity*, Oxford (2011)

  - Carlo Rovelli and Francesca Vidotto, *Covariant Loop Quantum Gravity*, Cambridge (2014); draft available online

- More advanced introductory/expository works:

  - Carlo Rovelli, *Quantum Gravity*, Cambridge University Press (2004); draft available online

  - Thomas Thiemann, *Introduction to modern canonical quantum general relativity*, e-print available as gr-qc/0110034

  - Thomas Thiemann, *Introduction to Modern Canonical Quantum General Relativity*, Cambridge University Press (2007)

  - Abhay Ashtekar, *New Perspectives in Canonical Gravity*, Bibliopolis (1988).

  - Abhay Ashtekar, *Lectures on Non-Perturbative Canonical Gravity*, World Scientific (1991)

  - Rodolfo Gambini and Jorge Pullin, *Loops, Knots, Gauge Theories and Quantum Gravity*, Cambridge University Press (1996)

  - Hermann Nicolai, Kasper Peeters, Marija Zamaklar, *Loop quantum gravity: an outside view*, e-print available as hep-th/0501114

  - H. Nicolai and K. Peeters, *Loop and Spin Foam Quantum Gravity: A Brief Guide for Beginners*, e-print available as hep-th/0601129

  - T. Thiemann The LQG – String: Loop Quantum Gravity Quantization of String Theory (2004)

- Conference proceedings:

  - John C. Baez (ed.), *Knots and Quantum Gravity*

- Fundamental research papers:

  - Ashtekar, Abhay (1986). "New variables for classical and quantum gravity". *Physical Review Letters* **57** (18): 2244–2247. Bibcode:1986PhRvL..57.2244A. doi:10.1103/PhysRevLett.57.2244. PMID 10033673

- Ashtekar, Abhay (1987). "New Hamiltonian formulation of general relativity". *Physical Review D* **36** (6): 1587–1602. Bibcode:1987PhRvD..36.1587A. doi:10.1103/PhysRevD.36.1587

- Roger Penrose, *Angular momentum: an approach to combinatorial space-time* in *Quantum Theory and Beyond*, ed. Ted Bastin, Cambridge University Press, 1971

- Rovelli, Carlo; Smolin, Lee (1988). "Knot theory and quantum gravity". *Physical Review Letters* **61** (10): 1155–1158. Bibcode:1988PhRvL..61.1155R. doi:10.1103/PhysRevLett.61.1155.

- Rovelli, Carlo; Smolin, Lee (1990). "Loop space representation of quantum general relativity". *Nuclear Physics* **B331**: 80–152.

- Carlo Rovelli and Lee Smolin, *Discreteness of area and volume in quantum gravity*, Nucl. Phys., **B442** (1995) 593-622, e-print available as gr-qc/9411005

- Kuchař, Karel (1973). "Canonical Quantization of Gravity". In Israel, Werner. *Relativity, Astrophysics and Cosmology*. D. Reidel. pp. 237–288. ISBN 90-277-0369-8.

- Thiemann, Thomas (2006). "Loop Quantum Gravity: An Inside View". *Approaches to Fundamental Physics*. Lecture Notes in Physics **721**: 185–263. arXiv:hep-th/0608210.Bibcode:2007LNP...721..185T.doi:17/978-3-540-71117-9_10. ISBN 978-3-540-71115-5.

29.15   External links

- "Loop Quantum Gravity" by Carlo Rovelli Physics World, November 2003

- Quantum Foam and Loop Quantum Gravity

- Abhay Ashtekar: Semi-Popular Articles . Some excellent popular articles suitable for beginners about space, time, GR, and LQG.

- Loop Quantum Gravity: Lee Smolin.

- Loop Quantum Gravity on arxiv.org

- A list of LQG references catered to fresh graduates

- Loop Quantum Gravity Lectures Online by Lee Smolin

- Spin networks, spin foams and loop quantum gravity

- Wired magazine, News: *Moving Beyond String Theory*

- April 2006 Scientific American Special Issue, *A Matter of Time*, has Lee Smolin LQG Article *Atoms of Space and Time*

- September 2006, The Economist, article *Looping the loop*

- Gamma-ray Large Area Space Telescope: http://glast.gsfc.nasa.gov/

- Zeno meets modern science. Article from Acta Physica Polonica B by Z.K. Silagadze.

- Did pre-big bang universe leave its mark on the sky? - According to a model based on "loop quantum gravity" theory, a parent universe that existed before ours may have left an imprint (*New Scientist*, 10 April 2008)

Chapter 30

Supersymmetric gauge theory

In theoretical physics, one often analyzes theories with supersymmetry which also have internal gauge symmetries. So, it is important to come up with a supersymmetric generalization of gauge theories.

We begin this article with a brief introduction to gauge theory. Then we will see what happens when gauge theories are formulated in the framework of supersymmetry (SUSY).

30.0.1 Gauge theory

A gauge theory is a mathematical framework for analysing gauge symmetries. There are two types of symmetries, viz., global and local. A global symmetry is the symmetry which remains invariant at each point of a manifold (manifold can be either of space-time coordinates or that of internal quantum numbers). A local symmetry is the symmetry which depends upon the space over which it is defined, and changes with the variation in coordinates. Thus such symmetry is invariant only locally (i.e., on a subset of the manifold).

Maxwell's equations and quantum electrodynamics are famous examples of gauge theories.

30.0.2 SUSY

According to the Particle Physics, there exist two kinds of particles in nature, namely, bosons and fermions. Bosons are integer spin particles. Their characteristic feature is that any amount of these can occupy a single place in space. Thus they are identified with the forces of nature. Fermions are half-integer spin particles. No two of which, bearing the same quantum number can occupy a single position in space-time. Thus they are identified with the matter. This is precisely the reason why SUSY is such an ideal candidate for the Unification of radiation and matter.

This mechanism works via an operator Q, known as supersymmetry generator, which acts as follows:

$Q|\text{boson}\rangle = \text{fermion}$
$Q|\text{fermion}\rangle = \text{boson}$

For instance, the supersymmetry generator can take a photon as an argument and transform it into a photino and vice versa. This happens through translation in the (parameter) space. This superspace is a $\mathbb{Z}_2$ graded vector space $\mathcal{W} - \mathcal{W}^0 \oplus \mathcal{W}^1$, where $\mathcal{W}^0$ is the bosonic Hilbert space and $\mathcal{W}^1$ is the fermionic Hilbert space.

30.0.3 SUSY gauge theory

The motivation for a supersymmetric version of gauge theory can be the fact that gauge invariance is consistent with supersymmetry.

Because both the half-integer spin fermions and the integer spin bosons can become gauge particles. Moreover the vector fields and the spinor fields both reside in the same representation of the internal symmetry group.

Suppose we have a gauge transformation $V_\mu \to V_\mu + \partial_\mu A$, where V_μ is a vector field and A is the gauge function. The main problem in construction of SUSY Gauge Theory is to extend the above transformation in a way that is consistent with SUSY transformations.

The Wess-Zumino gauge provides a successful solution to this problem. Once such suitable gauge is obtained, the dynamics of the SUSY gauge theory work as follows: we seek a lagrangian that is invariant under the Super-gauge transformations (these transformations are an important tool needed to develop supersymmetric version of a gauge theory). Then we can integrate the lagrangian using the Berezin integration rules and thus obtain the action. Which further leads to the equations of motion and hence can provide a complete analysis of the dynamics of the theory.

30.1 $N = 1$ SUSY in 4D (with 4 real generators)

In four dimensions, the minimal $N = 1$ supersymmetry may be written using a superspace. This superspace involves four extra fermionic coordinates $\theta^1, \theta^2, \bar{\theta}^1, \bar{\theta}^2$, transforming as a two-component spinor and its conjugate.

Every superfield, i.e. a field that depends on all coordinates of the superspace, may be expanded with respect to the new fermionic coordinates. There exists a special kind of superfields, the so-called chiral superfields, that only depend on the variables θ but not their conjugates (more precisely, $\overline{D} f = 0$). However, a vector superfield depends on all coordinates. It describes a gauge field and its superpartner, namely a Weyl fermion that obeys a Dirac equation.

$$V = C + i\theta\chi - i\bar{\theta}\bar{\chi} + \tfrac{i}{2}\theta^2(M+iN) - \tfrac{i}{2}\bar{\theta}^2(M-iN) - \theta\sigma^\mu\bar{\theta}v_\mu + i\theta^2\bar{\theta}\left(\bar{\lambda} + \tfrac{1}{2}\bar{\sigma}^\mu\partial_\mu\chi\right) - i\bar{\theta}^2\theta\left(\lambda + \tfrac{i}{2}\sigma^\mu\partial_\mu\bar{\chi}\right) + \tfrac{1}{2}\theta^2\bar{\theta}^2$$

$$\times \left(D + \tfrac{1}{2}\Box C\right)$$

V is the vector superfield (**prepotential**) and is real ($V = V$). The fields on the right hand side are component fields.

The gauge transformations act as

$$V \to V + \Lambda + \overline{\Lambda}$$

where Λ is any chiral superfield.

It's easy to check that the chiral superfield

$$W_\alpha \equiv -\tfrac{1}{4}\overline{D}^2 D_\alpha V$$

is gauge invariant. So is its complex conjugate $\overline{W}_{\dot{\alpha}}$.

A nonSUSY covariant gauge which is often used is the **Wess–Zumino gauge**. Here, C, χ, M and N are all set to zero. The residual gauge symmetries are gauge transformations of the traditional bosonic type.

A chiral superfield X with a charge of q transforms as

$$X \to e^{q\Lambda}X, \qquad \overline{X} \to e^{q\overline{\Lambda}}X$$

Therefore $Xe^{-qV}X$ is gauge invariant. Here e^{-qV} is called a **bridge** since it "bridges" a field which transforms under Λ only with a field which transforms under Λ only.

More generally, if we have a real gauge group G that we wish to supersymmetrize, we first have to complexify it to G^c. e^{-qV} then acts a **compensator** for the complex gauge transformations in effect absorbing them leaving only the real parts. This is what's being done in the Wess–Zumino gauge.

30.1.1 Differential superforms

Let's rephrase everything to look more like a conventional Yang–Mills gauge theory. We have a U(1) gauge symmetry acting upon full superspace with a 1-superform gauge connection A. In the analytic basis for the tangent space, the covariant derivative is given by $D_M = d_M + iqA_M$. Integrability conditions for chiral superfields with the chiral constraint

$$\overline{D}_{\dot\alpha} X = 0$$

leave us with

$$\left\{ \overline{D}_{\dot\alpha}, \overline{D}_{\dot\beta} \right\} = F_{\dot\alpha\dot\beta} = 0.$$

A similar constraint for antichiral superfields leaves us with $F\alpha\beta = 0$. This means that we can either gauge fix $A_{\dot\alpha} = 0$ or $A\alpha = 0$ but not both simultaneously. Call the two different gauge fixing schemes I and II respectively. In gauge I, $\overline{d}_{\dot\alpha} X = 0$ and in gauge II, $da\, X = 0$. Now, the trick is to use two different gauges simultaneously; gauge I for chiral superfields and gauge II for antichiral superfields. In order to **bridge** between the two different gauges, we need a gauge transformation. Call it e^{-V} (by convention). If we were using one gauge for all fields, XX would be gauge invariant. However, we need to convert gauge I to gauge II, transforming X to $(e^{-V})^q X$. So, the gauge invariant quantity is $Xe^{-qV}X$.

In gauge I, we still have the residual gauge e^Λ where $\overline{d}_{\dot\alpha}\Lambda = 0$ and in gauge II, we have the residual gauge e^Λ satisfying $da\,\Lambda = 0$. Under the residual gauges, the bridge transforms as

$$e^{-V} \to e^{-\overline{\Lambda} - V - \Lambda}.$$

Without any additional constraints, the bridge e^{-V} wouldn't give all the information about the gauge field. However, with the additional constraint $F_{\dot\alpha\beta}$, there's only one unique gauge field which is compatible with the bridge modulo gauge transformations. Now, the bridge gives exactly the same information content as the gauge field.

30.2   Theories with 8 or more SUSY generators ($N > 1$)

In theories with higher supersymmetry (and perhaps higher dimension), a vector superfield typically describes not only a gauge field and a Weyl fermion but also at least one complex scalar field.

30.3   See also

- super QCD

- superpotential

- D-term

- F-term

- current superfield

30.4   References

- Stephen P. Martin. *A Supersymmetry Primer*, arxiv.org/pdf/hep-ph/9709356v6.pdf .

- Prakash, Nirmala. *Mathematical Perspective on Theoretical Physics: A Journey from Black Holes to Superstrings*, World Scientific (2003).

Chapter 31

Wess–Zumino model

Not to be confused with Wess–Zumino–Witten model.

In theoretical physics, the **Wess–Zumino model** has become the first known example of an interacting four-dimensional quantum field theory with supersymmetry, at least in the Western world. In 1974, Julius Wess and Bruno Zumino studied, using modern terminology, dynamics of a single chiral superfield (composed of a complex scalar and a spinor fermion) whose cubic superpotential leads to a renormalizable theory.

The Lagrangian of the *free* massless Wess–Zumino model in four-dimensional spacetime with flat metric $\mathrm{diag}(-1,1,1,1)$ is

$$\mathcal{L} = -\frac{1}{2}(\partial S)^2 - \frac{1}{2}(\partial P)^2 - \frac{1}{2}\bar{\psi}\partial\!\!\!/\psi$$

with S a scalar field, P a pseudoscalar field and ψ a Majorana spinor field. The action is invariant under the transformations generated by the superalgebra. The infinitesimal form of these transformations is:

$$\delta_\epsilon S = \bar{\epsilon}\psi$$

$$\delta_\epsilon P = \bar{\epsilon}\gamma_5\psi$$

$$\delta_\epsilon \psi = \partial\!\!\!/(S + P\gamma_5)\epsilon$$

where ϵ is a Majorana spinor-valued transformation parameter and γ_5 is the chirality operator.

Invariance under a (modified) set of supersymmetry transformations remains if one adds mass terms for the fields, provided the masses are equal. It is also possible to add interaction terms under some algebraic conditions on the coupling constants, resulting from the fact that the interactions come from superpotential for the chiral superfield containing the fields S, P and ψ.

31.1   References

- Figueroa-O'Farrill, J. M. (2001). "Busstepp Lectures on Supersymmetry". arXiv:hep-th/0109172.

- Wess, J.; Zumino, B. (1974). "Supergauge transformations in four dimensions". *Nuclear Physics B* **70** (1): 39–50. Bibcode:1974NuPhB..70...39W. doi:10.1016/0550-3213(74)90355-1.

31.2 Text and image sources, contributors, and licenses

31.2.1 Text

- **Supersymmetry** *Source:* https://en.wikipedia.org/wiki/Supersymmetry?oldid=671662371 *Contributors:* Bryan Derksen, Taw, Andre Engels, Roadrunner, Maury Markowitz, Ewen, Stevertigo, Edward, Michael Hardy, Arpingstone, Theresa knott, IMSoP, Jeandré du Toit, Samw, Smack, Charles Matthews, Maximus Rex, Phys, Raul654, BenRG, Rursus, Mor~enwiki, Ancheta Wis, Giftlite, Mporter, Ferkelparade, Monedula, Fropuff, Xerxes314, Anville, Gus Polly, Moyogo, Unconcerned, DO'Neil, Maarten van Vliet, Pharotic, LiDaobing, Sam Hocevar, Lumidek, Deglr6328, Arivero, Rich Farmbrough, Roybb95~enwiki, Bender235, El C, Nornagon~enwiki, Duk, Tweet Tweet, LostLeviathan, Pearle, Gary, Francescog~enwiki, Wtmitchell, RJFJR, Reaverdrop, Blaxthos, Killing Vector, Jordan14, Ted BJ, MONGO, Mpatel, MFH, SeventyThree, Bodera, VermillionBird, Drbogdan, Rjwilmsi, Josiah Rowe, R.e.b., Bubba73, Maxim Razin, Drrngrvy, FlaBot, Cless Alvein, Nowhither, Itinerant1, Gparker, KFP, Lmatt, Chobot, Vyroglyph, YurikBot, Wavelength, RussBot, Ohwilleke, Bhny, Epolk, Maxim Leyenson, Chaos, Romanc19s, Bota47, Mgnbar, Closedmouth, Arthur Rubin, RG2, That Guy, From That Show!, A bit iffy, SmackBot, Mira, Kurochka, Wangjiaji, Gilliam, Bluebot, Cadmasteradam, Complexica, Bazonka, Colonies Chris, Can't sleep, clown will eat me, QFT, Ruff ilb, Robma, Solarapex, Radagast83, Jgwacker, TheMaster42, Martijn Hoekstra, Ligulembot, Acjohnson55, Yevgeny Kats, Charleswestbrook, TriTertButoxy, Lambiam, Tktktk, Xiaphias, JarahE, Mdanziger, Dan Gluck, Newone, Marysunshine, Tawkerbot2, Cydebot, Hydraton31, Bazzargh, David edwards, Michael C Price, Crum375, Koeplinger, Headbomb, J.christianson, Escarbot, Salgueiro~enwiki, Kborland, Jpod2, Cgingold, Maliz, TimidGuy, C9, Kostisl, R'n'B, Zentropa77, Natsirtguy, Maurice Carbonaro, Kevin Hickerson, Shawn in Montreal, Idioma-bot, Sheliak, Cuzkatzimhut, Nxavar, Kawakameha, Cuboidal, Ptrslv72, PhysPhD, Kbrose, SieBot, Nn123645, ClueBot, Jcpilman, Chessmaster7m, Kitsunegami, Rhododendrites, Mastertek, Mishas42, Scrabby~enwiki, TimothyRias, WikHead, MystBot, Addbot, DOI bot, Zahd, Barak Sh, F Notebook, Lightbot, Luckas-bot, Yobot, Ibayn, TaBOT-zerem, Amirobot, Nonnormalizable, AnomieBOT, Girl Scout cookie, Citation bot, ArthurBot, Plumpurple, Tomwsulcer, Omnipaedista, Gsard, CES1596, FrescoBot, HaloStereo1, Paine Ellsworth, Xmikywayx, Citation bot 1, Gil987, Kikeku, Jonesey95, Eddie Nixon, MondalorBot, Aknochel, Tom1661, Gagoga ju, TobeBot, Puzl bustr, Andraas, EmausBot, Djloststylez, Ddimensões, Arbnos, Susy is it, ChuispastonBot, Isocliff, ClueBot NG, KagakuKyouju, IJVin, Frietjes, Helpful Pixie Bot, Bibcode Bot, BG19bot, Teika kazura, JayBeeEye, Ninmacer20, ChrisGualtieri, Logosun, AHusain314, NA48, Rfassbind, Katherine Pendleton, Lioinnisfree, Liquidityinsta, TaiSakuma, Stamptrader, Kdmeaney, Qxxxxxq, Almaionescu, Monkbot, Janhaithabu, Mammoth2011, Jwill530, Stacie Croquet, Cuttlas1 and Anonymous: 172

- **Spacetime symmetries** *Source:* https://en.wikipedia.org/wiki/Spacetime_symmetries?oldid=62780 *Contributors:* Edward, Charles Matthews, Hooperbloob, Oleg Alexandrov, Mpatel, BD2412, Salix alba, Ligulem, BradBeattie, Bornhj, Hillman, Grafen, That Guy, From That Show!, SmackBot, Mgiganteus1, Myasuda, Michael H 34, Legobot, Yobot, Citation bot, RockSolidCosmo, Maschen, AHusain314 and Anonymous:1

- **Minimal Supersymmetric Standard Model** *Source:* https://en.wikipedia.org/wiki/Minimal_Supersymmetric_Standard_Model?oldid=6711 *Contributors:* Phys, Dmytro, Gandalf61, Rursus, Connelly, Marcika, Waltpohl, Pharotic, Carandol~enwiki, HorsePunchKid, Grunt, Pja-cobi, Jensbn,ElC, Jag123, JohnyDog, RJFJR, DV8 2XL, Woohookitty, Mpatel, VermillionBird, Rjwilmsi, Goudzovski, Bhny, JabberWok,Shawn81, SCZenz,Closedmouth, Caco de vidro, Tom Lougheed, Stepa, Dauto, Chris the speller, Bluebot, Colonies Chris, Sl1982, QFT,MBlume, Jgwacker, Pulu, CenozoicEra, NNemec, Waggers, Dan Gluck, Iridescent, Antonio Prates, Lottamiata, CmdrObot, Michael C Price,Dchristle, RoadMap, Headbomb, CannedhamX, Knotwork, Yill577, Paulnilsson, Maliz, Dr. Morbius, Andre.holzner, Wilsonge, Red Act,Pjoef, StewartMH, PipepBot, ArdClose, Mastertek, Chaosdruid, Rreagan007, SkyLined, Addbot, DOI bot, Mjamja, Tokikake, Luckas-bot,Yobot, Wireader, AnomieBOT, Archon 2488, Citation bot, GenQuest, GrouchoBot, Omnipaedista, Ernsts, Paine Ellsworth, Identitaamore,Citation bot 1, PigFlu Oink, Puzl bustr, RjwilmsiBot, Akrose, EmausBot, WCEngineer, Arbnos, Suslindisambiguator, AManWithNoPlan,Iscoliff, Zukertort, Bibcode Bot, BG19bot, ElphiBot, Physlad, ChrisGualtieri, Cinaro, Stamptrader and Anonymous: 49

- **Physics beyond the Standard Model** *Source:* https://en.wikipedia.org/wiki/Physics_beyond_the_Standard_Model?oldid=670089290 *Contributors:* David spector, Ewen, Michael Hardy, Andrewman327, Donarreiskoffer, Nurg, Rursus, David Gerard, Alison, David Schaich, RJHall, El C, Kwamikagami, I9Q79oL78KiL0QTFHgyc, Jeodesic, 4v4l0n42, Alinor, Count Iblis, Rjwilmsi, Strait, Eyu100, HappyCamper, Lmatt, BradBeattie, Ohwilleke, Bhny, SCZenz, CecilWard, Karl Andrews, Nlu, Dna-webmaster, Pawyilee, 2over0, Caco de vidro, Jaysbro, SmackBot, Mdj, Nickst, Chris the speller, Bluebot, Scwlong, QFT, Pepsidrinka, Jgwacker, Yevgeny Kats, Doug Bell, John, Dspitzle, RandomCritic, JarahE, Kurtan~enwiki, Headbomb, Peter Gulutzan, N shaji, Lenny Kaufman, VoABot II, Email4mobile, Maliz, R'n'B, HEL, Natsirtguy, Rod57, Tarotcards, DadaNeem, Goop Goop, Fences and windows, Michael H 34, Venny85, Wing gundam, Beast of traal, Bhuna71, Mild Bill Hiccup, Djr32, Excirial, RCalabraro, Brews ohare, Mastertek, TimothyRias, Truthnlove, Addbot, Luckas-bot, Zhitelew, Yobot, AnomieBOT, Citation bot, LilHelpa, Smk65536, Stevebow, Omnipaedista, Seeleschneider, A. di M., Kenneth Dawson, Steve Quinn, Citation bot 2, Aturen, Tom.Reding, ErgSlider, Physics therapist, Gistmass, Bj norge, Vstarsky, Serketan, ZéroBot, Galaktiker, Arbnos, Suslindisambiguator, Wiggles007, Smtchahal, ClaudeDes, Braincricket, Widr, Helpful Pixie Bot, Mike9110, DryRun, Bibcode Bot, BG19bot, Brainssturm, Qtom.masters, ThePeriodicTable123, M0532062613, Andyhowlett, Cinaro, I am One of Many, Kowtje, CtrlAltBackspace, 22merlin, Monkbot, Delbert7, Tetra quark, TQuentin, MauiPhoenix and Anonymous: 73

- **Hierarchy problem** *Source:* https://en.wikipedia.org/wiki/Hierarchy_problem?oldid=671659630 *Contributors:* The Anome, WhisperToMe, Phys, AnonMoos, Jni, Giftlite, Xerxes314, Thincat, Lumidek, Rich Farmbrough, FT2, Pt, Jag123, I9Q79oL78KiL0QTFHgyc, Mindmatrix, GregorB, VermillionBird, Coemgenus, Mattmartin, Strait, Salix alba, UkPaolo, Ugha, Bhny, Netrapt, Ephraim33, QFT, Jgwacker, NNemec, Ninjakannon, Shambolic Entity, Dr. Morbius, Drgnrave, X!, James Banogon, Megalekaitrane, D.scain.farenzena, Copyeditor42, Alexbot, Lalegria, Addbot, Mixen Dixon, TutterMouse, Debresser, Topquark22, Yobot, AnomieBOT, Yemibedu, Materialscientist, Citation bot, Neurolysis, ArthurBot, Pra1998, Omnipaedista, A. di M., Erik9bot, Banak, FrescoBot, Paine Ellsworth, Puzl bustr, Bj norge, Hauntedpz, RjwilmsiBot, EmausBot, AsceticRose, Arbnos, Suslindisambiguator, Quondum, Jbackroyd, Bibcode Bot, Ervin Goldfain, Drcooljoe, IluvatarBot, Ownedroad9, MSUGRA, Prokaryotes, Mfb and Anonymous: 44

- **Higgs mechanism** *Source:* https://en.wikipedia.org/wiki/Higgs_mechanism?oldid=668245280 *Contributors:* CYD, Roadrunner, Ubiquity, Michael Hardy, Julesd, Palfrey, Charles Matthews, Doradus, Tpbradbury, Phys, Sbisolo, David Gerard, Ancheta Wis, Giftlite, Herbee, Edcolins, Lumidek, Ukexpat, Benzh~enwiki, Chris Howard, FT2, Mat cross, David Schaich, Saintswithin, Mal~enwiki, Bender235, Viriditas, BDD, Oleg Alexandrov, Kelly Martin, Linas, BoLingua, Duncan.france, Christopher Thomas, BD2412, Rjwilmsi, Koavf, Strait, R.e.b., Jehochman,

FlaBot, Goudzovski, Markdroberts, Gareth E Kegg, Chobot, Algebraist, YurikBot, Bambaiah, Wester, Darsie, AVM, Bhny, Stephenb, Długosz, Dna-webmaster, Tetracube, Caco de vidro, Finell, Triple333, SmackBot, Maksim-e~enwiki, ZerodEgo, Chris the speller, Sbharris, Jmnbatista, Lambiam, JorisvS, Ckatz, Meco, Newone, Benabik, MarsRover, Myasuda, Xxanthippe, Michael C Price, Quibik, Ldussan, Difty, Thijs!bot, Epbr123, Headbomb, West Brom 4ever, Mattfiller, D.H, RogierBrussee, VoABot II, Bakken, Jpod2, RickyCayley, JohnWilliams, Hekerui, Rif Winfield, MartinBot, Haydarhan, Gillleke, Cuzkatzimhut, VolkovBot, Off-shell, LokiClock, TXiKiBoT, Calwiki, Moose-32, Ptrslv72, Coffee, Gerakibot, Likebox, JacquesPHI, Henry Delforn (old), Pac72, Mr. Stradivarius, LoserJoke, ClueBot, General Epitaph, Wwheaton, Drmies, Auntof6, Brews ohare, M.O.X, Crowsnest, XLinkBot, Scvblwxq, Addbot, Eric Drexler, SpBot, Bob K31416, Barak Sh, Tide rolls, Yoavd, Luckas-bot, Yobot, Ptbotgourou, Fraggle81, Galaxydraem, AnomieBOT, Ciphers, Rubinbot, ArthurBot, LilHelpa, Xqbot, TheAMmollusc, Capricorn42, DSisyphBot, RibotBOT, Waleswatcher, Benzen, FrescoBot, BenzolBot, XeBot, Citation bot 1, Benji1986, O.anatinus, RedBot, MastiBot, Aknochel, Beth Ann Lindstrom, Felix0411, Meier99, Mary at CERN, WildBot, EmausBot, WikitanvirBot, LHC Tommy, Slawekb, JSquish, Quondum, L Kensington, Cerlbar, Zueignung, BabbaQ, CBuiltother, PhysicsAboveAll, Giuseppe Vitiello, Jj1236, Parthdu, Curb Chain, Bibcode Bot, Tirebiter78, Ownedroad9, ChrisGualtieri, Abits52, Konbini, Ajsal.ea, Itchmean, Cjean42, Crigeos, Crbeals, Jwratner1, Atotalstranger, Jzampardi, KasparBot and Anonymous: 146

- **Quantum field theory** *Source:* https://en.wikipedia.org/wiki/Quantum_field_theory?oldid=671846854 *Contributors:* AxelBoldt, CYD, Mav, The Anome, XJaM, Roadrunner, Stevertigo, Michael Hardy, Tim Starling, IZAK, TakuyaMurata, SebastianHelm, Looix~enwiki, Ahoerstemeier, Cyp, Glenn, Rotem Dan, Stupidmoron, Charles Matthews, Timwi, Jitse Niesen, Kbk, Rudminjd, Wik, Phys, Bevo, BenRG, Northgrove, Robbot, Bkalafut, Gandalf61, Rursus, Fuelbottle, Tobias Bergemann, Ancheta Wis, Giftlite, Lethe, Dratman, Alison, St3vo, Mboverload, DefLog~enwiki, ConradPino, Amarvc, Pcarbonn, Karol Langner, APH, AmarChandra, D6, CALR, Urvabara, Discospinster, Guanabot, Igorivanov~enwiki, Masudr, Pjacobi, Vsmith, Nvj, MuDavid, Bender235, Pt, El C, Shanes, Sietse Snel, Physicistjedi, KarlHallowell, PWilkinson, Helix84, Thialfi, Varuna, Gcbirzan, Docboat, Count Iblis, Egg, Mpatel, Marudubshinki, Graham87, Opie, Vanderdecken, Rjwilmsi, MarSch, Earin, R.e.b., RE, Strobilomyces, Arnero, Itinerant1, Alfred Centauri, Srleffler, Chobot, UkPaolo, Wavelength, Bambaiah, Hairy Dude, Russ-Bot, TimNelson, Archelon, CambridgeBayWeather, SCZenz, Odddmonster, E2mb0t~enwiki, Semperf, Tetracube, Garion96, Erik J, Robert L, Banus, RG2, SmackBot, Stephan Schneider, Tom Lougheed, Melchoir, KocjoBot~enwiki, Mcld, Dauto, Chris the speller, Complexica, Threepounds, RuudVisser, QFT, Jmnbatista, Cybercobra, Rebooted, Victor Eremita, DJIndica, Lambiam, Mgiganteus1, Zarniwoot, Jim.belk, Stwalkerster, SirFozzie, Hu12, Dan Gluck, Iridescent, Joseph Solis in Australia, Albertod4, Van helsing, BeenAroundAWhile, Witten Is God, Cydebot, Jamie Lokier, Meno25, Michael C Price, The 80s chick, Mendicus~enwiki, AstroPig7, Msebast~enwiki, Mbell, Headbomb, Nick Number, Mentifisto, AntiVandalBot, Bt414, Bananan~enwiki, Martin Kostner, Moltrix, Kasimann, Kromatol, Puksik, Lerman, LLHolm, RogueNinja, Tlabshier, JEH, Nikolas Karalis, Storkk, JAnDbot, Igodard, Four Dog Night, N shaji, Bongwarrior, Andrea Allais, Soulbot, Etale, Maliz, Custos0, HEL, J.delanoy, Acalamari, Jeepday, Policron, Blckavnger, Juliancolton, Skou, Telecomtom, GrahamHardy, Sheliak, Cuzkatzimhut, VolkovBot, Bktennis2006, Marksr, HowardFrampton, The Original Wildbear, Dj thegreat, Markisgreen, TBond, Lejarrag, Moose-32, Raphtee, Sue Rangell, Neparis, Drschawrz, YohanN7, SieBot, TCO, Yintan, Likebox, Paolo.dL, Tugjob, Henry Delforn (old), Jecht (Final Fantasy X), OKBot, StewartMH, ClueBot, EoGuy, Wwheaton, The Wild West guy, Shvav~enwiki, Bob108, Brews ohare, Thingg, Count Truthstein, XLinkBot, PSimeon, SilvonenBot, Truthnlove, HexaChord, Addbot, ConCompS, Pinkgoanna, Leapold~enwiki, Dmhowarth26, Glane23, Hanish.polavarapu, Lightbot, Scientryst, R.ductor, Ettrig, Yndurain, Legobot, Luckas-bot, Yobot, Ht686rg90, Niout, Tamtamar, AnomieBOT, Ciphers, Palpher, IRP, Gjsreejith, Materialscientist, Citation bot, Bci2, ArthurBot, Northryde, LilHelpa, Caracolillo, Amareto2, MIRROR, Professor J Lawrence, Plasmon1248, Omnipaedista, RibotBOT, Spellage, JayJay, FrescoBot, Kenneth Dawson, D'ohBot, Knowandgive, N4tur4le, Hyqeom, Newt Winkler, Hickorybark, Lotje, Dinamik-bot, LilyKitty, Fortesque666, Reaper Eternal, Minimac, Marie Poise, Yaush, Dylan1946, EmausBot, Racerx11, GoingBatty, Carbosi, Thecheesykid, ZéroBot, Cogiati, Jjspinorfield1, Suslindisambiguator, Quondum, Maschen, Zueignung, Davidaedwards, Lom Konkreta, ClueBot NG, Gilderien, Iloveandrea, Vacation9, Heyheyheyhohoho, Fortune432, The ubik, Zak.estrada, Widr, Helpful Pixie Bot, Evanescent7, Ykentluo, Martin.uecker, Walterpfeifer, Pfeiferwalter, Klilidiplomus, W.D., CarrieVS, Khazar2, Momo1381, Dexbot, Cerabot~enwiki, Garuda0001, AHusain314, Thepalerider2012, A.entropy, Mark viking, Faizan, Aj7s6, संजीव कुमार, Lemnaminor, BerFinelli, Axel.P.Hedstrom, Kclongstocking, Mutley1989, I art a troler, Liquidityinsta, Prokaryotes, DemonThuum, Dingdong2680, Asherkirschbaum, Monkbot, Gjbayes, Thedarkcheese, BradNorton1979, UareNumber6, Teelaskeletor, YeOldeGentleman, Mret81, KasparBot and Anonymous: 293

- **Supersymmetry algebra** *Source:* https://en.wikipedia.org/wiki/Supersymmetry_algebra?oldid=670730267 *Contributors:* Michael Hardy, Dod1, R.e.b., Pred, Papa November, Mathsci, Myasuda, Michael C Price, Thijs!bot, Headbomb, Nilradical, MystBot, Addbot, Barak Sh, AnomieBOT, Omnipaedista, Charvest, Jonesey95, ClueBot NG, Bibcode Bot, Colbert Sesanker, Mark viking, TuxLibNit, AHusain3141 and Anonymous: 5

- **Supercommutative algebra** *Source:* https://en.wikipedia.org/wiki/Supercommutative_algebra?oldid=628427822 *Contributors:* Michael Hardy, TakuyaMurata, Silverfish, Phys, Fropuff, Oleg Alexandrov, Linas, R.e.b., Mathbot, SmackBot, David Farris, Addbot, AnomieBOT, Erik9bot, Brad7777, Mark viking and Anonymous: 5

- **Supersymmetric quantum mechanics** *Source:* https://en.wikipedia.org/wiki/Supersymmetric_quantum_mechanics?oldid=671923698 *Contributors:* Smack, Charles Matthews, Phys, Phil Boswell, Anville, CALR, Rjwilmsi, Myasuda, AnonyScientist, Addbot, Lightbot, Yobot, Wireader, Manoridius, Unara, Mstftsm, Citation bot 1, DavidCooperCS, RjwilmsiBot, Fh387dfyt87, Faolin42, ZéroBot, Ohyoungloo, Bibcode Bot, PHert, Paritto, AHusain314, Fredcath, Cmedlock, Anchenyao, TheGreatGauss and Anonymous: 13

- **Supersymmetry as a quantum group** *Source:* https://en.wikipedia.org/wiki/Supersymmetry_as_a_quantum_group?oldid=532071012 *Contributors:* Charles Matthews, Rich Farmbrough, Squids and Chips, Kawakameha, SchreiberBike, Addbot, Erik9bot, Maschen and Anonymous: 2

- **Superspace** *Source:* https://en.wikipedia.org/wiki/Superspace?oldid=599351404 *Contributors:* Michael Hardy, SebastianHelm, Angela, Charles Matthews, Phys, Chuunen Baka, Giftlite, Edcolins, Hidaspal, Gauge, Physicistjedi, BRW, Linas, Mpatel, BD2412, R.e.b., SmackBot, Jagged 85, Bluebot, Silly rabbit, JarahE, West Brom 4ever, JustAGal, RobHar, Noclevername, S tyler, Dougher, R'n'B, SchreiberBike, AnonyScientist, Addbot, Erik9bot, 777sms, Hhhippo, Maschen, Brad7777, Duplij, Vutshi, ZAKI1905 and Anonymous: 11

- **Supergeometry** *Source:* https://en.wikipedia.org/wiki/Supergeometry?oldid=552875458 *Contributors:* Michael Hardy, Jason Quinn, Helopticor, MZMcBride, SmackBot, Reedy Bot, Niceguyedc, Bte99, LilHelpa, Gsard, EmausBot, Helpful Pixie Bot and Brad7777

- **Supergroup (physics)** *Source:* https://en.wikipedia.org/wiki/Supergroup_(physics)?oldid=639050883 *Contributors:* Edward, Michael Hardy, Docu, Charles Matthews, Mwoolf, Phys, Fropuff, Brian0918, SmackBot, Commander Keane bot, FlyHigh, CmdrObot, RogierBrussee, Ludvikus, Michael H 34, SieBot, Hans Adler, Jovianeye, Addbot, Omnipaedista, Erik9bot, BoogityBang, GoingBatty and Anonymous: 9

- **Supergravity** *Source:* https://en.wikipedia.org/wiki/Supergravity?oldid=668440740 *Contributors:* AxelBoldt, Michael Hardy, TakuyaMurata, Angela, Charles Matthews, Phys, Bevo, Robbot, Gandalf61, Giftlite, Herbee, LeYaYa, Fropuff, Moyogo, Jeremy Henty, Leonard G., Urvabara, Arivero, Masudr, Dmr2, Srbauer, Markryherd, Physicistjedi, Axl, Wtmitchell, Japanese Searobin, Linas, Kzollman, Mpatel, GregorB, Canderson7, Marasama, Gurch, LeCire~enwiki, Chobot, Roboto de Ajvol, Hillman, Conscious, E. Menay, Wimt, Smoggyrob, QmunkE, Ilmari Karonen, Caco de vidro, SmackBot, Melchoir, FlashSheridan, Vald, Chris the speller, Colonies Chris, QFT, BWDuncan, TheST, Kuru, Jim.belk, JarahE, Michaelbusch, Zero sharp, CapitalR, Jorbesch, Crichigno, CmdrObot, Myasuda, Equendil, Phatom87, Pyro95819, Mbell, WVhybrid, West Brom 4ever, Icep, Shlomi Hillel, Yill577, David Eppstein, N.Nahber, Andre.holzner, Mschel, EdBever, Freeboson, Wesino, WJBscribe, Fuenfundachtzig, Signalhead, Cuzkatzimhut, Jickle, Robdunst, WereSpielChequers, Caltas, Wing gundam, Paolo.dL, Oxymoron83, Lightmouse, JL-Bot, EmanWilm, RS1900, ClueBot, ArdClose, Mild Bill Hiccup, JavierReynaldo, Vivio Testarossa, Mastertek, Pqnelson, AnonyScientist, Truthnlove, Addbot, Some jerk on the Internet, Wentuq, Luckas-bot, Yobot, Bility, AnomieBOT, ArthurBot, Omnipaedista, Gsard, Hep thinker, FrescoBot, Paine Ellsworth, Pxpt, Tom.Reding, Casimir9999, Wornsear, EmausBot, Slightsmile, Wikipelli, HiW-Bot, ZéroBot, Cogiati, Arbnos, Quantumor, Terraflorin, Bbeehvh, ClueBot NG, Joefromrandb, Helpful Pixie Bot, Bibcode Bot, BG19bot, Altaïr, BattyBot, Jeremy112233, M0532062613, Jamesx12345, Mamzypig99, Bitprior, Monkbot, KasparBot and Anonymous: 75
- **Supercharge** *Source:* https://en.wikipedia.org/wiki/Supercharge?oldid=552067788 *Contributors:* Stevertigo, Phys, Lumidek, Commander Keane, Bluemoose, Obey, Hydrogen Iodide, TheFarix, Kborland, Yobot, FrescoBot, Itebero, Twctinc1, KLBot2 and Anonymous: 7
- **Superfield***Source:*https://en.wikipedia.org/wiki/Superfield?oldid=613453761*Contributors:*Phys, Lumidek, BD2412, Conscious, GrinBoi,That Guy, From That Show!, Andstergiou, FlyHigh, Keitei, Michael C Price, Cuzkatzimhut, ArdClose, Addbot, 777sms and Anonymous: 3
- **Superpartner** *Source:* https://en.wikipedia.org/wiki/Superpartner?oldid=666820365 *Contributors:* Roadrunner, SimonP, Phys, Donarreiskoffer, Giftlite, Kocio, Alai, Duncan.france, Mpatel, Rjwilmsi, R.e.b., Drrngrvy, FlaBot, KFP, Conscious, SCZenz, SmackBot, Reedy, Dauto, Jgwacker, Thijs!bot, Headbomb, Maliz, Hans Dunkelberg, LovroZitnik, Agharo, Antixt, AlleborgoBot, Madacs, Bobathon71, Alexbot, SilvonenBot, SkyLined, Addbot, Barak Sh, Luckas-bot, ArthurBot, Xqbot, Erik9bot, Carlog3, Paine Ellsworth, Haeinous, Cracrunch, RedBot, EmausBot, Hydroxonium, Flloater, ClueBot NG, Bibcode Bot, Hrttu523, Rolf h nelson, Akro7 and Anonymous: 14
- **Graviton** *Source:* https://en.wikipedia.org/wiki/Graviton?oldid=671813129 *Contributors:* CYD, Bryan Derksen, Timo Honkasalo, XJaM, Fubar Obfusco, Maury Markowitz, Kaczor~enwiki, Jketola, TakuyaMurata, Eric119, Looxix~enwiki, Glenn, Cyan, Wooster, Charles Matthews, Timwi, Wik, BenRG, Donarreiskoffer, Scott McNay, Stephan Schulz, Arkuat, Chris Roy, Merovingian, Davidl9999, Giftlite, Xerxes314, Jason Quinn, Matt Crypto, CryptoDerk, RetiredUser2, Icairns, Zfr, Lumidek, Ukexpat, Urvabara, Discospinster, Pjacobi, Vapour, Brian0918, El C, Joanjoc~enwiki, Dalf, Army1987, Mpvdm, La goutte de pluie, Physicistjedi, Daniel Arteaga~enwiki, Zenosparadox, Dethtron5000, Keenan Pepper, Viridian, Falcorian, Skeejay, Simetrical, Dr Archeville, Mpatel, Kyleca, Tmassey, Christopher Thomas, Tevatron~enwiki, Kbdank71, Nightscream, Koavf, Mike Peel, Ems57fcva, FlaBot, RexNL, Chobot, DVdm, Roboto de Ajvol, Spacepotato, Anonymous editor, SnoopY~enwiki, Salsb, Bachrach44, Hyperbrand, NickBush24, Pnrj, RL0919, EEMIV, IslandGyrl, Bota47, C h fleming, Petri Krohn, Mario23, Alias Flood, Tim314, Teply, GrinBot~enwiki, SmackBot, Amcbride, Melchoir, Eskimbot, Gilliam, Skizzik, Timneu22, Complexica, Villarinho, Colonies Chris, V1adis1av, Chlewbot, Xyzzyplugh, Jmnbatista, Fuhghettaboutit, Sadi Carnot, Yevgeny Kats, TenPoundHammer, Lambiam, Zaphraud, JorisvS, Mr Stephen, Ramuman, Quasar Jarosz, Lottamiata, Firewall62, Kurtan~enwiki, CmdrObot, BeenAroundAWhile, WeggeBot, Shultz IV, UncleBubba, Michael C Price, Anthmoo, Thijs!bot, Epbr123, Headbomb, KevinS06, Opelio, Spartaz, JAnDbot, Xoneca, SHCarter, Pikazilla, Robin S, STBot, Kostisl, J.delanoy, Tarotcards, Coppertwig, Wesino, Sava ankit2006, Tygrrr, Idioma-bot, Sheliak, JoAnneThrax, TXiKiBoT, WilliamSommerwerck, Hqb, Anonymous Dissident, Antixt, SieBot, Flyer22, Henry Delforn (old), ClueBot, Ergn, Darkicebot, DenverRedhead, Addbot, Eric Drexler, Uruk2008, DOI bot, BrianBop, PJonDevelopment, F Notebook, Legobot, Picturesofnothing, Dov Henis, Alfredschrader, Eric-Wester, AnomieBOT, VanishedUser sdu9aya9fasdsopa, Jim1138, Materialscientist, Citation bot, Tomflaherty, ProtectionTaggingBot, Waleswatcher, FrescoBot, Juto20, LucienBOT, Paine Ellsworth, I dream of horses, Tom.Reding, RedBot, Omar.tigereyes, IVAN3MAN, Ashish.kotwal, Michael9422, D0wnfalle, EmausBot, Octaazacubane, 8digits, Slightsmile, K6ka, Thecheesykid, User10 5, Rcsprinter123, Orbjeeples, Puffin, Herk1955, ClueBot NG, Raidr, Helpful Pixie Bot, Bibcode Bot, BG19bot, Shapoopy178, ServiceAT, PhnomPencil, Trevayne08, Brainssturm, Tjamcclain2, ChrisGualtieri, Ariscod, TheUyulala, LightandDark2000, Jessybun, Makecatbot, Kryomaxim, JRYon, Andyhowlett, Mark viking, Yorsh07, CensoredScribe, WPratiwi, Monkbot, Bryan Paul Senior, Dr.Begich, Nompynuthead, Jacobflarsen and Anonymous: 196
- **Neutralino** *Source:* https://en.wikipedia.org/wiki/Neutralino?oldid=671536470 *Contributors:* Angela, Julesd, Schneelocke, Charles Matthews, Saltine, Donarreiskoffer, Robbot, Rursus, Moink, Awolf002, Herbee, Xerxes314, Waltpohl, Pharotic, Eequor, Icairns, Urvabara, Pjacobi, Slicky, Physicistjedi, JPFlip, SDC, Theofilatos, Kbdank71, FlaBot, Roboto de Ajvol, YurikBot, Conscious, SCZenz, AndrewWTaylor, SmackBot, Stepa, Nickst, V1adis1av, Pulu, Lester, Newone, Wadoli Itse, Thijs!bot, Barticus88, Headbomb, Stannered, Squantmuts, NicZ~enwiki, Choihei, Antixt, ClueBot, SkyLined, Addbot, Prim Ethics, Yobot, AnomieBOT, Citation bot, ArthurBot, Br77rino, Ernsts, A. di M., Erik9bot, Tom.Reding, ZéroBot, David C Bailey, Helpful Pixie Bot, Bibcode Bot, Halfb1t, Manar al Zraiy, Makecat-bot, Phseek and Anonymous: 24
- **Theory of everything** *Source:* https://en.wikipedia.org/wiki/Theory_of_everything?oldid=670253196 *Contributors:* AxelBoldt, Paul Drye, CYD, The Anome, Eclecticology, Toby Bartels, Roadrunner, Zippy, Stevertigo, Lorenzarius, Michael Hardy, Rojclague, Nixdorf, TakuyaMurata, Karada, Skysmith, Kosebamse, CesarB, Anders Feder, Angela, Julesd, Salsa Shark, Ugen64, Poor Yorick, Evercat, Schneelocke, Feedmecereal, Timwi, Dcoetzee, Dysprosia, Jitse Niesen, Wik, Jakenelson, Omegatron, Raul654, Nnh, Kevin M C Harkess, UninvitedCompany, Fredrik, Altenmann, Nurg, Naddy, Gandalf61, Mirv, Academic Challenger, Rursus, Blainster, Caknuck, Wereon, Diberri, Pengo, Tobias Bergemann, Hooloovoo, Ancheta Wis, Dbenbenn, Mporter, Jabra, Ferkelparade, Bfinn, Xerxes314, Curps, Alison, FeloniousMonk, McGravin, Behnam, Gzornenplatz, JRR Trollkien, Steuard, Andycjp, Sonjaaa, Antandrus, Kim54, Tomruen, Lumidek, Gscshoyru, WpZurp, TJSwoboda, Zondor, JimJast, Discospinster, Rich Farmbrough, H0riz0n, Pjacobi, Vsmith, Pluke, Autiger, Mal~enwiki, Pavel Vozenilek, Floorsheim, El C, Lycurgus, Sourcecode, Oldsoul, PhilHibbs, Sietse Snel, Jpgordon, Atraxani, Smalljim, Slicky, LostLeviathan, Matpitka, Juesch, Danski14, Alansohn, Gary, DariuszT, ShardPhoenix, Kocio, Pion, Hdeasy, Bart133, Schaefer, BanyanTree, ClockworkSoul, Tycho, Suruena, Count Iblis, DV8 2XL, Gene Nygaard, Euphrosyne, Squidwina, Ott, Siafu, Roylee, Woohookitty, Mindmatrix, RHaworth, TigerShark, Savantnavas, MrDarcy, Mpatel, GregorB, Athletec64, Christopher Thomas, Ashmoo, BD2412, Drbogdan, Rjwilmsi, Kinu, Strait, Lordsatri, Dennis Estenson II, HappyCamper, LjL, Bubba73, The wub, Yamamoto Ichiro, JohnDBuell, FayssalF, ColinJF, Wragge, Windchaser, Musical Linguist, Mindloss, RexNL, Gurch, Pete.Hurd, Lmatt, Diza, Zayani, Spencerk, Chobot, Sharkface217, DVdm, Hmonroe, Bgwhite, Ptah~enwiki, Ugha, Wavelength, Hillman, StuffOfInterest, Phantomsteve, John Smith's, Zigamorph, SpuriousQ, Jobe457, Stephenb, CambridgeBayWeather, Rsrikanth05, Vibritannia, Neilbeach, Salsb, Big Brother 1984, Anomalocaris, NawlinWiki, Joncolvin, ErkDemon, Trovatore, ETTan, Schrei, THB, Syrthiss, Wknight94, Richardcavell, FF2010, CWenger, Kevin, Caco de vidro, Katieh5584, Banus, Sbyrnes321,

Narkstraws, SmackBot, R.E. Freak, Kurochka, DuoDeathscyther 02, Bayardo, McGeddon, Delldot, Kintetsubuffalo, Portillo, Rmosler2100, Bluebot, Jjalexand, 7777777s, Silly rabbit, George Church, Colonies Chris, A. B., Calc rulz, Nicknitro71, Zsinj, TallyJoe, John Hyams, Jamse, Scott3, Jefffire, Serenity-Fr, Bilgrau, Avb, Rrburke, Addshore, DrL, Mr.LMNOP, Rassisi, Spanyard, Byelf2007, Nishkid64, Giovanni33, Soap, Cronholm144, Loadmaster, Stupid Corn, Benjaminlobato, FredrickS, SirFozzie, Waggers, Alexander Gieg, Gcavep, Abel Cavaşi, Newone, Courcelles, Tubezone, Esn, Dave Runger, Valoem, JRSpriggs, Kurtan~enwiki, 0-8, Duduong, Friendly Neighbour, CRGreathouse, Geremia, Tkoeppe, Ken Gallager, DepartedUser2, Cydebot, Vanished user 2340rujowierfj08234irjwfw4, Ninguém, Steel, Peterdjones, Hebrides, David edwards, Michael C Price, Raoul NK, Wortzman, Ulnevets, Konradek, Mojo Hand, Raymond Feilner, Headbomb, Marek69, Inve40, Twcjr, Duncan McB, KrakatoaKatie, Luna Santin, Gdo01, Byrgenwulf, Myanw, Knotwork, Len Raymond, JAnDbot, Barek, MER-C, Txomin, Inks.LWC, Matthew Fennell, Instinct, MoralMajority, Promking, Bongwarrior, VoABot II, JamesBWatson, JBKramer, DAGwyn, Theroadislong, Lenschulwitz, 28421u2232nfenfcenc, Peatbog, Allstarecho, Fang 23, Spellmaster, Philg88, Peter J Schoen, Denis tarasov, MartinBot, R'n'B, JCarlos, J.delanoy, Pharaoh of the Wizards, Maurice Carbonaro, LordAnubisBOT, Pyrospirit, AntiSpamBot, NewEnglandYankee, DadaNeem, Cometstyles, WJBscribe, Foofighter20x, Econofire, Squids and Chips, Germanium, Reelrt, ChaosCon343, Danwills, RingtailedFox, Jeff G., TXiKiBoT, Nxavar, Rei-bot, Vishal144, Pouya sh, Corvus cornix, Michael H 34, Martin451, Cheffoxx, Betanon, BotKung, Everything counts, Popopp, MrMelonhead, Stephenmolesey, James McBride, Deanlsinclair, Pageman~enwiki, Monty845, Logan, Kpa4941, PaddyLeahy, Dogah, SieBot, Tiddly Tom, Robdunst, Wing gundam, Gammanon, Bentogoa, Likebox, Tiptoety, SteakNShake, Momo san, Freeman501, BartekChom, Monkeyspangler, Lightmouse, Anakin101, Divinestuff, Carbogen, Ayleuss, Soporaeternus, ArepoEn, ClueBot, LAX, Cliff, Ian the Aussie, Monomath1, Boing! said Zebedee, Heldbacktheband, LonelyBeacon, Neverquick, Excirial, WikiZorro, Tamaratrouts, Wndl42, Brews ohare, PhySusie, Morel, Mastertek, Mikaey, 7, Crowsnest, Thinking Stone, TimothyRias, PatDunphey, JKeck, XLinkBot, Bvssvni, Ougner, Truthnlove, YeAaMsLtA, Thatguyflint, Tayste, Balungifrancis, Addbot, Proofreader77, Some jerk on the Internet, Uruk2008, DOI bot, Couchie, Johnchang6868, Discrepancy, Mjamja, Bobtron5000, Fluffernutter, KaityJoe, MrOllie, Favonian, Barak Sh, F Notebook, Tide rolls, Scientryst, WikiDreamer Bot, Meisam, Blah28948, Yobot, Finiter, Ptbotgourou, Ezequiels.90, Jgmoxness, Amble, Mirandamir, RDemelo, AnomieBOT, ^musaz, Girl Scout cookie, 9258fahsflkh917fas, Theunify, Anxfisa, Kanat Abildinov, Materialscientist, Citation bot, Subhajit Ganguly, Fleaman5000, Amareto2, Addihockey10, Smk65536, Mlpearc, GrouchoBot, Rwmeo, Omnipaedista, Shirik, RibotBOT, Fa.alt3r3g0, Fsdjfsdfk, Chaseroads, ⁇⁇, FrescoBot, Paine Ellsworth, Ribashka, Steven Avraham Rosten, PhysicsExplorer, Ottokar~enwiki, Tank hasmukh Khimjibhai, Tank theorist of everything, Hasmukh Khimjibhai Tank, DivineAlpha, Citation bot 1, Gil987, Three887, Tom.Reding, A8UDI, NarSakSasLee, Casimir9999, AndrewGrieder, Aknochel, IVAN3MAN, SchreyP, Noel Edward, Natwatchmaker, Weedwhacker128, Suffusion of Yellow, Koozedine, RjwilmsiBot, Specal ops, Afteread, DASHBot, Golumbo, EmausBot, Ikerus, Katherine, Dewritech, RA0808, K6ka, Zero939, Thecheesykid, Hhhippo, Traxs7, Arbnos, SporkBot, DanielBurnstein, FinalRapture, Aatu Koskensilta, Staszek Lem, Sridattadev, M00se1989, Wiggles007, Andrushkkutza, Vedoder, Donner60, GIAN PHIL, Davidaedwards, WHF Christie, Terra Novus, Matevz91, Isocliff, Sanno89, Cgt, Will Beback Auto, ClueBot NG, Stein Sivertsen, ClaudeDes, Lord God Almighty, Hindustanilanguage, Helpful Pixie Bot, Nightingale.zj, B21O303V3941W42371, Bibcode Bot, Wiki13, Akashankitjain, Neutral current, Aranea Mortem, Stimulieconomy, Steven.w.kowalski, MathewTownsend, Flyerbri, GroupT, Megajakeroo, La marts boys, Zofo, LightandDark2000, Josepht404, Nickhwee, Davidyevgeny, Kingcircle, Vith Nix, Illuusio, Davidyevgenyroven, QuantumNico, Vladimir Leonov, Friek555, HesterShaw, Sol1, Phaedrx, Jwratner1, Jmassion, HeymynamesJon, Bigfootrobert, Elitousson, Mdsheraj, Kdmeaney, JaconaFrere, Somecdnguy4, Monkbot, LollyBear12, StacyPoyPie, Mujii loving, Mayojohns, Gronk Oz, Yoyosami, Hakan tomaşoğlu, Pfpguy, Cirksena, Svm sudhan, 39Debangshu, Quantalogos, KasparBot, Patrickmantonio, Quackriot and Anonymous: 516

- **Superstring theory** *Source:* https://en.wikipedia.org/wiki/Superstring_theory?oldid=669906748 *Contributors:* Mav, Bryan Derksen, Stevertigo, Michael Hardy, Erik Zachte, Minesweeper, Looxix~enwiki, Ahoerstemeier, JWSchmidt, Cyan, Palfrey, Evercat, Schneelocke, Hashar, Charles Matthews, Tpbradbury, Motor, David Shay, Omegatron, Bevo, Bcorr, Robbot, Fredrik, Hadal, Vuara, Giftlite, Barbara Shack, Herbee, Fropuff, Anville, Maarten van Vliet, WalkinDownThirtyThree, Christopherlin, Steuard, Karol Langner, Lumidek, Prestonmarkstone, Rich Farmbrough, Igorivanov~enwiki, Autiger, Pavel Vozenilek, El C, Rgdboer, Shadow demon, Causa sui, Billymac00, BM, Gary, Pion, Tycho, Cal 1234, Redvers, Postrach, Supercool Dude, Mindmatrix, Mpatel, Joke137, Mandarax, Bill37212, Yamamoto Ichiro, Bubbleboys, Chobot, Ben Tibbetts, Wavelength, RussBot, Chris Capoccia, Chensiyuan, Cate, Chaos, NawlinWiki, Astral, Voidxor, TheMadBaron, Zerodamage, Allens, SmackBot, Android 93, Kurochka, McGeddon, Kintetsubuffalo, Cesoid, Silly rabbit, Stevage, Baronnet, Colonies Chris, Scwlong, Mesons, Kurrupt3d, Bjankuloski06en~enwiki, Makyen, MathStuf, Hu12, Iridescent, Kahalachan, Gatortpk, Mattbr, Neelix, Gregbard, Nauticashades, Cydebot, Davidanzaldua, ChKa, Headbomb, Escarbot, Jj137, Shlomi Hillel, Dougher, JAnDbot, 100110100, 28421u2232nfenfcenc, Aziz1005, Jean-Pierre Petit~enwiki, Hans Dunkelberg, Maurice Carbonaro, Bot-Schafter, SmilesALot, Student7, Cmichael, Sheliak, JayCo777, Calwiki, Andrius.v, Molinogi, Billinghurst, Lamro, Enviroboy, Seraphita~enwiki, Drschawrz, Henry Delforn (old), Lightmouse, Altzinn, Gratedparmesan, ClueBot, Arakunem, Vergil 577, Frdayeen, Vizzini101, Niceguyedc, Neverquick, Resoru, Mastertek, Kakofonous, Princess Janay, Alex123irish123, Madeinmexico567, Oldnoah, Madeinmexico566, Truthnlove, YeAaMsLtA, Addbot, Physicman123, CWatchman, Cuaxdon, Semdino, AnnaFrance, LinkFA-Bot, TaBOT-zerem, Evans1982, Gerixau, Eric-Wester, Magog the Ogre, AnomieBOT, DemocraticLuntz, ^musaz, Josh Guffin, Jim1138, Citation bot, Renaissancee, BLP-outrageous move logs, Omnipaedista, RibotBOT, Paine Ellsworth, Steve Quinn, Tom.Reding, Klavesin, Tkachyk, Dinamik-bot, Bj norge, Idh0854, Arbnos, Vramasub, L Kensington, Particle hep, Isocliff, ClueBot NG, Widr, Adminium, Delivernews, Bibcode Bot, Khanduras, Quarkgluonsoup, Flowerhat15, MythosMagic, OCCullens, Aldrich2122, Graphium, AHusain314, Jochen Burghardt, WorldWideJuan, Jakec, Liquidityinsta, E8xE8, Polytope24, FlaviusCorcoata, Cirksena, BakedLikaBiscuit, KasparBot, Rantonels and Anonymous: 171

- **Coleman–Mandula theorem** *Source:* https://en.wikipedia.org/wiki/Coleman%E2%80%93Mandula_theorem?oldid=571287679 *Contributors:* Michael Hardy, Charles Matthews, Phys, Giftlite, Lumidek, Nicobn~enwiki, Bender235, Jag123, Delius, Ricky81682, Marudubshinki, Rjwilmsi, Conscious, Bhny, Bluebot, QFT, Ligulembot, Myasuda, Cydebot, Headbomb, Intovarius, Masterpiece2000, DumZiBoT, MystBot, Addbot, DOI bot, Luckas-bot, Yobot, JackieBot, Xqbot, RedBot, Virtakuono, Afteread, EmausBot, Bibcode Bot, Brad7777 and Anonymous: 3

- **Haag–Lopuszanski–Sohnius theorem** *Source:* https://en.wikipedia.org/wiki/Haag%E2%80%93Lopuszanski%E2%80%93Sohnius_theo?oldid=622628714 *Contributors:* Charles Matthews, Giftlite, Trafton, Aronbeekman, MFH, R.e.b., QFT, Jim.belk, Dycedarg, ChKa, Head-bomb, Legoktm, Addbot, Lightbot, Yobot, Xqbot, Molitorppd22, 777sms, EmausBot, WikitanvirBot, Bibcode Bot and Anonymous: 5

- **Magnetic monopole** *Source:* https://en.wikipedia.org/wiki/Magnetic_monopole?oldid=671447955 *Contributors:* Bryan Derksen, The Anome, Ap, Andre Engels, Roadrunner, Maury Markowitz, Heron, Camembert, Patrick, Michael Hardy, Tim Starling, EddEdmondson, Dominus, Ixfd64, Skysmith, Looxix~enwiki, Mkweise, Ahoerstemeier, Stevenj, Aarchiba, Cyan, HolIgor, Charles Matthews, Timwi, Phys, Jerzy, BenRG,

Jeffq, Henrygb, Rasmus Faber, Pengo, Cutler, Enochlau, Giftlite, Mintleaf~enwiki, Xerxes314, Rapjo, Waltpohl, Pharotic, Peter Ellis, Nova77, ConradPino, Gzuckier, Beland, MFNickster, Anythingyouwant, Elektron, Icairns, Lumidek, Karl Dickman, Mike Rosoft, Urvabara, Jkl, Rich Farmbrough, TedPavlic, Pjacobi, ArnoldReinhold, MuDavid, Bender235, ESkog, Kjoonlee, El C, Sasquatch, Thuktun, Congruence, Alansohn, Anthony Appleyard, Cmprince, Pauli133, Nick Mks, Falcorian, Linas, JarlaxleArtemis, Ruud Koot, Mpatel, Tabletop, GregorB, CharlesC, TheAlphaWolf, Emerson7, Mandarax, Aarghdvaark, BD2412, Rjwilmsi, HonoluluMan, MarSch, Eyu100, Seraphimblade, DonSiano, Gareth McCaughan, R.e.b., Erkcan, Drrngrvy, Mathbot, Nihiltres, Tardis, Adarsh116098, Chobot, DVdm, Amaurea, YurikBot, Phmer, RussBot, Xihr, JabberWok, Gaius Cornelius, PoorLeno, DragonHawk, Wiki alf, Welsh, Długosz, Gillis, Dchoulette, Jstrater, Crasshopper, Tony1, Crumley, SamuelRiv, 2over0, Reyk, KingCarrot, Sbyrnes321, Mhardcastle, SmackBot, Michaelliv, Melchoir, Jonathan Karlsson, Octahedron80, Skatche, V1adis1av, QFT, Alex Fix, Ianmacm, Khukri, Mohseng, Skiminki, Yevgeny Kats, Nat2, JorisvS, Loadmaster, Rock4arolla, Stephen B Streater, Norm mit, Gorog, Courcelles, Piccor, Achoo5000, Chetvorno, Disambiguator, Randall Nortman, GRB, Capefeather, Moyerjax, Michael C Price, Quibik, Doug Weller, DumbBOT, Karl-H, Difty, Wikid77, Headbomb, Luna Santin, Thranduil, Fru1tbat, Spartaz, JAnDbot, Igodard, Catslash, Bakken, Sjanusz, David Eppstein, Stevvers, WLU, 2bithacker, C.R.Selvakumar, Nsande01, Nlalic, NerdyNSK, JA.Davidson, Rod57, Dawright12, Aoosten, Tarotcards, Plasticup, Loohcsnuf, Barraki, Ross Fraser, Ratfox, Dorftrottel, Trmatthe, VolkovBot, FDominec, Rei-bot, Lixo2, Mathfreak11235, Wingedsubmariner, Antixt, RaseaC, Stigin, SieBot, CatherS, Likebox, Pit-trout, Henke37, Lisatwo, Dickontoo, Skeptical scientist, Maxime.Debosschere, Martarius, Balashpersia~enwiki, Unbuttered Parsnip, Razimantv, Mild Bill Hiccup, LonelyBeacon, Wrsh11, Sfitzsi, SchreiberBike, DumZiBoT, XLinkBot, Oldnoah, Avoided, Addbot, Jacopo Werther, DOI bot, Мышка, Barak Sh, 84user, Qaswqaswgd, Skippy le Grand Gourou, Luckas-bot, Munkel Davidson, Yobot, KamikazeBot, AnomieBOT, Floquenbeam, Citation bot, Flying hazard, Xqbot, Renaissancee, Kbodouhi, Charvest, Nagualdesign, FrescoBot, Goodbye Galaxy, Citation bot 1, Relke, DrilBot, Cwedhrin, Q0k, MarcelB612, RedBot, Trappist the monk, YURI-21century, Morphotomy, Splartmaggot, FKLS, Deanmullen09, Wrotesolid, Waylah, Giscard2, John of Reading, WikitanvirBot, Wikipelli, ZéroBot, Prayerfortheworld, Cogiati, N0RND123, StringTheory11, Quondum, Maschen, Particle hep, Zoooooooooaa, Isocliff, David Thorne, ClueBot NG, Andrija radovic, MerlIwBot, Bibcode Bot, BG19bot, PearlSt82, Gorthian, F=q(E+v^B), Niqomi, BattyBot, ChrisGualtieri, Khazar2, MaxwellDecoherence, Enyokoyama, JRYon, Jaxcp3, Andyhowlett, GabeIglesia, Razibot, Consecutor, Monkbot, BoltNinja and Anonymous: 232

- **Atiyah–Singer index theorem** *Source:* https://en.wikipedia.org/wiki/Atiyah%E2%80%93Singer_index_theorem?oldid=640788336 *Contributors:* AxelBoldt, Rmhermen, Michael Hardy, TakuyaMurata, Angela, Schneelocke, Loren Rosen, Charles Matthews, David Newton, Psychonaut, Humus sapiens, Smb1001, Tosha, Giftlite, ShaunMacPherson, Lethe, Lupin, Fropuff, Steuard, Icairns, Oleg Alexandrov, GregorB, Rjwilmsi, R.e.b., Algebraist, Wavelength, RussBot, Netrapt, Fram, RDBury, BeteNoir, Chris the speller, Silly rabbit, Nbarth, Akriasas, RJBurkhart, Vina-iwbot~enwiki, Michael Kinyon, Mathsci, JarahE, Rastaco, Darklilac, Quarague, JanCK, Ling.Nut, David Eppstein, Ndokos, VolkovBot, Temurjin, Rei-bot, Saibod, Kawakameha, Dmoskovich, EGetzler, Marsupilamov, MystBot, Addbot, DOI bot, Lightbot, Yobot, Kilom691, Citation bot, Xqbot, Drilnoth, Point-set topologist, Citation bot 1, WQUlrich, Rausch, Meier99, RjwilmsiBot, EmausBot, Slawekb, Suslindisambiguator, D.Lazard, NTeleman, Harsimaja, Helpful Pixie Bot, Bibcode Bot, Teika kazura, Brad7777, Enyokoyama, Makecat-bot, Hamoudafg, SakeUPenn, Anrnusna, Pally blaga, Nicolae teleman and Anonymous: 38

- **Noncommutative geometry** *Source:* https://en.wikipedia.org/wiki/Noncommutative_geometry?oldid=670089379 *Contributors:* Youssefsan, Michael Hardy, TakuyaMurata, GTBacchus, William M. Connolley, Charles Matthews, Phys, Tobias Bergemann, Giftlite, Lupin, Jrdioko, Tristanreid, DefLog~enwiki, CSTAR, Chris Howard, JonL, Paul August, Gauge, Count Iblis, Ceyockey, Oleg Alexandrov, Japanese Searobin, Linas, -Ril-, Mpatel, Triddle, GregorB, Rjwilmsi, R.e.b., John Z, Bmicomp, Siddhant, RussBot, Michael Slone, Shell Kinney, Yserarau, Crasshopper, JonathanD, Caco de vidro, SmackBot, Unyoyega, Zoran.skoda, Scwlong, CRGreathouse, Ntsimp, Thijs!bot, Konradek, Headbomb, Magioladitis, Taborgate, Lantonov, Cuzkatzimhut, Schucker, JohnBlackburne (old), LokiClock, Henry Delforn (old), Sphilbrick, Alterationx10, Mild Bill Hiccup, DragonBot, ProfessorTarantoga, Pqnelson, Addbot, Krampma, Yobot, Wireader, Felipe Gonçalves Assis, 9258fahsflkh917fas, Xqbot, Omnipaedista, Gsard, Tonyxty, Stephan Spahn, ZéroBot, Booqorm, Anselrill, Davidaedwards, Helpful Pixie Bot, Bibcode Bot, Brad7777, Qtom.masters, Deltahedron, Enyokoyama, Mark viking, Foredit and Anonymous: 41

- **Quantum group** *Source:* https://en.wikipedia.org/wiki/Quantum_group?oldid=670469457 *Contributors:* Michael Hardy, TakuyaMurata, Charles Matthews, Phys, Giftlite, MSGJ, Anville, Almit39, Chris Howard, Michall~enwiki, Zaslav, Linas, Rjwilmsi, R.e.b., Mathbot, Itinerant1, Mas-n evets, Figaro, Reyk, SmackBot, Colonies Chris, Vina-iwbot~enwiki, FlyHigh, Khazar, Mets501, CBM, Ntsimp, Headbomb, OrenBochman,Ar turj, Efio, Arcfrk, Phe-bot, Henry Delforn (old), Gpap.gpap, Addbot, Yobot, Udoh, Sz-iwbot, LilHelpa, Charvest, Sławomir Biały, Citationbot 1, Tkuvho, Jonesey95, Trappist the monk, Tinfoilcat, Crpytozoic, Korepin, RjwilmsiBot, Jowa fan, ZéroBot, Sameenahmedkhan, BibcodeBot, BG19bot, Simon Lentner, CripesBatey, Garuda0001, Mark viking, CsDix, Escspeed, 314Username, Elysion, Monkbot and Anonymous:21

- **Loop quantum gravity** *Source:* https://en.wikipedia.org/wiki/Loop_quantum_gravity?oldid=668889779 *Contributors:* Bryan Derksen, The Anome, AstroNomer~enwiki, RK, Toby Bartels, Miguel~enwiki, Schewek, Ewen, Michael Hardy, TakuyaMurata, Islandboy99, GTBacchus, Mcarling, Looxix~enwiki, Ahoerstemeier, Cyp, Kimiko, Palfrey, Jordi Burguet Castell, Mxn, Charles Matthews, Sanxiyn, Maximus Rex, Phys, Omegatron, Finlay McWalter, Dmytro, Sdedeo, Astronautics~enwiki, Peak, Chris Roy, Mirv, Sverdrup, Kn1kda, Hadal, Jheise, Clementi, Connelly, Giftlite, Sj, Fastfission, Herbee, Anville, Dratman, Curps, JeffBobFrank, Jason Quinn, Gzornenplatz, C17GMaster, DÅ,ugosz, PhiloVivero, DefLog~enwiki, Gadfium, HorsePunchKid, Sam Hocevar, Lumidek, Tdent, Joyous!, M1ss1ontomars2k4, Eep², Poccil, Rich Farmbrough, Avriette, Pjacobi, Vsmith, MuDavid, Pavel Vozenilek, Bender235, ESkog, Clement Cherlin, Peter M Gerdes, Drhex, John Vandenberg, C S, Cmdrjameson, GTubio, Tweet Tweet, Slicky, Ral315, Lysdexia, Arthena, Xaphan9966, Wtmitchell, Greg Kuperberg, Count Iblis, Egg, Lee-Anne, Kazvorpal, Killing Vector, Linas, Merlinme, HFarmer, Sympleko, Hfarmer, Mpatel, GregorB, J M Rice, Ae7flux, Tjbk tjb, Alienus, Fleisher, Sjö, Rjwilmsi, Nightscream, Zbxgscqf, Bubba73, FlaBot, John Baez, Don Gosiewski, Smithbrenon, Chobot, Spasemunki, Bgwhite, Roboto de Ajvol, YurikBot, Wavelength, RobotE, Rt66lt, Hillman, DanMS, Chaos, Salsb, Welsh, Schmock, Crasshopper, Beanyk, Akashmitra, Bota47, JonathanD, Endomion, Modify, Petri Krohn, Ilmari Karonen, Caco de vidro, Benandorsqueaks, SmackBot, Bayardo, FlashSheridan, Unyoyega, Vald, JMiall, Chris the speller, IvanAndreevich, DHN-bot~enwiki, Colonies Chris, Chlewbot, Pepsidrinka, Chrylis, MegaHasher, TriTertButoxy, Lambiam, Vincenzo.romano, Loadmaster, Konklone, K, G-W, Kurtan~enwiki, Harold f, Will314159, Friendly Neighbour, Vyznev Xnebara, Ian Beynon, Myasuda, Gmusser, Rjm656s, Fournax, Headbomb, Nick Number, MichaelMaggs, Edokter, Byrgenwulf, Knotwork, Arch dude, Igodard, Yill577, WolfmanSF, Tonyfaull, Skylights76, Rickard Vogelberg, Gwern, AltiusBimm, Melamed katz, Vanished user 47736712, WJBscribe, Izno, KittyHawker, Sheliak, AlnoktaBOT, Nxavar, Jackfork, Carlorovelli, Anotherak, SieBot, Keskival, AS, Robdunst, Hugh16, Senderista~enwiki, Bnsreenath, Caidh, Oxymoron83, Dcattell, Swiebodzice, Sk8hack, Danthewhale, Martarius, Sfan00 IMG, Shaded0, Djr32, CohesionBot, JavierReynaldo, Arjayay, SchreiberBike, Pqnelson, Mjaniec, DumZiBoT, Ianbay, Neuralwarp,

XLinkBot, Fastily, Tenner47, Arthur chos, Avoided, Tenderbuttons, Benplusnumber, Balungifrancis, Addbot, DOI bot, 15lsoucy, Tarosic, Debresser, SamatBot, Yobot, Ibayn, 4th-otaku, AnomieBOT, VanishedUser sdu9aya9fasdsopa, Archon 2488, Francois33, Citation bot, Xqbot, Imushfiq, MIRROR, Pra1998, Dumontierc, Omnipaedista, Franco3450, Rr2000, FrescoBot, Paine Ellsworth, Nunc aut numquam, Martlet1215, Citation bot 1, Jonesey95, Tom.Reding, Schiefesfragezeichen, ROMVLVS, Casimir9999, RobinK, Meier99, Dinamik-bot, Bj norge, ElPeste, Afteread, EmausBot, Detogain, John of Reading, Racerx11, GoingBatty, XinaNicole, Ensabah6, Uploadvirus, ZéroBot, Arbnos, Zueignung, WaterCrane, Crown Prince, LaurentRDC, Isocliff, Vodkacannon, Raidr, Helpful Pixie Bot, Titodutta, Bibcode Bot, BG19bot, Spaligo, KateWishing, PhnomPencil, Sylvain.maurin, Kecchina, Halfb1t, Brad7777, Fylbecatulous, Jimw338, MyTuppence, Mogism, LT-Woods, Andyhowlett, Jawa0, &reasNink, SomeFreakOnTheInternet, Tentinator, EvergreenFir, DimReg, Pedarkwa, Db9199 24, Anrnusna, Notspelly, Ntomlin1996, Monkbot, Isbromberg, Dsprc, Tetra quark and Anonymous: 328

- **Supersymmetric gauge theory** *Source:* https://en.wikipedia.org/wiki/Supersymmetric_gauge_theory?oldid=671555895 *Contributors:* Phys, Lumidek, Conscious, SmackBot, JarahE, Aruton, AnonyScientist, Addbot, Luckas-bot, Omnipaedista, Erik9bot, Imran Parvez, AHusain314, Benlarcc and Anonymous: 6

- **Wess–Zumino model** *Source:* https://en.wikipedia.org/wiki/Wess%E2%80%93Zumino_model?oldid=551278243 *Contributors:* Phys, Giftlite, Lumidek, Bender235, Conscious, ChKa, Headbomb, Cuzkatzimhut, Klappspatier, Addbot, Lightbot, Legobot, Luckas-bot, Yobot, Anne Bauval, J04n, Omnipaedista, RedBot and Anonymous: 5

31.2.2   Images

- **File:1e0657_scale.jpg** *Source:* https://upload.wikimedia.org/wikipedia/commons/a/a8/1e0657_scale.jpg *License:* Public domain *Contributors:* Chandra X-Ray Observatory: 1E 0657-56 *Original artist:* NASA/CXC/M. Weiss

- **File:AIP-Sakurai-best.JPG** *Source:* https://upload.wikimedia.org/wikipedia/commons/2/2b/AIP-Sakurai-best.JPG *License:* Public domain *Contributors:* Own work *Original artist:* self

- **File:Ambox_important.svg** *Source:* https://upload.wikimedia.org/wikipedia/commons/b/b4/Ambox_important.svg *License:* Public domain *Contributors:* Own work, based off of Image:Ambox scales.svg *Original artist:* Dsmurat (talk · contribs)

- **File:Black_Hole_Merger.jpg** *Source:* https://upload.wikimedia.org/wikipedia/commons/d/d1/Black_Hole_Merger.jpg *License:* Public domain *Contributors:* Taken from http://www.space.com/imageoftheday/image_of_day_060203.html credit is listed to NASA. *Original artist:* NASA

- **File:CERN_LHC_Tunnel1.jpg** *Source:* https://upload.wikimedia.org/wikipedia/commons/f/fc/CERN_LHC_Tunnel1.jpg *License:* CC BY-SA 3.0 *Contributors:* Own work *Original artist:* Julian Herzog (website)

- **File:CuttingABarMagnet.svg** *Source:* https://upload.wikimedia.org/wikipedia/commons/4/43/CuttingABarMagnet.svg *License:* CC0 *Contributors:* Own work *Original artist:* Sbyrnes321

- **File:Cyclic_group.svg** *Source:* https://upload.wikimedia.org/wikipedia/commons/5/5f/Cyclic_group.svg *License:* CC BY-SA 3.0 *Contributors:*

- Cyclic_group.png *Original artist:*

- derivative work: Pbroks13 (talk)

- **File:Dynkin4A3lift.png** *Source:* https://upload.wikimedia.org/wikipedia/commons/d/db/Dynkin4A3lift.png *License:* CC BY-SA 3.0 *Contributors:* Own work *Original artist:* Pacman 2.0

- **File:Dynkin_Diagram_Triangle.jpg** *Source:* https://upload.wikimedia.org/wikipedia/commons/b/b3/Dynkin_Diagram_Triangle.jpg *License:* CC BY-SA 3.0 *Contributors:* Own work *Original artist:* Pacman 2.0

- **File:Edit-clear.svg** *Source:* https://upload.wikimedia.org/wikipedia/en/f/f2/Edit-clear.svg *License:* Public domain *Contributors:* The *Tango! Desktop Project. Original artist:*
The people from the Tango! project. And according to the meta-data in the file, specifically: "Andreas Nilsson, and Jakub Steiner (although minimally)."

- **File:Em_dipoles.svg** *Source:* https://upload.wikimedia.org/wikipedia/commons/f/f0/Em_dipoles.svg *License:* CC0 *Contributors:* Own work *Original artist:* Maschen

- **File:Em_monopoles.svg** *Source:* https://upload.wikimedia.org/wikipedia/commons/2/2f/Em_monopoles.svg *License:* CC0 *Contributors:* Own work *Original artist:* Maschen

- **File:Higgs,_Peter_(1929).jpg** *Source:* https://upload.wikimedia.org/wikipedia/commons/0/0d/Higgs%2C_Peter_%281929%29.jpg *License:* CC BY-SA 2.0 de *Contributors:* Mathematisches Institut Oberwolfach (MFO), http://owpdb.mfo.de/detail?photo_id=12812 *Original artist:* Gert-Martin Greuel

- **File:Hqmc-vector.svg** *Source:* https://upload.wikimedia.org/wikipedia/commons/6/68/Hqmc-vector.svg *License:* CC BY 3.0 *Contributors:* Own work *Original artist:* VermillionBird

- **File:Hydrogen300.png** *Source:* https://upload.wikimedia.org/wikipedia/commons/a/ad/Hydrogen300.png *License:* Public domain *Contributors:* Transferred from en.wikipedia; transferred to Commons by User:OverlordQ using CommonsHelper.
Original artist: PoorLeno (talk) Original uploader was PoorLeno at en.wikipedia

- **File:LQG_black_hole_Horizon.jpg** *Source:* https://upload.wikimedia.org/wikipedia/en/9/9d/LQG_black_hole_Horizon.jpg *License:* CC-BY-SA-3.0 *Contributors:*
created on xfig
Previously published: 2007-09-01
Original artist:
Ibayn

31.2.3 Content license